Shurong Wang, Zhongyang Luo
Pyrolysis of Biomass
GREEN Alternative Energy Resources

GREEN Alternative Energy Resources

Volume 1

Shurong Wang, Zhongyang Luo

Pyrolysis of Biomass

DE GRUYTER

Science Press
Beijing

Authors
Shurong Wang
State Key Laboratory of Clean Energy Utilization
Zhejiang University
Hangzhou, 310027
China
srwang@zju.edu.cn

Zhongyang Luo
State Key Laboratory of Clean Energy Utilization
Zhejiang University
Hangzhou, 310027
China

ISBN 978-3-11-037457-5
e-ISBN (PDF) 978-3-11-036963-2
e-ISBN (EPUB) 978-3-11-038612-7
Set-ISBN 978-3-11-036966-3

Library of Congress Cataloging-in-Publication Data
A CIP catalog record for this book has been applied for at the Library of Congress.

Bibliographic information published by the Deutsche Nationalbibliothek
The Deutsche Nationalbibliothek lists this publication in the Deutsche Nationalbibliografie; detailed bibliographic data are available on the Internet at http://dnb.dnb.de.

Cover image: vencavolrab/iStock
Typesetting: PTP-Berlin, Protago-TEX-Production GmbH, Berlin
Printing and binding: CPI books GmbH, Leck
♾ Printed on acid-free paper
Printed in Germany

www.degruyter.com

Preface

Energy is a fundamental resource for the survival of human society. Significant progress in human civilization is bound to be accompanied by the improvement and adjustment of energy structures. Nowadays, energy is one of the core issues in the world, and the demand for energy intensifies every day with the expansion of the global economy. However, problems due to the lack of primary energy, such as coal, oil and natural gas, and further potential risks still exist, which may add to the hazards of energy safety. Moreover, environmental pollution resulting from the utilization of fossil fuels cannot be ignored. The harm caused by increasing greenhouse gas emissions, acid rain and PM2.5 is raising more and more concern.

Biomass is a clean renewable energy resource, and it is much more environmentally friendly than fossil fuels like coal, because neutral emission of CO_2 is achieved when utilizing biomass. Furthermore, biomass is the only renewable carbon-containing energy resource that can be used for large-scale production of liquid fuels. If the abundant biomass resources in the world can be converted to liquid fuels, the tense and unbalanced status of the oil supply will be greatly alleviated. Biomass fast pyrolysis liquefaction is a mainstream technology that can efficiently convert biomass into high-energy-density bio-oil, which is a type of liquid fuel that is easy to store and transport. Bio-oil can be directly used as fuel in a boiler or be further modified for high-grade liquid motor fuel applications. However, raw bio-oil has some disadvantages such as high water content, high oxygen content, low heating value, low pH value, complicated composition and instability, which have greatly limited its application. There are two key points for solving these problems: the first is to fundamentally analyze the mechanisms involved in biomass pyrolysis so as to understand the laws of pyrolysis, which can be helpful for improving the quality of bio-oil during the preparation process; the second point is to promote research on bio-oil upgrading. For example, some bio-oil upgrading technologies, such as catalytic hydrogenation, catalytic cracking and catalytic esterification, can combine distributed biomass pyrolysis liquefaction with centralized bio-oil upgrading, so that high-grade vehicle fuels and even aviation fuels can be obtained. I have devoted myself to research on the biomass pyrolysis mechanisms and the production of high-grade liquid fuels through biomass fast pyrolysis liquefaction since 1998. And it is my great pleasure to have cooperated with Prof. Zhongyang Luo in writing this book with the aim of introducing the pyrolysis of biomass.

The book is divided into seven chapters. It follows the thread of the three major components of biomass and systematically illustrates the theories involved in biomass pyrolysis liquefaction. This book begins with the structure of lignocellulosic biomass. The basic structure, physical and chemical properties of the three major components of biomass, namely cellulose, hemicellulose and lignin, are introduced in detail, and common model compounds of the three major components as well as their proper-

DOI 10.1515/9783110369632-001

ties and corresponding extraction methods are also expounded in the first chapter. Then from Chapter 2 to Chapter 4 we give an in-depth illustration of the pyrolysis mechanism of the three major components of biomass. Beginning with the basic process of pyrolysis, the influence of different factors on biomass component pyrolysis is elaborated, and the biomass component pyrolysis mechanism is discussed based on kinetics research, product formation processes and theoretical simulation at molecular level. Chapters 5 and 6 introduce the influence of the interaction of components, catalysts and inorganic salts on the pyrolysis behavior of biomass. Finally, pyrolysis behavior and pyrolysis liquefaction characteristics of different biomass species are illustrated, and research about bio-oil graded upgrading based on the separation of bio-oil is presented in Chapter 7.

I have been engaged in researching pyrolysis mechanisms and biomass pyrolysis liquefaction for almost 20 years. None of the achievements could have been obtained without support from the State Key Laboratory of Clean Energy Utilization at Zhejiang University and my colleagues, especially the long-term guidance from Academician Kefa Cen, Prof. Mingjiang Ni and Prof. Zhongyang Luo. I am also deeply grateful for the project funding for the biomass energy research field from the National Natural Science Foundation of China, Ministry of Science and Technology of China, Ministry of Education of China, and Science and Technology Department of Zhejiang Province, and some of the research results are also reflected in corresponding chapters. I sincerely offer my special thanks to all of my postgraduate students. Their theses greatly enrich this book. It is because of their hard work that the group's biomass pyrolysis liquefaction research has become much more systematic and thorough over time. Special thanks also to Bin Ru, Xiujuan Guo, Qinjie Cai, Lingjun Zhu, Qi Wang, Yurong Wang, Xinbao Li, Qianqian Yin and Wenwen Guo who have already graduated, and Haizhou Lin, Gongxin Dai, Fan Zhang, Yuan Zhao, Junhao Chen, Xiaoliu Wang and Shi Yin who are still studying, for their great help in collecting literature, arranging illustrations and text proofreading. In addition, I would like to express my sincere thanks to the publisher De Gruyter for their excellent work at every stage of this project.

Finally, many thanks to my wife Ruoqun Luo and my son Ziyang Wang for their full support and sincere understanding during the years of writing this book.

Although the authors have made every effort to accurately reflect related work on biomass pyrolysis and its results, some improper renderings or omissions may unavoidably exist. Every comment from readers and relevant experts in this field is deeply appreciated, and will be used to further improve future versions.

Shurong Wang
August 2016

Acknowledgement to financial support

National Natural Science Foundation of China (51276166, 51476142, 50676085, 50476057, 50176046, 29976039, 90610035, 50025618)
National Basic Research Program of China (2007CB210200, 2013CB228100)
National High Technology Research and Development Program of China (2009AA05Z407)
International Science & Technology Cooperation Program of China (2009DFA61050)
Program for New Century Excellent Talents in University (NCET-10-0741)
Research Fund for the Doctoral Program of Higher Education of China (20090101110034)
Zhejiang Provincial Natural Science Foundation (R1110089)
Zhejiang Provincial Key Science and Technology Innovation Team Project (2009R50012)
Zhejiang Provincial New Century 151 Talent Program (12-1-040)

DOI 10.1515/9783110369632-002

Contents

1 Biomass components and characteristics

1.1 Biomass components

Generally, biomass is organic matter derived from living organisms. Its formation is closely connected with photosynthesis, directly or indirectly, and it includes all animals, plants and microbes, as well as organic matter derived, excreted or metabolized from or by these living organisms. Biomass can be used as a source of energy and it most often refers to plants or plant-based materials which are not used for food or feed, specifically called lignocellulosic biomass, especially focusing on agricultural and forestry wastes, such as woodchip and rice straw.

1.1.1 Composition analysis of biomass

Biomass is composed primarily of three elements: carbon, hydrogen and oxygen, whose total content generally reaches above 95 %. It also contains small amounts of nitrogen and sulfur, which are generally recognized as deriving from the residual proteins in protoplasm at the early growth stage of a cell [1]. Tab. 1.1 lists the composition analysis of several typical types of biomass, such as Mongolian Scots pine and rosewood in forestry biomass, rice straw and husk in agricultural biomass, bamboo and grassiness in herbaceous biomass, and algae in aquatic biomass. The fixed carbon and volatiles content are similar for different biomass species, while the ash content varies markedly. As typical forestry biomass, Mongolian Scots pine and rosewood have the lowest ash content, while agricultural biomass has relatively higher ash content. Except for carbon, hydrogen, oxygen, nitrogen and sulfur, biomass still contains some inorganic metal elements like potassium, calcium, sodium, magnesium, aluminum, iron and copper etc., in the form of inorganic compounds. Furthermore, potassium and calcium content are significantly higher than those of

Tab. 1.1: Proximate and ultimate analysis of various kinds of biomass.

Biomass	M_{ad} (%)	A_{ad} (%)	V_{ad} (%)	FC_{ad} (%)	$Q_{b,ad}$ (MJ/kg)	C_{ad} (%)	H_{ad} (%)	N_{ad} (%)	S_{ad} (%)	O_{ad} (%)
Mongolian Scots pine	13.90	0.30	73.74	12.06	18.84	45.92	4.41	0.10	0.03	35.34
Rosewood	13.45	0.35	71.07	15.13	17.07	44.32	4.88	0.16	0	36.84
Rice straw	11.21	16.12	61.36	11.31	13.87	36.89	3.44	1.19	0.20	30.95
Rice husk	12.30	12.26	60.98	14.46	14.57	40.0	3.66	0.53	0.13	31.12
Bamboo	5.40	3.68	75.70	15.22	17.54	45.32	2.51	0.82	0.04	42.23
Grassiness	8.21	2.44	73.09	16.26	16.65	44.45	4.68	0.31	0.16	39.75
Algae	16.30	10.09	60.39	13.22	12.65	34.17	3.84	2.16	1.04	32.40

DOI 10.1515/9783110369632-003

other elements. Potassium content is relatively higher in herbaceous biomass like straw, while wood contains more calcium. In addition, biomass may still differ in appearance and reaction characteristics, even with similar chemical compositions.

1.1.2 Distribution of biomass components

Cellulose, hemicellulose and lignin are the three main components of lignocellulosic biomass, constituting the cell wall of plants. As displayed in Fig. 1.1, the main components are distributed unevenly in the cell wall, forming the skeleton, linking material and hard solids. Cellulose molecules regularly gather as pipes forming tough microfibers that function as the skeleton material of a plant's cell wall, and the inner space is filled with amorphous linking material, namely hemicellulose and hard-solid lignin [2]. The connection between cellulose and hemicellulose or lignin molecules is mainly by hydrogen bonds, while that between hemicellulose and lignin contains both hydrogen and covalent bonds, resulting in small amounts of carbohydrates adhered to the lignin extracted from lignocellulosic biomass [3]. Furthermore, biomass also contains free compounds, called extractives, which can be extracted by polar or nonpolar organic solvents. The extractives are nonstructural compounds, mainly including paraffin, fat, resin, tannin, starch and pigment etc. [4]. In addition, there are a few inorganic metal salts, mainly composed of alkali and alkaline earth metal salts. Overall, the physical and chemical behaviors of biomass are closely related to its composition and the relative content of cellulose, hemicellulose and lignin, as well as the extractives and inorganic salts.

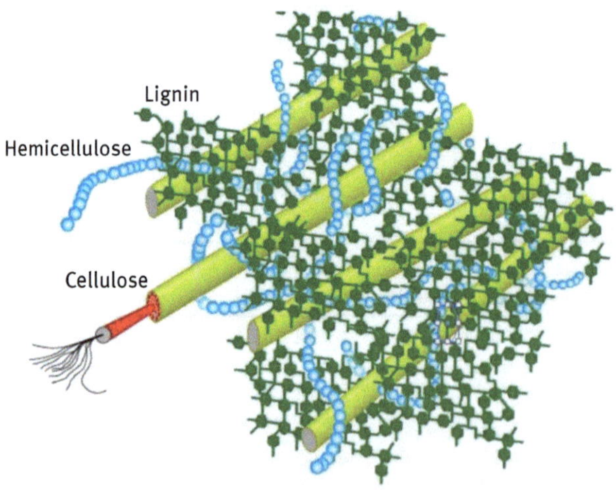

Fig. 1.1: Biomass components.

The cellulose, hemicellulose, and lignin content in biomass are greatly influenced by the biomass species. Generally, the cellulose content can reach 40–80 %, and the hemicellulose content ranges from 15 % to 30 %, while lignin content is around 10–25 % [5]. Acidic or basic solvents are often used to remove a certain component step by step, so as to calculate the exact mass of each component quantitatively. Van Soest analysis is the most widely used of the various quantification methods [6].

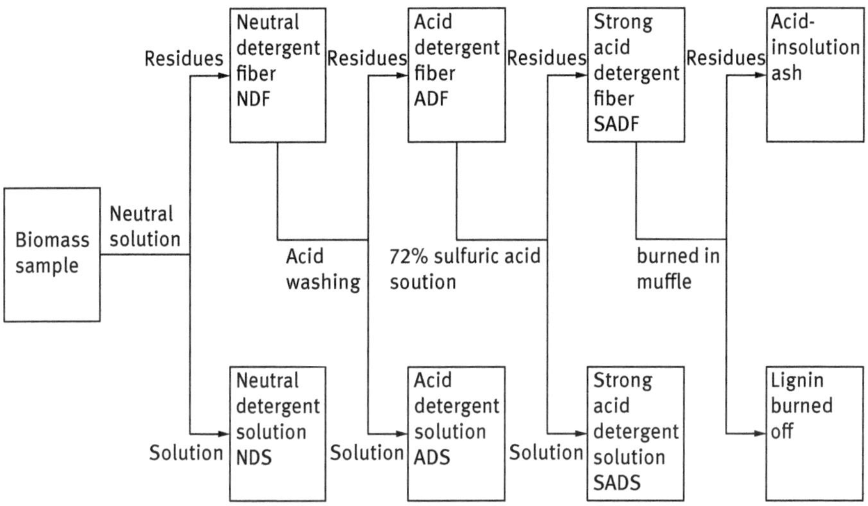

Fig. 1.2: Van Soest analysis of biomass sample.

A biomass sample with particle sizes within a given range is obtained after grinding and sieving and subsequent extraction according to the operation processes presented in Fig. 1.2. Firstly, the biomass is washed with neutral solution, dissolving and removing the detergent solution (NDS) to get the neutral detergent fiber (NDF) that mainly contains cellulose, hemicellulose, lignin and ash. Then, by washing with acid solution, the acid detergent fiber (ADF) that is mainly composed of cellulose, lignin and ash is obtained after the removal of hemicellulose. After that, 72 % sulfuric acid solution is used to dissolve and remove the cellulose so as to obtain strong acid detergent fiber (SADF), whose main components are lignin and a little insoluble ash. Finally, the residues are burned in a muffle furnace to remove lignin, leaving acid-insoluble ash. By weighing the mass of residues after each procedure, the distribution of the three main components, as well as the extractives content and acid-insoluble ash content in biomass can be calculated.

Tab. 1.2 displays the cellulose, hemicellulose, lignin and acid-insoluble ash content in several typical forestry, agricultural, herbaceous and algae biomass measured with Van Soest analysis. It can be seen that cellulose, hemicellulose and lignin are the majority components in forestry biomass with total content of more than 90 %,

Tab. 1.2: Component analysis of biomass samples (unit: %).

Samples	Van Soest analysis			Calculated component distribution				
	NDF	ADF	SADF	Extractives	Cellulose	Hemicellulose	Lignin	Acid-insoluble ash
Cedar wood	92.3	80.36	31.46	7.7	48.9	11.94	31.21	0.25
Mongolian Scots pine	91.3	73.6	25.8	8.7	47.8	17.7	25.5	0.3
Fast growing poplar	97.06	78	17.3	2.94	60.7	19.06	14.8	2.5
Rosewood	91.57	74.77	21.15	8.43	53.62	16.8	21.08	0.07
Manchurian ash	95.67	70.03	13.61	4.33	56.42	25.64	13.47	0.14
Betula alnoides	91.08	70.69	17.56	8.92	53.13	20.39	17.21	0.35
Rice straw	78.19	37.08	7.55	21.81	29.53	41.11	5.07	2.48
Rice husk	81.42	44.27	20.4	18.59	23.87	37.15	12.84	7.56
Bamboo	79.21	33.37	13.37	20.79	20	45.84	12.83	0.54
Grassiness	72.32	39.11	9.12	27.69	29.99	33.21	3.13	5.99
Chlamydomonas [8]				65.5	/	/	/	/
Green alga [8]				70.4	/	/	/	/
Chlorella[9]				67.21	/	/	/	/
Pine				3	46–50	19–22	21–29	0.3
Pine [10]				2.88	44.55	21.9	27.67	0.32
Pine [11]				2.7	41.7	22.5	25.8	0.3
Beech [12]				2.6	44.2	33.5	21.8	0.5
Rice husk [13]				8.4	31.3	24.3	14.3	23.5
Rice straw [13]				13.1	37	22.7	13.6	19.8
Rice straw [14]				13.1	37	16.5	13.6	19.8

of which cellulose has the highest amount. The distribution of the three major components in agricultural and herbaceous biomass is similar, with relatively higher cellulose and hemicellulose content and lower lignin content. In addition, the acid-insoluble ash content is a little high. The main reason for low lignin content was likely the easy removal of hard rind where lignin is concentrated, which was caused by drying, string and grinding operations during pretreatment [7]. Algae mainly contains more than 50 % of extractives, while the quantities of the three major conventional components are too small to be analyzed. It is worth noting that the overall component distribution is similar, even though there are some differences in some detailed composition values obtained by different researchers. These can likely be attributed to the structural differences for various biomass species. The cellulose content in pine and beech, belonging to forestry biomass, are high. For rice husk and rice straw in agricultural biomass, the cellulose and lignin content are relatively lower, but their ash and extractives content are obviously higher than those in forestry biomass.

1.2 Cellulose

Cellulose is the macromolecular polysaccharide that is most widely distributed and most abundant in nature. About 50 % of the carbon in plants exists in cellulose. For example, 40–45 % of dry material exists in the form of cellulose in wood. The cellulose content even reaches above 90 % in cotton, thus cotton can be used to produce microfiber filters [15]. Usually cellulose is packed with hemicellulose and lignin, so it is hard to be obtained by direct removal.

1.2.1 Structure of cellulose

1.2.1.1 Chemical structure of cellulose

Cellulose mainly contains three elements: carbon, hydrogen and oxygen in the proportions 44.2 %, 6.3 % and 49.5 %, respectively, which can be expressed as $C_6H_{10}O_5$ in chemical formula. The chemical structure is shown in Fig. 1.3. As a kind of macromolecular polysaccharide, cellulose is often indicated as $(C_6H_{10}O_5)_n$, where n is the degree of polymerization. For instance, n is about 6,000–8,000 for woody fiber and 14,000 for cotton fiber [16]. The X-ray diffraction pattern of cellulose crystal indicates that the crystallographic repeating units are two anhydroglucose linked with a helical axis. The left and right anhydroglucose rotate around each other by 180° along with the helical axis, so it is cellobiose that constitutes the basic unit of cellulose instead of glucose [17]. The glucose in cellulose belongs to D-pyran glucose, and the basic glucose units are linked by β-1,4-glycosidic bonds. Glycosidic bonds are not strong and tend to degrade under acid or high temperature conditions. Each glucose unit has three alcoholic hydroxyl groups, and the hydrogen in hydroxyl groups can form a

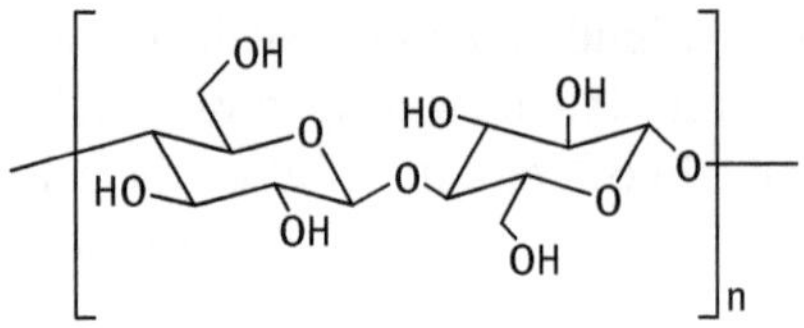

Fig. 1.3: Basic monomer structure of cellulose.

hydrogen bond with the oxygen in the side chain, forming a strong fibrillar structure with gathered molecular chains [18–20]. Although the monomer is cellobiose in cellulose instead of glucose, there are strong correlations between cellulose and glucose. More than 90 % glucose could be obtained when cellulose was hydrolyzed under dilute phosphoric acid solution in a microwave reactor. It indicates that the glycosidic bonds in the cellulose structure could be broken as a priority under certain hydrolysis conditions, forming large amounts of glucose monomers [21].

1.2.1.2 Crystalline structure of cellulose

Most chain molecules of cellulose integrate in a linear beam, usually called microfiber, which is the final form of the cellulose skeleton. According to the arrangement format of chain molecules, the cellulose ultrastructure can be divided into crystalline and amorphous regions. In the crystalline region, molecules are arranged uniformly, while in the amorphous region, molecular arrangement is loose. But there is no obvious boundary between the two regions. The crystallinity of cellulose is an important index to evaluate the properties of the cellulosic skeleton structure. It is positively correlated with the length of the microfibers. Cellulosic crystal appears in polymorphism, and can be divided into five crystalline modifications (type I, II, III, IV, and V) [22–24] based on the microstructure and arrangement of cellulosic chains. Type I(natural cellulose) and II(regenerated cellulose or mercerized cellulose) are the most typical and often selected as the representatives for cellulose. It is commonly believed that chain arrangement in type I cellulose is parallel while that in type II is antiparallel. When treated with liquid ammonia, these two types will produce type III cellulose. After further heat treatment, type III may convert into type IV cellulose. Among all the forms, cellobiose is the most basic translation unit, and its existence gives the molecular chain of cellulose a repeatable length of 10.3 Å.

Atalla and Vanderhart [25] studied natural cellulosic crystals through ^{13}C-CP/MAS-NMR and found that cellulose in nature could be divided into I_α and I_β by crystal configuration. The former corresponds to triclinic crystal with one chain while the latter is a monoclinic crystal containing two chains, and the density of the latter is also higher than that of I_α. In addition, the thermodynamic stability of I_β cellulose is higher than that of I_α cellulose. Under certain conditions (such as heat treatment in basic solution), I_α cellulose can be gradually converted to I_β cellulose. In nature, I_α cellulose mainly exists in lower plants, while in higher plants, cellulose is mainly in the form of I_β. Further study by Atalla and Vanderhart [25] showed that I_α cellulose was the main type in algae and bacteria cellulose, and constituted more than 60 wt%, and the I_β cellulose content in higher plants like cotton could reach 80 wt%. They also found that the I_β cellulose content increased after the heat treatment of biomass.

1.2.1.3 Apparent structure of cellulose

The structural information can be specified according to the shape of the crystal, which can be used to confirm and optimize cellulose geometry in quantum chemistry simulation. Another widely adopted method is macroscopic classification. By analyzing the apparent structure of cellulose through scanning electron microscopy (SEM) and transmission electron microscopy (TEM), cellulose is divided into bacterial cellulose (BC), microfibrillated cellulose (MFC), microcrystalline cellulose (MCC), networked cellulose (NC), nanocrystalline cellulose (NCC) and their corresponding complexes [26]. BC mainly comes from the production process of bacterial strains. When cellulose undergoes shearing treatment, total disruption of the cell wall material is induced and microfibrillated cellulose (MFC) is released. MCC is a kind of hydrolyzed cellulose particle consisting of large cellulosic microcrystal and some amorphous regions. NC is regenerated in ethanol after MCC has been dissolved by sulfuric acid, and its chains are randomly bundled together. NCC is a typical rigid rod-shaped monocrystalline cellulose, and it can be easily isolated from cellulose fibers through acid hydrolysis.

1.2.2 Characteristics of cellulose

Natural cellulose is a white filament that is odorless and colorless. The density of crystalline cellulose is around 1,600 kg/m^3, while that of wood is about 1,500 kg/m^3.

Cellulose is hydrophilic and easy to expand due to the polar hydroxyl groups, and the strength of water adsorption depends on cellulosic structure and capillarity. At a temperature of 20 °C and relative air humidity of 60 %, cellulose will adsorb 6–12 % water [27]. The absorbed water can be divided into bound water and free water. The former refers to the water entering the cellulosic amorphous region and forming hydrogen bonds by combining with cellulosic hydroxyl. When the absorbed water reaches saturation, water molecules keep going into the cell cavity and the interspace of cellulose and this is the free water. Cellulosic solubility is poor in water, but good in sulfuric and hydrochloric acid. The dissolving process must be kept at a low temperature, otherwise cellulose will decompose quickly.

The glucose unit in cellulose has three active hydroxyls with different reaction abilities that are distributed in the pyranose ring of 2, 3, 6 carbon positions [28, 29]. Hydroxyl at C_6 has the least steric hindrance, thus its reaction activity is higher than that of the other two hydroxyl groups. Generally, the esterification rate of hydroxyl at C_6 is 10 times faster than that of the other hydroxyl groups, and the etherification rate of hydroxyl at C_2 is twice as fast as that of the hydroxyl at C_3 [30]. Apart from hydroxyl, almost all reactions relating to cellulose are associated with the cleavage of glycosidic bonds between two adjacent glucose monomers. Glycosidic bonds in cellulose have the features of acetal bonds and are very sensitive to acids. When reacting with acids or acid solutions, glycosidic bonds will rupture leading to a decrease in the degree of polymerization [31]. In addition, cellulose can undergo esterification, etherification, oxidation and photocatalysis reactions under certain conditions.

1.2.3 Isolation of cellulose and its model compounds

The structure of cellulose is simple, and the properties of cellulose isolated from different biomass are stable, show good repeatability and do not change much with type of biomass. As the main component of lignocellulosic biomass, cellulose and its model compounds are usually selected to represent biomass. Qualitative filter paper and microcrystalline cellulose are often selected as model compounds of cellulose. The former belongs to the natural celluloses and is derived from cotton fiber with high cellulose content after simple treatment. The latter is powdered cellulose from the hydrolysis of pure cotton cellulose. The elemental composition of the two model compounds is basically the same (Tab. 1.3), mainly including carbon, hydrogen and oxygen. The calculated molecular formula of microcrystalline cellulose is $C_6H_{9.6}O_{4.9}$, and that of qualitative filter paper is $C_6H_{9.6}O_{4.8}$. Both of them are highly coincident with that of cellulose ($C_6H_{10}O_5$).

Tab. 1.3: Ultimate analysis of qualitative filter paper and microcrystalline cellulose.

Raw material	C (%)	H (%)	O (%)	N (%)	S (%)
Qualitative filter paper	44.98	6.20	48.71	0.09	0.02
Microcrystalline cellulose	43.26	5.78	46.57	0.45	0.05

Both qualitative filter paper and microcrystalline cellulose are the ideal cellulose model compounds, and their SEM and FTIR spectra are presented in Fig. 1.4. Although obvious morphological differences appear in the SEM picture, the FTIR spectra are very similar for these two celluloses with typical polysaccharide features. Cellulose is a linear chain molecule without branches, and strong polar hydroxyl exists on its glucose units. Adjacent molecular chains could form macromolecules in twister through the linking of hydrogen bonds. In FTIR spectra, the peak at 3,342 cm^{-1} is assigned to –OH vibration, and the peaks located at 1,635 cm^{-1} and 1,161 cm^{-1} correspond to C=O and C–O–C absorption, while the peaks in the range from 1,368 to 1,279 cm^{-1} correspond to C–H and $-CH_2-$ absorption.

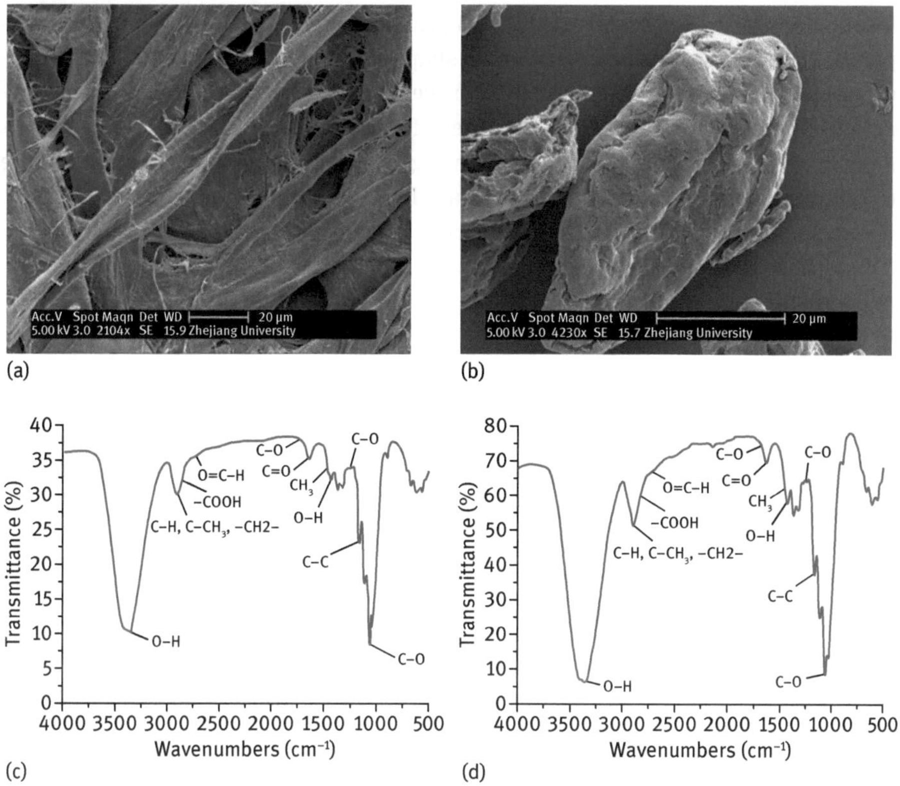

Fig. 1.4: SEM and FTIR spectra of cellulose; (1) SEM: qualitative filter paper, (2) SEM: microcrystalline cellulose; (a) FTIR spectrum: qualitative filter paper, (b) FTIR spectrum: microcrystalline cellulose.

Cellulose can be extracted by many methods. The appropriate method to extract cellulose should take the species and characteristics of biomass into careful consideration. In the whole biomass structure, cellulose is often linked with the short chains of hemicellulose. Thus during the extraction process, some extra sugar units, like mannose and xylose, will remain as a residue. In order to obtain cellulose with high purity, biomass rich in cellulose is often used. The commonly used methods for cellulose extraction are acid-alkaline hydrolysis [32, 33], synthesis-solvent [34] and steam explosion [35].

1.3 Hemicellulose

Cellulose and lignin are closely combined in a plant's cell wall by a mixture of glycan, called hemicellulose, distributing crosswise in various cell wall layers. The content and structure of hemicellulose vary across different biomass species. The hemicellulose content is 10–15 % in softwood, 18–23 % in hardwood and 20–25 % in herbaceous plants. Hemicellulose appears in a chain shape similar to cellulose, with a lower degree of polymerization of less than 200 on average. So hemicellulose molecule aggregation exists in powder rather than in thread form, but they are still short linear polymers in nature. Compared with the uniform glycan and single linear chain structure of cellulose, hemicellulose can form glycan that is either homogeneous or heterogeneous. In addition, various monosaccharides can be connected in different ways to construct different glycan structures. As mentioned above, hemicellulose is the general term for various glycans [36].

1.3.1 Structure of hemicellulose

Hemicellulose has a complex structure, and its fundamental sugar units are mainly xylose, mannose, galactose and arabinose etc., which are shown in Fig. 1.5. These sugar units form the principal basic structural units of hemicellulose, such as glucuronoxylan, galactoglucomannan, arabinoglucuronoxylan, xylanglucose, xyloglucan and arabinoxylan [37, 38].

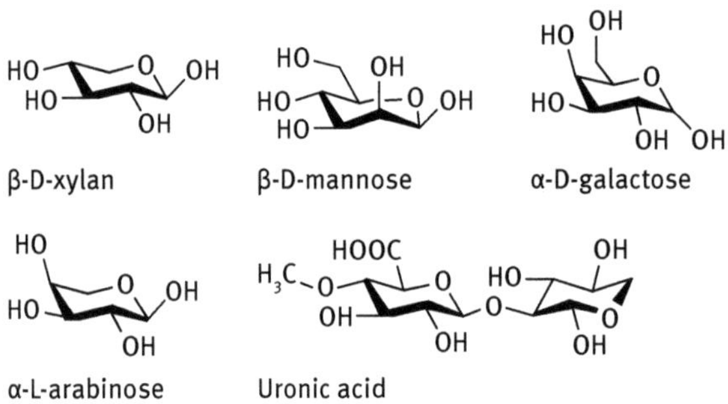

Fig. 1.5: Fundamental chemical structure of sugar units in hemicellulose.

Glucuronoxylan, also known as O-acetyl-4-O-methylglucuronoxylan, is the most common chemical structure in softwood hemicellulose, constituting 15–30 %. It mainly consists of β-D-xylose backbones connected through β-1,4-glycosidic bonds. As C_2 and C_3 atoms in xylose may be acetylated, a uronic acid unit occurs in one tenth of xylose units. The proportion of acetyl ranges from 8–17 %, corresponding to 3.5–7 acetyl groups per ten xylose units [37]. Side chains in 4-O-methylglucuronoxylan are relatively strong and hard to dissolve in acid conditions. Some rhamnose and galacturonic acid are also included in this structural unit.

Galactoglucomannan is the typical xylose structure in hardwood hemicellulose, constituting 20 %–25 %, while its content is less than 5 % in softwood hemicellulose. It is mainly a chain skeleton structure consisting of β-D-xylose and β-D-rhamnose units connected by β-1,4-glycosidic bonds. C_2 and C_3 atoms in xylose are partly acetylated and the α-D-galactose in substituted units is connected to glucose and mannose units through α-1,6 bonds, wherein the acetyl content in galactoglucomannan is around 6 % and 3–4 hexose units correspond to one acetyl on average. In addition, the proportion of mannose groups is far higher than that of glucan. According to previous research, the degree of galactosylation in galactoglucomannan influences the connection of cellulose and hemicellulose, as well as the extractability of hemicellulose indirectly [39].

Arabinoglucuronoxylan is the typical structure in non-wood (mainly agricultural biomass) hemicellulose, and also exists in small amounts in soft wood. It mainly consists of a linear β-1,4-D-xylose skeleton structure, including 4-O-methyl-α-D-pyran glucuronic acid connected by α-1,2-glycosidic bonds and α-L-arabinofuranose connected by α-1,3-glycosidic bonds [40].

Xyloglucan is the hemicellulose glycan existing largely in hardwood plants, mainly in dicotyledons, some in monocotyledons and also found in herbaceous plants [41]. It is mainly composed of a D-glucose skeleton connected by β-1,4-glycosidic bonds, including D-xylose, L-arabinose and D-galactose units. The structural units interact with microfibers in cellulose through the connection of hydrogen bonds, making a significant contribution to the integrality of cellulose structure [42].

Arabinoxylan mainly exists in cereal hemicellulose, similar to xylan in hardwood, but with higher L-arabinose content. Linear β-1,4-D-xylopyranose is replaced with an α-L-arabinofuranose unit at 2-O or an α-D-pyran glucuronic acid unit at 3-O position [43]. Acetyl substitution can also exist [44]. Esterified phenolic dimer will lead to intramolecular and intermolecular crosslink effects of xylan and increase the crosslink degree of xylan and cellulose structure. In addition, during the lignification process of biomass, xylan tends to be connected to the lignin structure through esterification of aldehyde acid side chains. All of these factors increase the difficulty of xylan extraction [45, 46].

Hemicellulose is often named according to the glycosyl in the backbone, which is beneficial to the analysis of structure and properties. The backbone can be composed of either one or more glycosyls, and is a nonuniform high polysaccharide. When there is only D-xylose in the backbone, the hemicellulose is called xylan-based hemicellulose. When both glucose and mannose exist in the backbone, it is called glucomannan-based hemicellulose.

An important function of hemicellulose is to crosslink the cellulose and lignin together in the whole biomass structure. The connection between hemicellulose and cellulose depends on the hydrogen bonds and Van der Waals force. The length of a glycan chain is normally longer than the distance between two cellulose chains, so the former can cover the cellulose surface and connect to multi-cellulose chains via hydrogen bonds. In addition, the connection of hemicellulose and lignin is based on covalent bonds.

1.3.2 Characteristics of hemicellulose

The average density of hemicellulose is 1,500 kg/m^3, large differences exist in the hemicelluloses obtained by various methods due to their structural instability. The degree of polymerization of hemicellulose is relatively low, only 150–200 for natural hemicellulose, and is far lower than that of cellulose. The degree of polymerization for softwood hemicellulose is the lowest, only around 100, while it is about 200 for hardwood hemicellulose. The abundant carboxyl side chains in hemicellulose lead to its fine hydrophilia and swelling of the cell wall, which makes it fibroelastic. The degrees of hydroscopicity and swelling are higher than those of cellulose, as it is difficult for hemicellulose to form a crystalline region and it is easy for water to enter. A small part of hemicellulose is soluble while most parts are not. For example, arabinogalactan dissolves easily in water. In general, the lower the degree of polymerization and the more branches it has, the more easily hemicellulose will dissolve in water.

The hemicellulose structure has abundant side chains, a low degree of polymerization and an amorphous form, so its reactivity is somewhat higher than that of cellulose. Hemicellulose is easily decomposed into monomer components, such as mannose, xylose and some glucuronic acid etc. The glycosidic bonds in hemicellulose tend to be cleaved under the effect of acids, resulting in hemicellulose degradation.

1.3.3 Isolation of hemicellulose and its model compounds

Hemicellulose is difficult to extract completely from biomass, so model compounds with the typical structure of hemicellulose are used in most research. The overall reaction characteristics are studied by investigating the specific structural units. The common model compound is xylan, but mannan and galactose are also used.

1.3.3.1 Xylan

Xylan is widely distributed in nature and is the main component of hemicellulose in the cell walls of many plants such as corn stover, cottonseed hull and sawdust. It is a complex polysaccharide, having the backbone of D-xylan connected by β-1,4 glycosidic bonds, also attached by many substituent groups, such as glucuronic acid, acetyl, arabinose, ferulic acid and cumaric acid. The composition and structure of xylan vary in different biomass, as well as the same biomass when different extraction methods are employed [47]. The typical structure is presented in Fig. 1.6. The main chain is composed of β-D-xylopyranose units connected by β-1, 4-glycosidic bonds, and the C_2 site is replaced by a 4-O-methy-α-D-glucuronic acid group, accounting for two groups in ten xylose units on average.

The structure of xylan largely depends on its source. In birch xylan, xylose, arabinose and glucose account for 89.3 %, 1 % and 1.4 %, respectively [48]. Rice bran

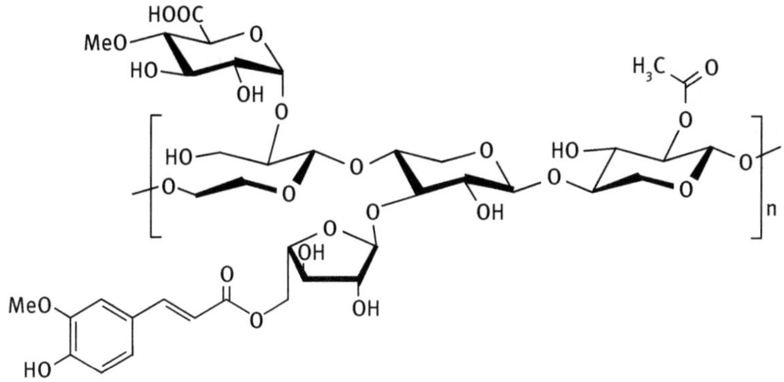

Fig. 1.6: Typical structure of xylan.

xylan has 46 % xylose, 44.9 % arabinose, 6.1 % galactose and 1.9 % glucose [43]. Wheat arabinoxylan is composed of 65.8 % xylose, 33.5 % arabinose, 0.1 % mannose, 0.1 % galactose and 0.3 % glucose [49]. Corncob xylan comprises 48–54 % xylose, 33–35 % arabinose, 5–11 % galactose and 3–6 % glucuronic acid [50]. Most xylan main chains are replaced by arabinose or glucuronic acid through xylose units connected at O-2 or O-3 [51]. Beech xylan produced by Sigma Company is mainly composed of C, H, O, and a little N and S. D-xylose is around 90 % glycosyl and it has abundant methylglucuronic acid side chains.

The FTIR spectra of xylan are displayed in Fig. 1.7, showing several characteristic absorption peaks of C–O–C, –OH, O–Ac, –COOH and $-OCH_3$, all of which exist in the structural model of xylan (Fig. 1.6). The phenylpropane structure that connects to the end of the glucose unit in the xylan branched chain is not observed, mainly due to its small quantity. It is derived from the lignin structure crosslinked to hemicellulose and might be removed completely in a sever extraction, which makes it hard to detect.

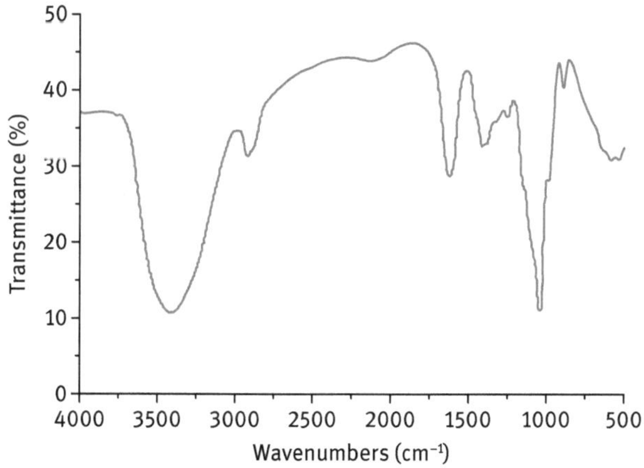

Fig. 1.7: FTIR spectrum of xylan.

1.3.3.2 Mannan

Mannan is widely distributed in plants, especially in softwood. It is a glycan with low molecular weight, mainly in the form of glactoglucomannan. The typical chemical structure is shown in Fig. 1.8. As shown in its FTIR spectrum (Fig. 1.9), mannan does not have acetyl or phenylpropane side chains compared with xylan, and the rest of the skeleton structure is similar.

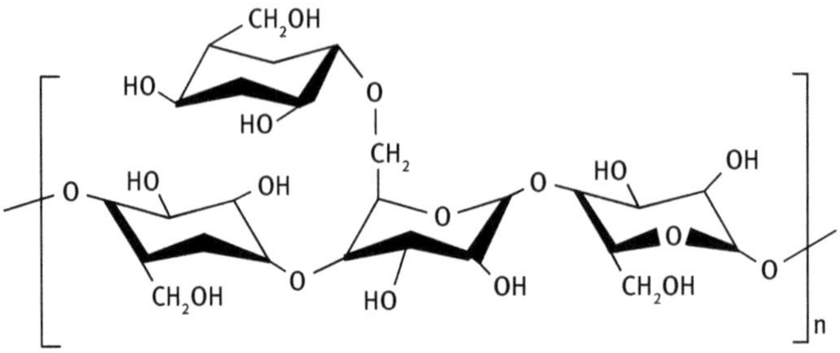

Fig. 1.8: Typical structure of mannan.

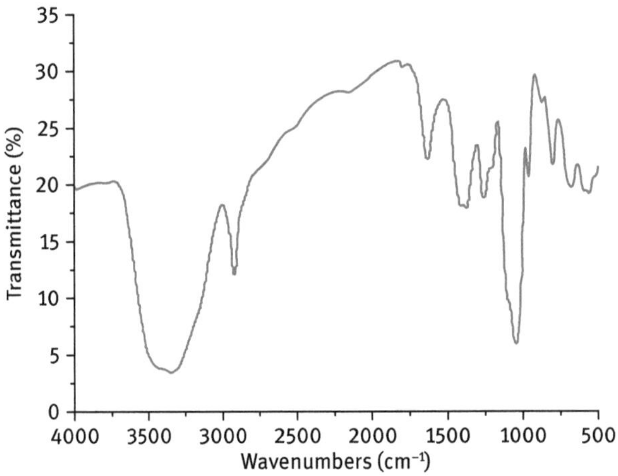

Fig. 1.9: FTIR spectrum of mannan.

1.3.3.3 Isolation of hemicellulose

Among the three components of biomass, hemicellulose is the most unstable and can be separated from wood, holocellulose and pulp using extraction methods. Since hemicellulose is an amorphous structure consisting of various glycans, it is impossible to separate these glycans totally. The extraction process for hemicellulose is similar to that for cellulose. Cellulose exits in the form of a residue while hemicellulose is soluble in the basic solution. That is, hemicellulose can be obtained through precipitation from the basic solution during the extraction process. In addition, hemicellulose can also be separated from holocellulose derived from biomass by delignification. When hemicellulose is separated from holocellulose, its structure varies along with the solvents used. In some effective neutral solvents, dimethyl sulfoxide is suitable for the extraction of xylan from holocellulose, as there are no chemical reactions and therefore little influence on the structure of xylan. Although more xylan can be extracted by basic solution, its chemical structure is obviously influenced.

1.4 Lignin

Lignin, as a filling and bonding material, can increase wood strength and prevent microbial contamination. Lignin content varies with plant type, it also varies even within different branches of the same plant. Lignin is also a polymer, but unlike the linear chain structure of cellulose, lignin chains distribute in every direction and form a three-dimensional macromolecular structure. Lignin content is 25–35 % in softwood, 20–25 % in hardwood, and 15–20 % in graminaceous monocotyledon, respectively. Unlike cellulose or hemicellulose, lignin belongs to the nonsaccharide macromolecular materials with very complex structure.

1.4.1 Structure of lignin

1.4.1.1 Structural units of lignin

Lignin mainly consists of phenylpropanoid units in a nonlinear and random form. Three basic structures are shown in Fig. 1.10, including the hydroxy-phenyl lignin (H-lignin) polymerized by *p*-hydroxy phenylpropanoid units, guaiacyl lignin (G-lignin) polymerized by guaiacyl phenylpropanoid units and syringyl lignin (S-lignin) polymerized by syringyl phenylpropanoid units [52]. Though lignin only has three basic structural units, there are different functional groups on the benzene rings, leading to the complexity of lignin structure. The distribution of the three basic structural units in lignin largely depends on the biomass type. In hardwood there are mainly S- and G-lignin units as well as a small amount of H-lignin units, while softwood mainly consists of G-lignin units and herbaceous biomass mainly has H-lignin units [53].

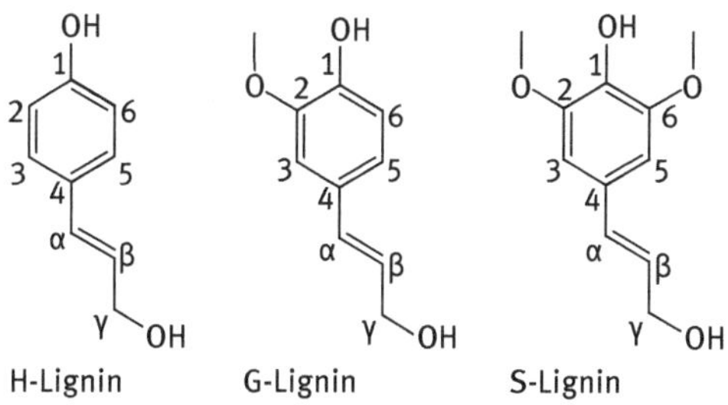

Fig. 1.10: Three basic structural units in lignin.

There are many kinds of functional groups in lignin, such as methoxyl and phenolic hydroxyl groups on the benzene ring, as well as carbonyl, alcoholic hydroxyl groups and carbon-carbon double bonds on side chains. Methoxyl groups are the most basic characteristic functional group in lignin. The methoxyl groups content is 14–16 % in softwood, 19–22 % in hardwood, and 14–15 % in herbaceous biomass, respectively [27]. The methoxyl groups on the benzene ring are stable and can only be removed by a strong oxidant. Alcoholic hydroxyl and phenolic hydroxyl groups attached on the propane side chains or the benzene rings also influence the characteristics of lignin. Of these, the phenolic hydroxyl groups are the main parameter for evaluating the degree of etherification and condensation, solubility and reaction activity of lignin.

The carbonyls, located on the side chains of lignin, can be divided into the conjugated and the unconjugated carbonyls according to the conjugate relations with the benzene ring. The carbon-carbon double bonds are often located on the side chains, such as cinnamyl aldehyde and cinnamyl alcohol structures, with only a low content, especially in hardwood lignin. The carbonyls are regarded as the key functional groups to determine whether the polymerization reaction of lignin can occur or not. In addition, some carboxylic acid groups exist because of the differences in extraction methods, especially for organic lignin.

1.4.1.2 Side chains and their linkages

The side chains of lignin are more complex than the aromatic ring skeleton structure, and have an important influence on product formation. As shown in Fig. 1.11, the side chains of lignin are basically divided into several structures: ethylene glycol (Fig. 1.11 (a)), glycerol (Fig. 1.11 (b)–(d)), coniferyl alcohols (Fig. 1.11 (e)) and coniferyl aldehyde (Fig. 1.11 (f)) on C_1 site. In addition, there are also alcoholic hydroxyl groups and ether structures on C_1 site, as well as the ester side-chain structures on C_1 or C_3 sites. Although lignin has only three basic structures, different functional groups are attached on their benzene rings, and the three structural units can be connected in a variety of random ways, leading to the complex structure of lignin.

There are many kinds of complex connections (Fig. 1.12) in lignin that can be divided into three classes: (1) ether bonds (comprising 60–70 %), mainly including those between a phenylpropane side chain and a benzene ring (β-O-4, α-O-4, γ-O-4), those between benzene rings (5-O-4), those between phenylpropane side chains (α-O-β′, α-O-γ′); (2) carbon-carbon bonds (accounting for 30–40 %), primarily containing 5-5, β-1, and β-5, as well as some β-6, α-6, α-β′, etc.; (3) ester bonds are not often observed and mainly exist in some herbaceous plants. Carbon-carbon bonds are the strongest and highly stable during thermal treatment; the strength of ether and ester bonds is much lower, especially ester bonds which are very likely to be cracked under basic conditions.

β-O-4 bonds are the most common linkages in lignin, with a proportion of 49–51 % in softwood spruce lignin and up to 65 % in hardwood beech [27]. β-O-4 bonds rup-

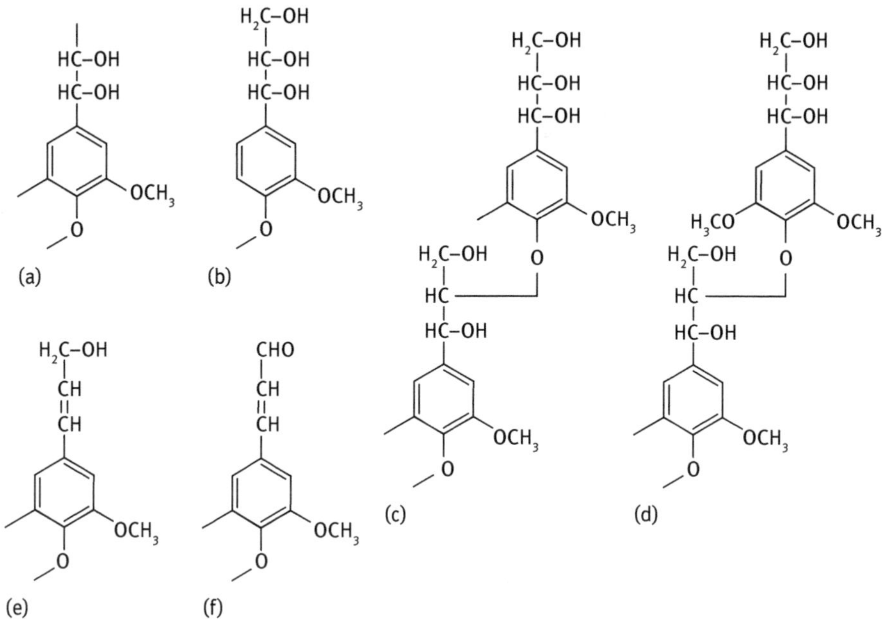

Fig. 1.11: Several side-chain structures of lignin.

ture first during the process of chemical treatment, resulting in the decomposition of lignin, hence playing an important role in the dissolution and degradation processes of lignin.

The various connection forms between the basic structures of lignin lead to its complex three-dimensional network structure without any regularity, so lignin is commonly expressed as a chemical structural model containing various basic structures and interunit linkages. But this structural model only describes a part of the representative molecules or an assumption of lignin.

1.4.2 Characteristics of lignin

Lignin is a white or almost colorless material. Some lignin shows color due to the separation and preparation processes. The relative density of lignin is 1,330–1,450 kg/m^3.

Lignin is a natural macromolecular polymer, whose molecular weight shows variation. The relative molecular weight of original lignin reaches from hundreds of thousands up to millions, but that of separated lignin is much less, generally varying from several thousand to tens of thousands with the highest number from 200,000 to 300,000. The probable reason for the large differences in molecular weight is ascribed to lignin separation, during which lignin undergoes degradation or denaturation in solution etc.

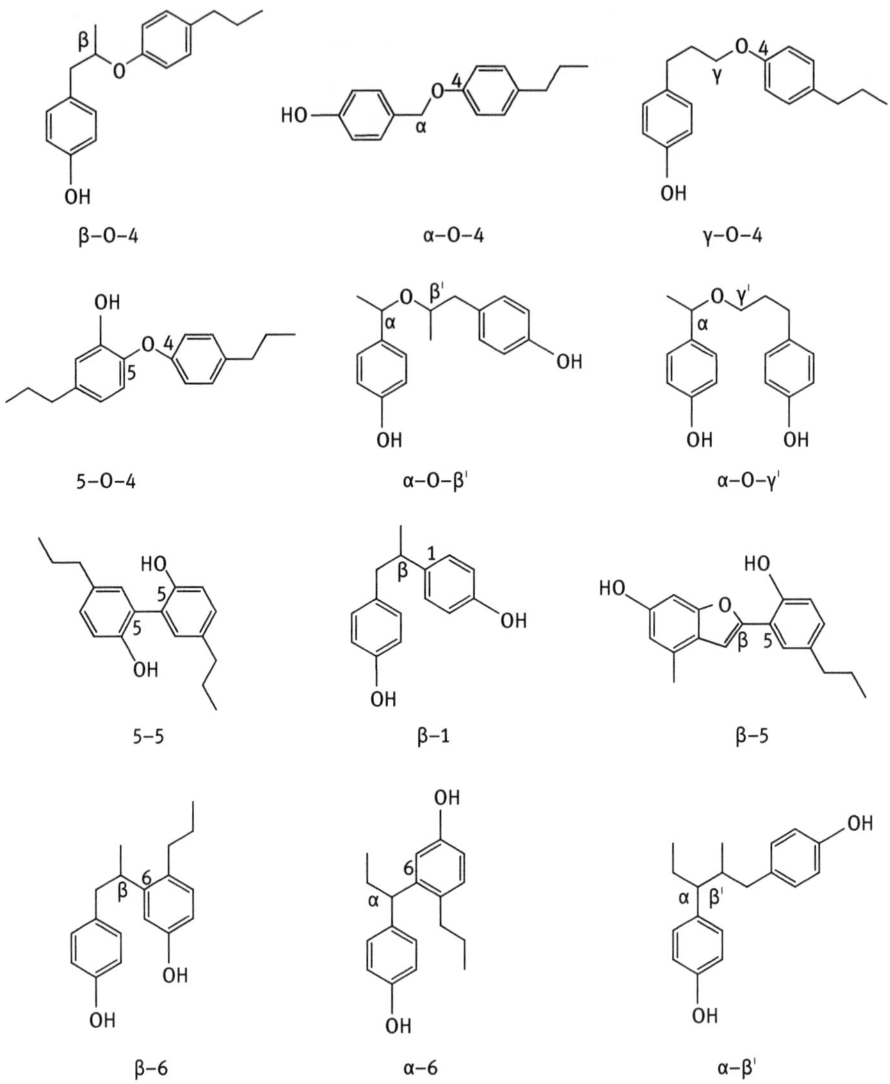

Fig. 1.12: Various connection forms in lignin structure.

There are many polar groups in lignin structures, as well as strong intramolecular and intermolecular hydrogen bonds, making it hard to dissolve it. However, for some separated lignin, its solubility is altered due to the changed physical properties of lignin caused by the condensation and degradation reactions during separation processes.

Lignin and majority separated lignin are amorphous thermoplastic polymers. They show slightly fragility at room temperature and have uncertain melting points and glass transition temperatures. Below glass transition temperature, lignin appears in glassy state; above this temperature, molecular chains begin to move and lignin intenerates resulting in its stickiness cohesion. Plant types, separation methods, and relative molecular weight influence the glass transition temperature. At the same time, the softening temperature will sharply decrease with an increase in water content, as water functions as a plasticizer in lignin. In addition, the softening temperature will also increase along with the molecular weight of lignin.

There are many active groups such as aryl, phenolic hydroxyl, alcoholic hydroxyl, carbonyl, methoxyl, carboxyl groups, and conjugated double bonds. As a result, many chemical reactions can occur, including oxidation, reduction, hydrolysis, alcoholysis, acidolysis, photolysis, acylation, sulfonation, alkylation, halogenation, nitration, polycondensation, or grafting copolymerization [54].

1.4.3 Isolation of lignin and its model compounds

It is difficult to obtain original lignin due to its complexity, and therefore, lignin extracted from biomass is often used as a substitute for original lignin. As listed in Tab. 1.4, there are many methods to separate lignin and they can be roughly divided into two major classes. One is removing the constituents other than lignin from the plant, then lignin will be filtered and reserved as the insoluble component. Another is taking lignin as a soluble component in the plant and dissolving it, while other

Tab. 1.4: Separation methods and characteristics of lignin.

<table>
<tr><th colspan="2">Separation method</th><th>Name of separated lignin</th><th>Degree of chemical change</th></tr>
<tr><td rowspan="3">Lignin was dissolved and then separated after precipitation and refinement (soluble lignin)</td><td>Organic solvent, precipitate out at neutral condition</td><td>Brauns lignin
Nard lignin
Milled wood lignin (MWL)
Cellulolytic enzyme lignin (CEL)</td><td>Very few chemical changes</td></tr>
<tr><td>Organic solvent, precipitate out at acid condition</td><td>Ethanol extracted lignin
Dioxane extracted lignin
Phenol extracted lignin
Water-soluble lignin</td><td>Coupled with chemical changes</td></tr>
<tr><td>Inorganic reagent</td><td>Sulfonate lignin
Alkali lignin
Sulfurated lignin
Chlorinated lignin</td><td>Coupled with chemical changes</td></tr>
<tr><td colspan="2" rowspan="2">Lignin was filtered as residue (insoluble lignin)</td><td>Sulfuric acid lignin (Klason)</td><td>Large chemical changes</td></tr>
<tr><td>Hydrochloric acid lignin
Cuprammonium lignin
Periodate lignin</td><td>Few chemical changes</td></tr>
</table>

components like cellulose are insoluble, and then it can be separated. There are few chemical changes in Brauns natural lignin and milled wood lignin and they can be regarded as the ideal representatives of original lignin. Hydrolysis lignin, alkali lignin, sulfonated lignin and organic solvent lignin are also often selected as the model compounds of lignin. In addition, black liquor from the pulping process also contains a lot of lignin which can be separated by filtration.

1.4.3.1 Milled Wood Lignin (MWL)

Lignin extracted using the Bjokman method, also called milled wood lignin, is a lignin model compound that to a large extent retains the chemical structural of lignin. The extraction process utilizes neutral solvent which will not cause a swelling effect under the condition of no added acid and no thermal treatment. First, the selected biomass raw material is ground and sieved to get particles at 40–60 mesh, degreased at room temperature, and then it is ground for more than 48 h to destroy the wood cell structure while suspended in a non-swelling solution, such as toluene. After that, the material is extracted several times using a mixed solution of dioxane and water, condensed and dried to get rough milled wood lignin, which is then dissolved in a mixed solution of 1,2-dichloroethane and ethanol (2 : 1). Next, diethyl ether is added to precipitate the sample, followed by washing and drying to get the final purified milled wood lignin.

The basic structural characterization of the milled wood lignin obtained by this process is an essential procedure to demonstrate its lignin representativeness. The FTIR spectra of milled wood lignins are shown in Fig. 1.13. The spectra of two kinds of milled wood lignin are apparently similar, but obvious differences can be found in the characteristic bands and the related intensity.

The spectrum of Mongolian Scots pine displays typical guaiacyl lignin features: three peaks with similar intensity at 1,600 cm^{-1}, 1,508 cm^{-1}, and 1,459 cm^{-1}, an enhanced peak between 1,270 cm^{-1} and 1,223 cm^{-1}, two strong peaks at 1,140 cm^{-1} and 1,031 cm^{-1}, two weak peaks at 856 cm^{-1} and 815 cm^{-1}, which are in accordance with the parameters for lignin classification by Faix [56]. In the spectrum of Manchurian ash, syringyl vibration appears and there also exists vibration for guaiacyl bonds. Compared with the spectrum of Mongolian Scots pine, that of Manchurian ash milled wood lignin displays relatively weaker absorption peaks at 1,508 cm^{-1}, 1,270 cm^{-1}, and 1,031 cm^{-1} with a new peak at 1,329 cm^{-1}. Furthermore, the peaks at 856 cm^{-1} and 815 cm^{-1} are combined to form a single peak at 834 cm^{-1}. In addition, a peak emerges at 1,125 cm^{-1}, attributing to guaiacyl and syringyl units appearing at the same time. However, the most apparent difference is in the peaks at 1,125 cm^{-1} and 1,329 cm^{-1}, corresponding to C–H deformation and C–O stretching vibration respectively, which can be used to distinguish softwood and hardwood lignin structures.

Quantitative analysis of methoxyl and phenolic hydroxyl groups in lignin can be conducted with ^{1}H-NMR spectra. Pyridine and acetic anhydride reagents are applied to acetylate lignin in order to dissolve lignin and enhance H signals, thus increasing the

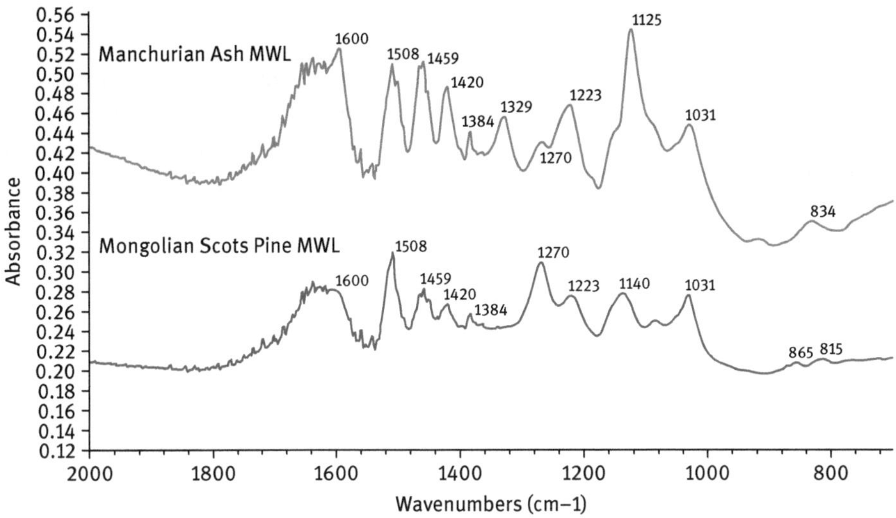

Fig. 1.13: FTIR spectra of Mongolian Scots pine and Manchurian ash MWL. Reprinted with permission from [55], © 2009 Elsevier.

accuracy of hydroxyl group analysis of lignin. The acetylated samples are dissolved in 0.6 ml of $CDCl_3$ and detected with DMX-500 apparatus, and tetramethylsilane (TMS) was used as an internal standard. The spectra are divided into several parts, according to the chemical shifts of lignin function groups and structural units. Therefore, the specific proton ratio is obtained from dividing the total integrated intensity by the regional integrated intensity. Based on ultimate analysis results, the H content is 3.96 % and 5.05 % in Mongolian Scots pine and Manchurian ash MWL, respectively. Furthermore, the proton is 3.96 and 5.05 per 100 g MWL from Mongolian Scots pine and Manchurian ash, respectively. The relative content of each chemical structural group is deduced from the proton based on the proton ratio in a specific area. As listed in Tab. 1.5, the aromatic ring comes from the guaiacyl and syringyl structural units in Manchurian ash MWL and the total proton is 0.70. Therefore, it can be calculated that guaiacyl units occupy 41.4 % and syringyl units account for 58.6 % of the aromatic rings. Typical hardwood features are observed in Manchurian ash MWL with the appearance of both guaiacyl and syringyl units. But guaiacyl units contribute to all the aromatic rings in Mongolian Scots pine MWL without syringyl units. That is, the methoxyl groups content of Manchurian ash MWL is far higher than that of Mongolian Scots pine MWL.

To understand the elemental composition and bond features of lignin macromolecules, the standard C_9 molecular formula could be obtained according to ultimate analysis and ^1H-NMR spectra. By calculation, the formula is $C_9H_{5.234}O_{2.935}(OCH_3)_{0.630}$ for Mongolian Scots pine MWL and $C_9H_{6.751}O_{3.007}(OCH_3)_{1.035}$ for Manchurian ash MWL. In detail, 1.620 aliphatic hydroxyls and 1.026 free phenolic hydroxyls exist in a C_9 unit

Tab. 1.5: ^{1}H-NMR spectra of acetylated MWL and proton per 100 g MWL.

Chemical shift Δ (ppm)	Assignment	Mongolian Scots pine MWL		Manchurian ash MWL	
		Peak area	Proton	Peak area	Proton
7.25–6.80	Aromatic ring proton on guaiacyl unit	0.55	0.58	0.18	0.29
6.80–6.25	Aromatic ring proton on syringyl unit	0.00	0.00	0.26	0.41
6.25–5.75	H_α in β-O-4 and β-1 structure	0.12	0.13	0.12	0.19
5.75–5.24	H_α in β-5	0.14	0.15	0.11	0.17
5.24–4.90	H in residual xylan	0.07	0.07	0.06	0.09
4.90–4.30	H_α and H_β in β-O-4	0.25	0.26	0.22	0.35
4.30–4.00	H in residual xylan and H_α in β-β structure	0.16	0.17	0.17	0.27
4.00–3.48	H on methoxy group	1.00	1.05	1.00	1.59
2.50–2.20	H in aromatic acetyl	0.54	0.57	0.49	0.78
2.20–1.60	H in fat acetate	0.85	0.90	0.56	0.89
<1.60	Highly blocked aliphatic protons	0.08	0.08	0.01	0.02
Sum		3.76	3.96	3.18	5.05

of Mongolian Scots pine MWL, while in Manchurian ash MWL, the quantities are 1.738 and 1.526, respectively. That is, more alcoholic and free phenolic hydroxyls exist in Manchurian ash MWL than in Mongolian Scots pine MWL.

1.4.3.2 Sulfuric acid lignin (Klason lignin)

Biomass was ground to powder and sieved in a 60–80 mesh, degreased in a Soxhlet extractor using benzene-ethanol mixture (2 : 1 in volume), then dried and added to 72 % sulfuric acid solution with intensive stirring. The solution was then boiled under a reflux condenser using distilled water and the filtered residue was washed in hot water repeatedly until acid-free. After drying, the obtained residue is called Klason lignin.

Tab. 1.6 lists the composition of Klason lignin, the sulfur content in Klason lignin is 0.22 % from Mongolian Scots pine and 0.35 % from Manchurian ash, but it is hardly found in the corresponding MWLs. Sulfur might be introduced by the use of sulfuric acid in the process of lignin extraction.

Tab. 1.6: Ultimate analysis of Klason lignin.

Sample	C (%)	H (%)	S (%)	N (%)	O (%)	O/C	H/C
Mongolian Scots pine Kalson lignin	58.28	5.65	0.22	0.40	35.45	0.46	1.16
Manchurian ash Kalson lignin	54.72	6.10	0.35	0.20	38.63	0.53	1.34

The characteristic peaks are basically the same between Klason lignin and MWL (Fig. 1.14), with some major peaks appearing. However, there are large differences in detailed intensity. Taking Manchurian ash as an example, most absorption peaks of Klason lignin are lower than those of MWL, and peaks at 1,270 cm^{-1} and 1,031 cm^{-1} are hardly observed. The intensity of characteristic peaks reflects the reaction activity of the structure to a certain extent. In this case, more changes are found in the chemical structure of Klason lignin than in MWL.

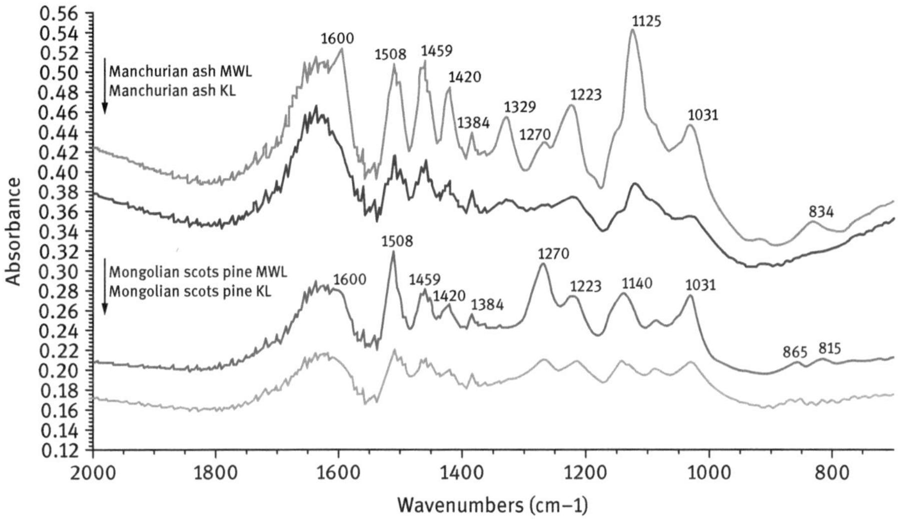

Fig. 1.14: FTIR spectra of Klason lignin. Reprinted with permission from [55], © 2009 Elsevier.

1.4.3.3 Sulfate lignin (Kraft lignin)

MWL and Klason lignin are the refined model compounds suitable for production in small quantities, but a large amount of industrial residues are produced in a pulping factory. The sulfate process, also called the Kraft process, is the most widely used process in pulping factories, having the advantages of a wide range of sources for raw materials and that chemical reagents can be re-utilized [57]. The black liquors can be divided into straw-based and wood-based black liquors. Lignin can be extracted from wood-based black liquor by the sulfate process, it can also be obtained from straw-based black liquor by alkaline pulping process. In the field of lignin extraction from black liquors, the emphasis is on how to achieve high-efficiency extraction.

1.4.3.4 Organic solvent lignin (organosolv lignin)

Organosolv lignin can be obtained by dissolving lignin in biomass with organic solvents such as ethanol and ketone at a certain temperature. Compared with Kraft lignin, the extraction of organosolv lignin is rather simple, and the structure contains more reactive groups without any sulfur impurities, hence it is more commonly used in macromolecular synthesis or other chemical modification. Ni et al. [58] studied the effects of different ethanol concentrations on lignin solubility. They found that the ethanol with a concentration of 70 % was most suitable for lignin extraction, and ascribed this result to the similar polarity of solvent and lignin. There are large differences between organosolv lignins extracted by using different organic solvents. The acetone-extracted lignin contains lower ash [59], while ethanol-extracted lignin results in rich carboxyl groups [60] and acetic acid-extracted lignin leads to a higher carbonyl content [61].

1.5 Extractives

Biomass also contains a small amount of extractives in addition to the three major components. They are a group of low molecular compounds that do not constitute cell walls or cell layers and can be extracted by polar or nonpolar organic solvents, water and water vapor. Extractives belong to the nonstructural components, including waxes, fats, resins, tannins, sugars, starches, pigments, etc. The extractives content and composition vary a lot depending on biomass categories, even different parts of the same biomass can influence its composition. On the whole, the components can be divided into three subfamilies: aliphatic compounds (mainly fats and waxes), terpenes and terpenoids, and phenolic compounds [62].

Generally, extractives are soluble in many solvents. Extractives in wood can be obtained by using a mixture of ethanol and benzene. Extractives are not structural materials in biomass, and their types, structures, and amounts depend on the biomass categories and extracts as well as the extraction time and methods. Most herbaceous biomass contains relatively more extractives than woody biomass [63]. The quality and quantity of extractives in wood are different for each tree species, even in the different parts in the same tree. Usually, extractives content increases gradually in the radial direction from sapwood inwards [15]. In addition, the corresponding extractive components are also different. Sugars, starches, fats are present in sapwood, while phenols concentrate in heartwood and here the molecular weight of polyphenol is higher in the medial part compared with the lateral part [15].

1.6 Inorganic salts

The principal inorganic salts in plants mainly exist in the ash of biomass, which has a different content and composition for various biomass. Low inorganic salt content is found in wood, but it is higher in agricultural biomass. Especially rice husk contains relatively more ash. In addition, inorganic salts content in different parts of the same plant are not the same. Generally, the content in sapwood is higher than that in heartwood, in branches it is higher than in bole, and in bark it is much higher than in the trunk.

1.6.1 Composition of inorganic salts

Inorganic elements in biomass are mainly potassium, calcium, sodium, magnesium, silicon, phosphorus, sulfur, chlorine, etc., and also include traces of aluminum, titanium, vanadium, manganese, iron, cobalt, nickel, copper, zinc, molybdenum, silver, barium, lead, etc. The quantities of various inorganic elements fluctuate within a certain range, but no obvious rules are found. Studies have shown that ash in the bark of wood biomass is mainly calcium oxide, comprising more than 70 %, followed by potassium oxide, magnesium oxide and sodium oxide [64]. The inorganic elements in biomass can be divided into detrital, authigenic and technogenic ones [65]. Detrital inorganic elements exist stably in biomass with their relatively higher melting points; authigenic inorganic elements are not stable, and the decomposition and melting temperatures are low during the process of biomass utilization; technogenic inorganic elements include many minerals with unique features that are highly different from natural materials. In biomass, authigenic inorganic elements are too closely combined with organic components to be separated, while for detrital minerals it is somewhat easier. Commonly, authigenic inorganic elements are the major factor influencing biomass thermochemical conversion. Inorganic salts influence the thermochemical conversion of biomass in different catalytic ways. In order to study the influence mechanisms of inorganic salts on biomass thermochemical conversion, it is necessary to be aware of the forms and correlated characteristics of inorganic elements present in raw materials.

In biomass raw materials, potassium is present in the form of water soluble salts, partly absorbed on carboxyl groups and other functional groups as ions. In almost all biomass materials, 90 % of the potassium is water soluble salt or exchangeable ions resulting in high mobility, and it tends to enter into the volatile phase during the pyrolysis process. Potassium release into the gaseous phase can be divided into two stages: when the temperature is less than 500 °C, organic potassium in biomass is released due to its thermal instability; when the temperature is over 500 °C, inorganic potassium is volatilized into gaseous phase due to an increase in steam pressure [66]. Potassium is generally considered to be released in the form of KCl if biomass contains chlorine. In the absence of chlorine, alkali metal hydroxide is the most stable gas compound (under oxidation atmosphere). According to the results of on-line mass spectrometry detection in biomass pyrolysis, potassium is most likely to be released in the forms of KCl and KOH [67]. In addition, there is still some potassium ion steam reacting with silicate substances to form insoluble residues during the pyrolysis process.

There is much less sodium than potassium, and it is generally not regarded as a necessary element for plant growth. But for some specific crops, sodium in low concentration plays an active role in place of potassium to a certain extent. How sodium exists in biomass and behaves during pyrolysis is very similar to potassium. Sodium has a similar behavior to potassium during the pyrolysis process. Furthermore, some insoluble sodium can be detected in raw biomass, ascribed to external impurities. For example, soil, ash and other sodium-containing impurities will be mixed in during the collection and treatment of biomass; clay is added as filler to the glossy printing surface of non-recyclable paper, in which the sodium is very stable, and generally does not participate in the pyrolysis process.

Calcium is an organic part of plant cell walls and other cell structures, and its main function is to strengthen cell wall hardness and the integral structure of the plant. Biomass, especially some fast-growing trees such as wicker or straw, contains a lot of calcium. Calcium in biomass raw material basically exists in the ion exchangeable and acid soluble substances. Unlike alkali metals tending to enter into the gaseous phase, almost all calcium has high stability and is not easily volatilized in the pyrolysis process, it is more likely to remain in the solid residues after pyrolysis. Experimental results show that all calcium in biomass feedstock can be found in solid products [68].

A small quantity of the element ferrum exists in biomass, mainly in acid soluble substances and some in ion exchange substances. Ferrum has two important functions in plants: one is to form coordination compounds with the plant organism; another depends on its activity during transmission and reversible redox reactions, and can be converted from Fe^{2+} to Fe^{3+}. Ferrum in plants is mainly concentrated in leaves, and 80 % of ferrum in plants appears in chlorophyll, playing an important role in photosynthesis.

Chlorine is another important element in biomass, which mainly exists in the form of chlorine ions, and plays a role in material balance during plant growth. The chlorine content in plants is more easily affected by soil conditions. Chlorine serves the transmission of inorganic matter in the biomass pyrolysis process, especially alkali metals, which react with chlorine to form highly volatile alkali metal chloride. Chlorine is an element of high volatility and is almost completely transformed into gas phase during the pyrolysis process. According to the chemical equilibrium, chlorine will react with potassium and sodium to form the stable and easy volatilized alkali metal chlorides. Over 600 °C, chlorine can be released in the form of alkali metal chloride [69]. In addition to the combination with alkali metals, hydrogen chloride is also an important way for chlorine to be released.

Tab. 1.7: Metallic ion content in rice husk and white pine after acid pickling.

Sample	Ca^{2+}	Mg^{2+}	K^+	Fe^{3+}	Na^+
Rice husk	1896.35	1791.6	10190.78	627	156
7 % HCl	439.42	235.61	1265.26	96.5	35.6
7 % H_3PO_4	286.57	361.36	2267.95	254.3	37.5
7 % H_2SO_4	367.56	296.31	2069.76	125.6	22.6
White pine	3480.9	296.64	984.97	762.43	295.88
7 % HCl	1090.6	65.58	116.79	139.65	126.78
7 % H_3PO_4	1121.63	117.95	120.98	135.68	109.76
7 % H_2SO_4	1132.64	132.96	132.35	126.43	106.79

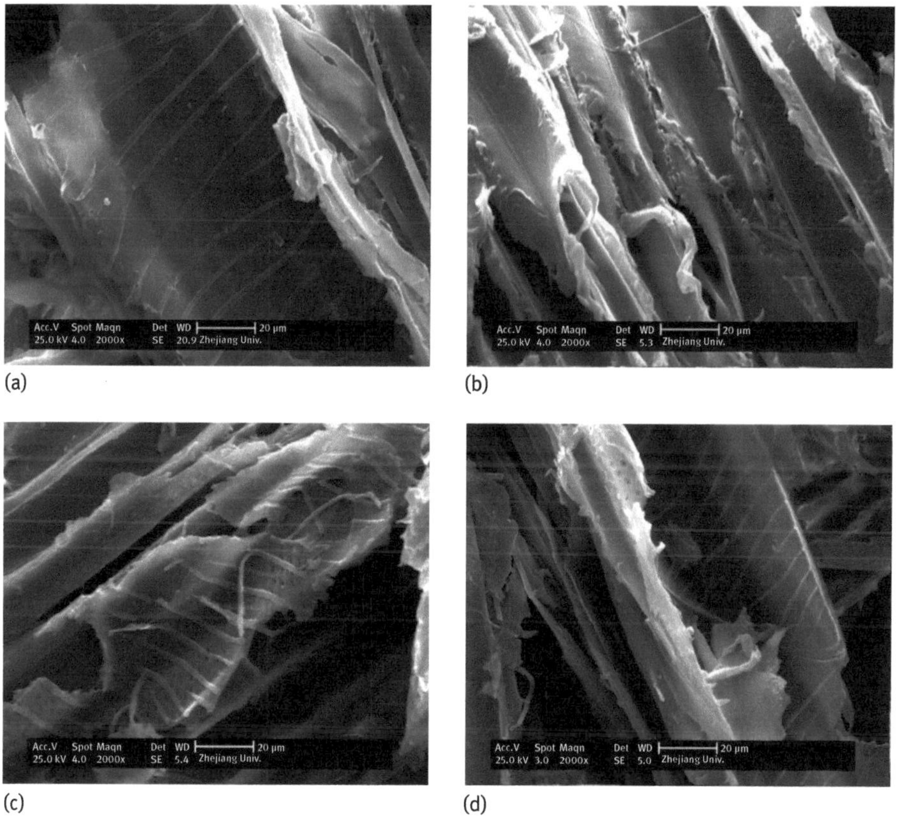

Fig. 1.15: SEM pictures of biomass before and after acid washing; (a) original sample, (b) hydrochloric acid washing, (c) sulfuric acid washing, (d) phosphoric acid washing.

1.6.2 Removal of inorganic salts

Inorganic salts can be removed to some extent by acid pickling or water washing, but the latter shows a poor effect, so acid pickling pretreatment is one of the most common methods to remove inorganic salts in biomass.

As listed in Tab. 1.7, ICP-AES analysis showed that the main metallic ions decreased obviously during the acid pickling process of rice husk. Different kinds of acids have a certain selectivity for removing metallic ions: hydrochloric acid is the best one for potassium ion removal, while phosphoric acid and sulfuric acid are suitable for the removal of calcium and sodium ions, respectively. Compared with rice husk, the inorganic metallic ions of white pine apparently decrease after acid picking but not so obviously, indicating that the small amount of inorganic salts in wood are not easy to remove.

Acid pickling can largely influence the structure and degree of polymerization. As shown in Fig. 1.15, sulfuric acid has the most intensive influence on structural change, leading to the destruction of some fiber structures and then some pores, which can be observed after the treatment. Although hydrochloric and phosphoric acid cause some decomposition of biomass, they have less influence than sulfuric acid on the biomass structure; after phosphoric acid treatment in particular, the samples almost retain their original structure. That means that acid pickling not only removes the ash in biomass, but also affects the structure and physical characteristics of biomass. In particular, partial hemicellulose tends to decompose in acid solution, leading to an increase in the relative content of cellulose and lignin [70].

1.7 Water in biomass

Water is presented in biomass in two types: (1) free water, existing in biomass cell cavities and intercellular spaces and influencing the density, combustion characteristics and permeability of biomass; (2) bound water, existing in cell walls and comprising about 30 % of dry biomass mass, largely affects the physical-mechanical properties of wood. The water content in biomass varies with the type of raw materials. Even for one tree, the water content is not same everywhere due to different growing seasons as well as different parts. When the water content is below 30 %, wood will tend to dry shrink due to water loss and wet expand due to water absorption. In general, on the basis of their water content, biomass raw materials are divided into absolute dry wood (after drying, having constant weight and no water), green wood (average water content 50–100 % of absolutely dry weight), moist wood (after water or wet transport, commonly water content greater than 100 %), air-dried wood (air drying, water content is 12–18 %), kiln-dried wood (normal artificial drying, water content is 7–15 %).

Biomass raw materials have some hygroscopicity, so their water content varies over a large range, leading to changes of volume and density of biomass. Zhou et al.

[71] investigated the bulk density change of sawdust (mixture of Douglas fir and lauan dust), and found that bulk density decreased with increasing water when the water content of sawdust was below 33.4 %, but increased along with increasing water when the water content was between 33.4–95.0 %. They consider that wood fiber has a saturated water content of 30 %, and wood would dry shrink once the water content dropped below this value.

References

[1] Hell R. Molecular physiology of plant sulfur metabolism. Planta. 1997;202(2):138–148.
[2] Sun R. Cereal straw as a resource for sustainable biomaterials and biofuels: chemistry, extractives, lignins, hemicelluloses and cellulose. Access Online via Elsevier. 2010.
[3] Heredia A, Jiménez A, Guillén R. Composition of plant cell walls. Zeitschrift für Lebensmittel-Untersuchung und Forschung. 1995;200(1):24–31.
[4] Chen JC, Liu WX, Yang GH, et al. The resource chemical of papermaking plants. Science Press, Beijing; 2012.
[5] Huber GW, Iborra S, Corma A. Synthesis of transportation fuels from biomass: chemistry, catalysts, and engineering. Chemical Reviews. 2006;106(9):4044–4098.
[6] Sharma HSS. Compositional analysis of neutral detergent, acid detergent, lignin and humus fractions of mushroom compost. Thermochimica Acta. 1996;285(2):211–220.
[7] Donaldson L, Hague J, Snell R. Lignin distribution in coppice poplar, linseed and wheat straw. Holzforschung. 2001;55(4):379–385.
[8] Kebelmann K, Hornung A, Karsten U, et al. Intermediate pyrolysis and product identification by TGA and Py-GC/MS of green microalgae and their extracted protein and lipid components. Biomass and Bioenergy. 2013;49:38–48.
[9] Miao X, Wu Q, Yang C. Fast pyrolysis of microalgae to produce renewable fuels. Journal of Analytical & Applied Pyrolysis. 2004;71(2):855–863.
[10] Hamelinck CN, Hooijdonk GV, Faaij AP. Ethanol from lignocellulosic biomass: techno-economic performance in short-, middle-and long-term. Biomass and Bioenergy. 2005;28(4):384–410.
[11] Kim P, Johnson A, Edmunds CW, et al. Surface functionality and carbon structures in lignocellulosic-derived biochars produced by fast pyrolysis. Energy & Fuels. 2011;25(10):4693–4703.
[12] Demirbaş A. Thermochemical conversion of biomass to liquid products in the aqueous medium. Energy Sources. 2005;27(13):1235–1243.
[13] Raveendran K, Ganesh A, Khilar KC. Influence of mineral matter on biomass pyrolysis characteristics. Fuel. 1995;74(12):1812–1822.
[14] Watanabe A, Katoh K, Kimura M. Effect of rice straw application on CH_4 emission from paddy fields. II. Contribution of organic-constituents in rice straw. Soil Science & Plant Nutrition. 1993;39(4):707–712.
[15] Junzo Nakano, Takayoshi H, Masashi S, et al. Wood chemistry. Chinese Forestry Publishing House, Beijing; 1989.
[16] Song J, Hou YF. The properties and utilizations of microcrystalline cellulose. Journal of Cellulose Science and Technology. 1995;3(3):1–10.
[17] Jing ZY. Carbohydrate chemistry. Chemical Industry Press, Beijing; 2008.
[18] Pastorova I, Arisz PW, Boon JJ. Preservation of d-glucose-oligosaccharides in cellulose chars. Carbohydrate Research. 1993;248(93):151–165.

[19] Northolt MG, Boerstoel H, Maatman H, et al. The structure and properties of cellulose fibres spun from an anisotropic phosphoric acid solution. Polymer. 2001;42(19):8249–8264.
[20] Zhbankov RG, Firsov SP, et al. Structural physico-chemistry of cellulose macromolecules. Vibrational spectra and structure of cellulose. Journal of Molecular Structure. 2002;614(1):117–125.
[21] Orozco A, Ahmad M, Rooney D, et al. Dilute acid hydrolysis of cellulose and cellulosic biowaste using a microwave reactor system. Process Safety and Environmental Protection. 2007;85(5):446–449.
[22] Su MY. The structure and research progress of cellulose crystal variant. Guangdong Chemical Fiber Technology Communication. 1980(01):26–39.
[23] Gao J, Tang LG. Journal of cellulose science. Science Press, Beijing; 1996.
[24] Pérez S, Samain D. Structure and engineering of celluloses. Advances in Carbohydrate Chemistry & Biochemistry. 2010;64:25–116.
[25] Atalla RH, Vanderhart DL. Native cellulose: a composite of two distinct crystalline forms. Science. 1984;223(4633):283–285.
[26] Krishnamachari P, Hashaikeh R, Tiner M. Modified cellulose morphologies and its composites; SEM and TEM analysis. Micron. 2011;42(8):751–761.
[27] Yang SH. Lignocellulosic chemistry. China Light Industry Press, Beijing; 2001.
[28] Hebeish A, Guthrie JT. The chemistry and technology of cellulosic copolymers. Springer Berlin, Heidelberg; 1981.
[29] Klemm DO. Regiocontrol in cellulose chemistry: principles and examples of etherification and esterification//ACS Symposium Series. ACS Publications, Washington, DC; 1998. pp. 19–37.
[30] Chen HZ. Biotechnology of lignocellulose. Chemical Industry Press, Beijing; 2005.
[31] Rinaldi R, Schüth F. Acid hydrolysis of cellulose as the entry point into biorefinery schemes. Chemsuschem. 2009;2(12):1096–1107.
[32] Fang JM, Sun RC, Tomkinson J. Isolation and characterization of hemicelluloses and cellulose from rye straw by alkaline peroxide extraction. Cellulose. 2000;7(1):87–107.
[33] Sun JX, Sun XF, Zhao H, et al. Isolation and characterization of cellulose from sugarcane bagasse. Polymer Degradation and Stability. 2004;84(2):331–339.
[34] Sun XF, Sun RC, Fowler P, et al. Isolation and characterisation of cellulose obtained by a two-stage treatment with organosolv and cyanamide activated hydrogen peroxide from wheat straw. Carbohydrate Polymers. 2004;55(4):379–391.
[35] Jiang M, Zhao M, Zhou Z, et al. Isolation of cellulose with ionic liquid from steam exploded rice straw. Industrial Crops and Products. 2011;33(3):734–738.
[36] Bendahou A, Dufresne A, Kaddami H, et al. Isolation and structural characterization of hemicelluloses from palm of Phoenix dactylifera L. Carbohydrate Polymers. 2007;68(3):601–608.
[37] Alén R. Structure and chemical composition of wood. Forest Products Chemistry. 2000:35.
[38] Pereira H, Graca J, Rodrigues JC. Wood chemistry in relation to quality. Cheminform. 2004;35(46):53–86.
[39] Ebringerova A, Hromadkova Z, Heinze T. Hemicellulose. In: Heinze T, editor. Polysaccharides I. Springer, Berlin; 2005. pp. 1–67.
[40] Timell TE. Wood hemicelluloses: Part II. Advances in Carbohydrate Chemistry. 1965;20:409–483.
[41] de Vries RP, Visser J. Aspergillus enzymes involved in degradation of plant cell wall polysaccharides. Microbiology and Molecular Biology Reviews. 2001;65(4):497–522.
[42] Carpita NC, Gibeaut DM. Structural models of primary cell walls in flowering plants: consistency of molecular structure with the physical properties of the walls during growth. The Plant Journal. 1993;3(1):1–30.
[43] Shibuya N, Iwasaki T. Structural features of rice bran hemicellulose. Phytochemistry. 1985;24(2):285–289.

[44] Wende G, Fry SC. O-feruloylated, O-acetylated oligosaccharides as side-chains of grass xylans. Phytochemistry. 1997;44(6):1011.
[45] Thammasouk K, Tandjo D, Penner MH. Influence of extractives on the analysis of herbaceous biomass. Journal of Agricultural and Food Chemistry. 1997;45(2):437–443.
[46] Gírio FM, Fonseca C, Carvalheiro F, et al. Hemicelluloses for fuel ethanol: A review. Bioresource Technology. 2010;101(13):4775–4800.
[47] Wang H, Li LT, Shi B. The research on composition and structure of corncob xylan and birch wood glycan. Food Science. 2004;25(z1):36–42.
[48] Kormelink F, Voragen A. Degradation of different glucurono arabino xylans by a combination of purified xylan-degrading enzymes. Applied Microbiology and Biotechnology. 1993;38(5):688–695.
[49] Gruppen H, Hamer RJ, Voragen AGJ. Water-unextractable cell wall material from wheat flour. 2. Fractionation of alkali-extracted polymers and comparison with water-extractable arabinoxylans. Journal of Cereal Science. 1992;16(1):53–67.
[50] Doner LW, Hicks KB. Isolation of hemicellulose from corn fiber by alkaline hydrogen peroxide extraction. Cereal Chemistry. 1997;74(2):176–181.
[51] Saha BC. Hemicellulose bioconversion. Journal of Industrial Microbiology and Biotechnology. 2003;30(5):279–291.
[52] Faravelli T, Frassoldati A, Migliavacca G, et al. Detailed kinetic modeling of the thermal degradation of lignins. Biomass and Bioenergy. 2010;34(3):290–301.
[53] Butler E, Devlin G, Meier D, et al. Characterisation of spruce, salix, miscanthus and wheat straw for pyrolysis applications. Bioresource Technology. 2013;131(3):202–209.
[54] Sheu DD, Chiu CH. Evaluation of cellulose extraction procedures for stable carbon isotope measurement in tree ring research. International Journal of Environmental Analytical Chemistry. 1995;59(1):59–67.
[55] Wang S, Wang K, Liu Q, et al. Comparison of the pyrolysis behavior of lignins from different tree species. Biotechnology Advances. 2009;27(5):562-567.
[56] Faix O, Jakab E, Till F, et al. Study on low mass thermal degradation products of milled wood lignins by thermogravimetry-mass-spectrometry. Wood Science and Technology. 1988;22(4):323–334.
[57] Ohra-aho T, Tenkanen M, Tamminen T. Direct analysis of lignin and lignin-like components from softwood kraft pulp by Py-GC/MS techniques. Journal of Analytical and Applied Pyrolysis. 2005;74(1):123–128.
[58] Ni Y, Hu QJ. Alcell® lignin solubility in ethanol–water mixtures. Journal of Applied Polymer Science. 1995;57(12):1441–1446.
[59] Evtuguin DV, Andreolety JP, Gandini A. Polyurethanes based on oxygen-organosolv lignin. European Polymer Journal. 1998;34(8):1163–1169.
[60] Hage RE, Brosse N, Chrusciel L, et al. Characterization of milled wood lignin and ethanol organosolv lignin from miscanthus. Polymer Degradation & Stability. 2009;94(10):1632–1638.
[61] Barros AM, Dhanabalan A, Constantino CJL, et al. Langmuir monolayers of lignins obtained with different isolation methods. Thin Solid Films. 1999;354(1):215–221.
[62] Sjostrom E. Wood Chemistry. China Forestry Publishing, Beijing; 1985.
[63] Sluiter JB, Ruiz RO, Scarlata CJ, et al. Compositional analysis of lignocellulosic feedstocks. 1. Review and description of methods. J Agric Food Chem. 2010;58(16):9043–9053.
[64] Vassilev SV, Baxter D, Andersen LK, et al. An overview of the chemical composition of biomass. Fuel. 2010;89(5):913–933.
[65] Vassilev SV, Baxter D, Andersen LK, et al. An overview of the organic and inorganic phase composition of biomass. Fuel. 2012;94:1–33.

[66] Yu CJ, Luo ZY, Zhang WN, et al. Inorganic material emission during biomass pyrolysis. Journal of Fuel Chemistry & Technology. 2000;(5):420–425.

[67] Bryers RW. Fireside slagging, fouling, and high-temperature corrosion of heat-transfer surface due to impurities in steam-raising fuels. Progress in Energy and Combustion Science. 1996;22(1):29–120.

[68] Thy P, Jenkins BM, Grundvig S, et al. High temperature elemental losses and mineralogical changes in common biomass ashes. Fuel. 2006;85(5):783–795.

[69] Olsson JG, Jaglid U, Pettersson J, et al. Alkali metal emission during pyrolysis of biomass. Energy & Fuels. 1997;11(4):779–784.

[70] Wang XH, Chen HP, Wang J, et al. Influences of mineral matters on biomass pyrolysis characteristics. Journal of Fuel Chemistry and Technology. 2008;36(6):679–683.

[71] Zhou HD, Xu CY, Ding HM. Studies on the basic characteristics of wooden bulk materials. Wood Processing Machinery. 2002;6(13):9–12, 19.

2 Pyrolysis of cellulose

Biomass pyrolysis, i.e. breaking chemical bonds in biomass macromolecules using heat energy under inert atmosphere, can convert biomass into low-polymerization products or even small molecular compounds through a series of complex reactions, such as depolymerization, ring-opening, and cleavage. Biomass pyrolysis products can be classified into gas, tar and char according to their phases at room temperature. The chemical reactions are extremely complicated during biomass pyrolysis owing to the diverse distribution of biomass components. Therefore, based on biomass component analysis, the whole of biomass pyrolysis behavior can be substantially understood by studying the individual pyrolysis behavior of the three main components and their corresponding interaction influence on biomass pyrolysis. As the main component of lignocellulosic biomass, cellulose and its model compounds usually receive more attention in biomass pyrolysis research due to their accessibility in nature. Furthermore, compared with hemicellulose and lignin, the structural and chemical properties of cellulose, isolated from different biomass under comparable conditions, show good repeatability as a result of its simple structure. Thus, cellulose pyrolysis studies conducted by different researchers can present quite valuable reference points, achieving widespread attention and abundant research results.

2.1 Fundamental process of cellulose pyrolysis

2.1.1 Introduction to cellulose pyrolysis

Studies of the mechanism of cellulose pyrolysis mainly focus on two aspects: (1) cellulose pyrolysis kinetics, which has been broadly concerned with the Broido–Shafizadeh model that involves the formation of tar, gas and char during cellulose pyrolysis [1]. According to this model, when cellulose is heated, a series of complex initial reactions involving dehydration, condensation and depolymerization take place, leading to the formation of active cellulose, an active intermediate product with relatively simple components and low degree of polymerization, followed by further competing reactions to form final products, mainly gas, bio-oil and char. In view of the possible applications, the purpose of kinetic studies is to obtain a relatively simple model to illuminate the thermogravimetric behavior of cellulose. (2) Chemical reactions in pyrolysis, especially the formation mechanism of the main products. Recently, the production of high value-added chemicals from cellulose by pyrolysis has become more and more popular in biomass utilization. As the main products of cellulose pyrolysis, studies on the formation mechanisms and evolution behaviors of levoglucosan, hydroxyacetaldehyde, hydroxyacetone, 5-hydroxymethylfurfural, furfural and other compounds have become hot topics in this research field. From a theoretical view, in-

DOI 10.1515/9783110369632-004

depth studies of cellulose pyrolysis reactions are of great significance for optimizing reaction conditions, which can maximize target products yields and improve product quality.

To qualitatively and quantitatively describe the pyrolysis behavior of biomass and its components usually requires the aid of analytical instruments or mechanism experimental apparatus with adjustable conditions. Among them, thermal gravimetric analysis is the technique to obtain the relationship between the sample weight change and reaction temperature on a thermal balance, which is usually performed at a constant heating rate under temperature programmed control. As an important method for thermal kinetic studies, thermal gravimetric analysis has the advantage of small dosage of test samples and short running time. Additionally, it is also convenient to study the reactions of the samples heated in the required temperature range [2]. Using thermal gravimetric (TG) analysis, the TG curve under a certain heating rate can be obtained, which records the relationship between the sample weight and its corresponding temperature during the heating process. Derivative thermal gravimetric (DTG) curves can be further obtained by first order differentiation of the TG curve. The DTG curve reflects the relationship between the weight loss rate of the sample and temperature. TG and DTG curves are essentially equivalent, except for the different data processing method. The TG curve can clearly show the weight loss over the whole pyrolysis process and the final amount of solid residue, and the DTG curve can identify the temperature corresponding to the maximum reaction rate and distinguish the different stages in the pyrolysis process. The peak height of the DTG curve is directly equal to the reaction rate at the corresponding temperature. Fig. 2.1 shows the typical characteristics of cellulose thermal decomposition. Weight loss is mainly comprised of several stages including physical water desorption, glass transition, main decomposition, char formation etc. Dehydration is the first major reaction occurring in cellulose pyrolysis. The anhydrocellulose formed by dehydration is the important intermediate for char formation, which is largely influenced by reaction parameters. In the main weight loss stage, cellulose first decomposes into oligosaccharide and then into small molecular gases and macromolecular condensable volatiles. And finally, cellulose pyrolysis enters into a stable char formation stage, which occurs at a relatively slow rate.

Fourier transform infrared spectroscopy (FTIR) is commonly used to identify chemical compounds. When a sample is irradiated by a beam of infrared light, the molecules of irradiated material will absorb a part of the light energy by transferring to vibrational and rotational energy of the functional groups in molecules, which elevates the excited molecules from the inherent energy level to some higher energy level. As a result, the spectrum appears as absorption bands. By performing thermal gravimetric analysis coupled with Fourier transform infrared spectroscopy (TG-FTIR), the thermal weight loss behavior of biomass can be understood. More importantly, the release characteristics of several important volatiles throughout the whole pyrolysis process can be obtained. According to the spectra from TG-FTIR analyses, the distribution of functional groups in cellulose pyrolysis products can

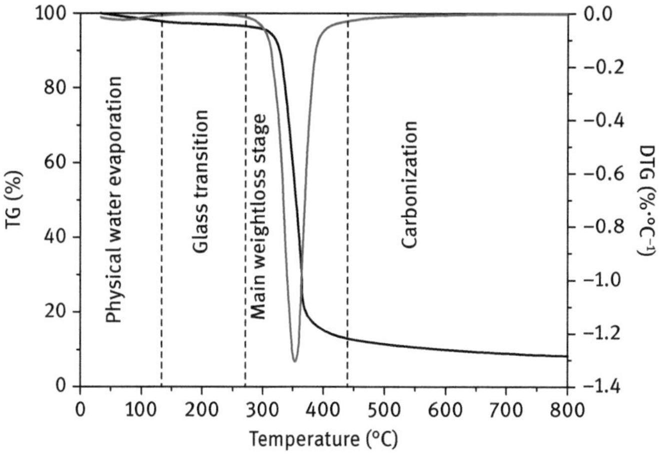

Fig. 2.1: TG/DTG curves of cellulose pyrolysis.

be determined. Subsequently, the evolution of the pyrolysis products can also be deduced. In the main reaction stage of cellulose pyrolysis, as shown in Fig. 2.2, the absorptions at 2,180 cm^{-1} and 2,110 cm^{-1} indicate the formation of CO. The bands at 2,395–2,235 cm^{-1} and 720–570 cm^{-1} can be ascribed to CO_2. The relatively intense bands at 3,100–2,650 cm^{-1}, 1,850–1,640 cm^{-1}, and 1,600–800 cm^{-1} are assigned to C–H stretching vibrations, C=O stretching vibrations and C–H bending vibration in plane, C–O and C–C skeleton vibrations, respectively. They correspond to various compounds including alkanes, aldehydes, ketones, carboxylic acids and alcohols. Similar results are also confirmed in other references. Maciel et al. [3] observed that the main stage of cellulose pyrolysis was in the temperature range of 270 to 420 °C

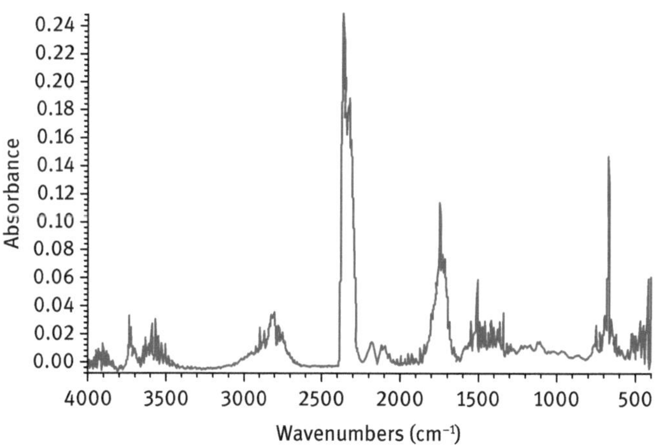

Fig. 2.2: FTIR spectrum of cellulose pyrolysis products released in the main reaction stage.

with the maximum weight loss rate at 335 °C. In the corresponding FTIR spectra, a band at 4,000–3,600 cm^{-1} (O–H stretching vibrations) was very obvious, which was ascribed to the release of water. The absorptions at 2,350 cm^{-1} and 2,167 cm^{-1} showed the presence of CO_2 and CO, respectively. Moreover, the signals occurring at 1,700–1,500 cm^{-1} (C=O vibrations), 1,470–1,430 cm^{-1} (C–H deformation in methoxyl group) and 1,170 cm^{-1} (C–O–C stretching vibrations) presented obviously. Biagini et al. [4] found that the maximum weight loss rate of cellulose pyrolysis occurred at 335 °C, and identified several pyrolysis products such as H_2O, CO, CO_2, hydrocarbon, and compounds containing C=O from the FTIR spectra.

Considering that only a part of the small molecular gases and volatile compounds with characteristic functional groups can be identified by the FTIR technique, gas chromatography (GC) coupled with mass spectrometry (GC/MS) or high performance liquid chromatography (HPLC) are favorably used to analyze the large molecular products from pyrolysis. In GC/MS, GC serves as a separation device and MS is the detector for the compounds from GC. GC analysis has the advantages of high separation performance, sensitivity and fast analysis, hence it is widely used for complex mixture analyses. However, GC analysis is mainly based on the comparison of retention time of the standard compounds, making it difficult for the qualitative identification of an unknown complex mixture. So GC is incapable of analyzing a bio-oil that contains hundreds of compounds. Conversely, MS analysis has a strong ability to identify a pure compound, however, it has no strong separation capability and then cannot be directly used for analyzing complex mixtures. Consequently, GC/MS, combining the complementary advantages of GC and MS, is very suitable for efficient analysis of complex mixtures, such as bio-oil.

Pyrolysis-gas chromatography/mass spectrometry analysis (Py-GC/MS) is an effective fast analytical method for biomass pyrolysis. Compared with the TG-FTIR method, which usually has a low heating rate, Py-GC/MS can be performed at a very high heating rate, which means that it can simulate the practical process of biomass fast pyrolysis. Therefore, identifying the volatile species and determining their contents by Py-GC/MS will be beneficial for understanding the mechanism of cellulose pyrolysis in depth. Fig. 2.3 shows the chromatogram of cellulose pyrolysis products at typical reaction condition by Py-GC/MS, in which each peak corresponds to one compound. Currently, hundreds of condensable volatile compounds from cellulose pyrolysis have been detected, including a variety of oxygenated compounds like acids, aldehydes, ketones, esters, ethers and phenols. According to the structural difference, the oxygenated compounds with relatively high content can be divided into three categories, namely linear small molecular compounds (hydroxyacetone, hydroxyacetaldehyde and acetaldehyde etc.), furans (5-hydroxymethylfurfural and furfural etc.), and pyrans (levoglucosan, anhydro-altrose levoglucosenone and anhydro-pyranose).

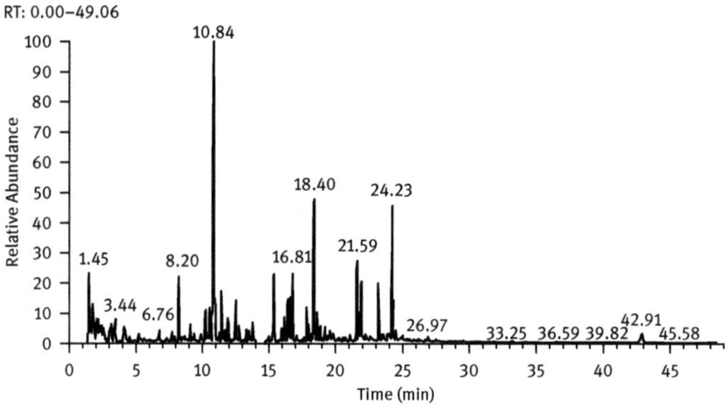

Fig. 2.3: Chromatogram of cellulose pyrolysis products by Py-GC/MS.

Levoglucosenone, levoglucosan, dianhydro-D-glucopyranose, furfural, 5-hydroxymethylfurfural, hydroxyacetone, acetaldehyde and CO_2 are relatively abundant in the products of cellulose pyrolysis. In addition, there are also some acids such as acetic acid and propionic acid, and cyclopentenones such as 1,2-cyclopentanedione. Lu et al. [5] investigated the distribution of products of cellulose pyrolysis at 500 °C, 600 °C, 700 °C and 800 °C by Py-GC/MS and found that more small molecular compounds were formed by the decomposition of oligomer with increasing temperature. Wu et al. [6] adopted the same method to simulate the pyrolysis of cellulose, and divided the products into two categories: one was the light volatiles with carbon atom number less than three, such as acrylic acid and hydroxyacetaldehyde; the other was the heavy volatiles containing more than three carbon atom, such as toluene, furanone, furfural, 5-methyfurfural, 5-hydroxymethylfurfural and glucopyranose. Qu et al. [7] studied cellulose pyrolysis in a fixed bed reactor, and found that cellulose pyrolysis products included acids, ketones, aldehydes, esters, carbohydrates, aromatics and so forth, among which carbohydrates were predominant. Zhou et al. [8] quantitatively analyzed the distribution of products of cellulose pyrolysis at 400–600 °C using Py-GC-MS-FID and found that levoglucosan was the prominent product with a yield of more than 50 % in the whole temperature range, followed by hydroxyacetaldehyde, dianhydro glucopyranose, water and CO_2 with a yield around 5 %, and the yield of 5-hydroxymethylfurfural was only about 1 %.

2.1.2 Pyrolysis of cellulose model compounds

Glucose and cellobiose are commonly used as the model compounds of cellulose. Because of their simple structure, they are convenient for deducing the reaction mechanism from the product distribution, which may be beneficial for understanding the pyrolysis behavior of cellulose. It has been found that the distribution of the pyrolysis products of glucose and cellobiose is similar to that of cellulose using Py-GC/MS, but quite different in the relative content. Especially the relative content of furans increases significantly while that of pyrans and linear small molecular compounds decreases. In the pyrolysis products of cellobiose, furfural and 5-hydroxymethylfurfural had a high relative content up to 30.97 % and 19.35 %, respectively, which are higher than that of cellulose under the same reaction conditions. The relative content of pyrans such as levoglucosan, levoglucosenone and dianhydro-D-glucopyranose from cellobiose pyrolysis, ranging from 1 % to 3 %, are lower than that from cellulose pyrolysis with the range of 2–9 %. Among the linear small molecular products from cellobiose pyrolysis, the relative content of CO_2 was the highest, followed by acetaldehyde, and both relative contents were lower than that from cellulose pyrolysis. And the relative content of acetic acid and propionic acid from cellobiose was close to that from cellulose. In the pyrolysis products of glucose, furfural and 5-hydroxymethylfurfural were prominent with a relative content of 31.73 % and 13.34 %, respectively. Moreover, alkyl-furans, alky-furfural, furan-ketones and furfur alcohol were also observed, indicating that furans are the primary products of glucose pyrolysis. The CO_2 and acetaldehyde content was also higher, reaching 5.08 % and 3.99 %, respectively. Acetic acid, propionic acid, and butyric acid were also present in the linear small molecular products and the pyrans content was quite low. Comparing the products from pyrolysis of three carbohydrates, it was found that furans are prominent for glucose and least for cellulose, while pyrans present an opposite trend. Patwardhan et al. [9] analyzed the pyrolysis products of cellobiose, and found that furfural and 5-hydroxymethylfurfural were quite abundant, while the yields of levoglucosan, levoglucosenone and other small molecules were lower and the point out that it might be due to the short residence time in the reactor. The intermediate macromolecule that was produced was difficult to crack into linear small molecules during short residence times. Mettler et al. [10] investigated the pyrolysis product distributions of cellulose and six glycosyl oligomers with the degree of polymerization (DP) ranging from 1 to 6. The results showed that the product categories of the seven carbohydrates were almost the same but varied in content, and the abundant products included levoglucosan, 5-hydroxymethylfurfural, furfural, methylglyoxal, hydroxyacetaldehyde and acetic acid. A higher DP enhanced the yield of levoglucosan, while the yields of other products changed with increasing DP. Zhou et al. [8] compared the pyrolysis product distributions of four cellulosic model compounds: glucose, cellobiose, maltohexaose and cellulose at 500 °C, and also found that the yield of levoglucosan increased from 8.10 % to 54.50 % with increasing DP, further suggesting that the formation of levoglucosan had a significant correlation with the cleavage of glycosidic bonds. It was generally believed that the cations formed from the cleavage of glycosidic bonds easily bind to the hydroxyl group at C_6 position of the glucose residue to form levoglucosan; while the yield of 5-hydroxymethylfurfural decreased as DP increased, indicating that 5-hydroxymethylfurfural was mainly generated from the stepwise dehydration of monosaccharide units.

2.2 Effect of reaction parameters on the pyrolysis behavior of cellulose

During the process of cellulose pyrolysis, reaction temperature, residence time, reaction pressure, particle size of raw material and reaction atmosphere will affect the pyrolysis process and the final product distribution. Furthermore, torrefaction before pyrolysis to remove moisture in cellulose and acid pretreatment will also have a significant impact on pyrolysis behavior. Fig. 2.4 shows the infrared radiation reactor used to study the effect of reaction parameters on the pyrolysis behavior of cellulose. The heat required for cellulose pyrolysis is provided by high-power silicon carbide pipes that are located outside of the quartz glass tube. In the pyrolysis process, the released volatiles are quickly swept through the filter by pure nitrogen, followed by cooling and collection in the multi-condensation system. Cellulose filter paper is used as feedstock, and its ultimate analysis shows that the molecular formula is the same as that of cellulose. Due to the fast rate of cellulose pyrolysis, before the heat transfer between sample and reactor reaches balance, pyrolysis is already completed, thus it is not appropriate to describe the pyrolysis process by using the changing temperature of the sample. Furthermore, it is difficult to measure the temperature of the sample directly due to shrinkage deformation during the pyrolysis process. Consequently, it is more reasonable to use the temperature of the radiation source (TRS) to study the effect of reaction temperature on the pyrolysis process. It needs to be pointed out that there is a temperature gradient between TRS and sample. Fig. 2.5 shows the simulated temperature curves of the cellulose outside surface at the TRS temperature of 610 °C and 800 °C, respectively.

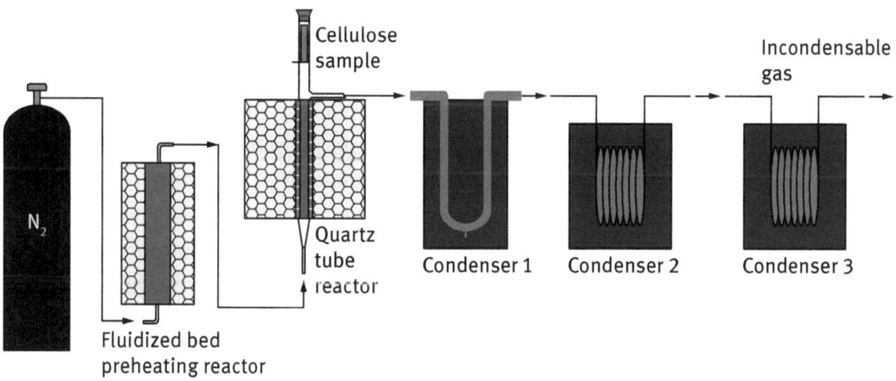

Fig. 2.4: Cellulose rapid pyrolysis system. Reprinted with permission from [11], © 2004 American Chemical Society.

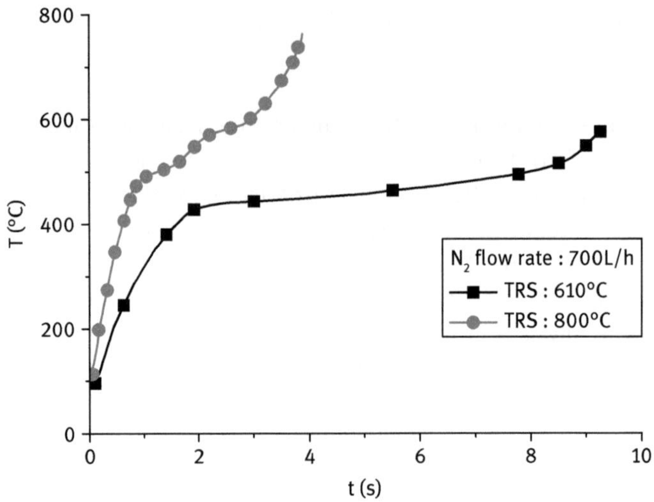

Fig. 2.5: Temperature change of cellulose outside surface.

2.2.1 Effect of reaction temperature

Reaction temperature plays a dominant role in the pyrolysis process. When other conditions remain unchanged, the change of the product distribution from cellulose pyrolysis with TRS temperature is shown in Fig. 2.6. Due to deficient heating at the low temperature stage, cellulose undergoes dehydration and releases a small amount of volatiles, and solid char is the main product. The heating rate increases with rising TRS temperature, leading to the formation of levoglucosan by the cracking of the glycosidic bonds, accompanied by the ring rupture of cellulose monomers and further cracking of ring-opened intermediates to produce small molecular gases and condensable volatiles [12], thus the yield of bio-oil increases and reaches the maximum at TRS temperature of 610 °C. The yield of gases varies smoothly, the char production decreases obviously with the elevated temperature, resulting from the overall exothermic reaction of char formation [13]. Increasing the temperature further accelerates the cleavage of cellulose monomer rings and the release of gas products; meanwhile, the secondary reaction of primary products occurs since the unstable bonds and groups of the primary volatiles are easily cracked at high temperature, promoting the formation of smaller molecular gases. Therefore, the yield of bio-oil decreases while the production of gases increases, and the char yield remains almost unchanged with a low value. However, due to the different heating methods and temperature measurements, the reaction temperature corresponding to the maximum bio-oil yield in different reactors varies. Luo et al. [14] obtained the maximum bio-oil yield of 58.6 % in cellulose pyrolysis at about 450 °C in a fixed bed reactor. Kojima et al. [15] performed cellulose pyrolysis experiments in a fluidized bed reactor and got the maximum bio-oil yield of

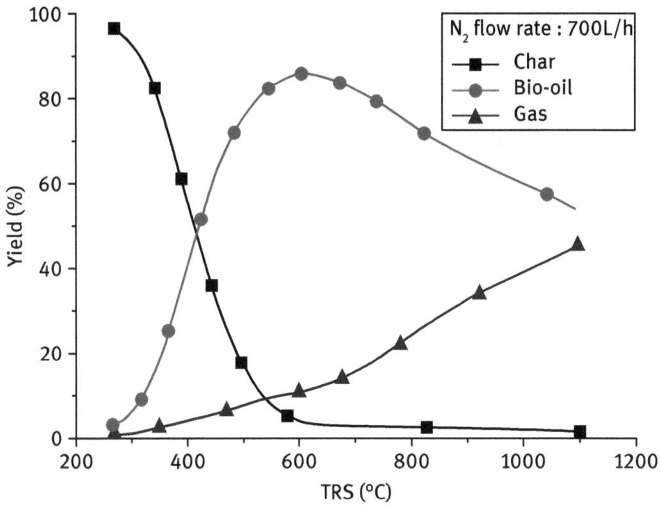

Fig. 2.6: Effect of temperature on the products distribution in cellulose pyrolysis. Reprinted with permission from [11], © 2004 American Chemical Society.

63 % at about 400 °C. Rutkowski et al. [16] studied cellulose pyrolysis in an infrared heating tube reactor, and the results showed that the maximum bio-oil yield was up to 84 % at about 500 °C.

Reaction temperature also has a significant impact on the composition of pyrolytic products, and the influence of reaction temperature on the yield of the main components of bio-oil, namely levoglucosan, hydroxyacetaldehyde and hydroxyacetone, is shown in Fig. 2.7. With increasing TRS temperature, the levoglucosan content in bio-oil decreases continually, while the overall yield presents a trend that first increases and then decreases with a maximum of 53.3 % at about 580 °C. In addition, the hydroxyacetaldehyde and hydroxyacetone content in bio-oil and their yields keep increasing. Further experimental research of levoglucosan pyrolysis reveals that levoglucosan is quite stable at a low temperature, indicating that hydroxyacetaldehyde and hydroxyacetone are mainly generated from the decomposition of active cellulose; and levoglucosan is easily decomposed into hydroxyacetaldehyde and hydroxyacetone at a high temperature by secondary reaction. Piskorz et al. [17] also observed similar results in their experiments and ascribed this phenomenon to the further dehydration and rearrangement when levoglucosan is heated, which led to the formation of hydroxyacetaldehyde and hydroxyacetone. Norinaga et al. [18] investigated the product distributions of cellulose pyrolysis at higher temperatures (700 °C, 750 °C, 800 °C), and the results showed that the yield of levoglucosan decreased sharply from 36.1 % at 700 °C to 1.96 % at 800 °C, with a dramatic increase in gas yield, especially the yield of CO increasing from 30.0 % to 56.0 %. They proposed that the radicals derived from the breakage of branched aliphatic hydrocarbon at high temperature are further cy-

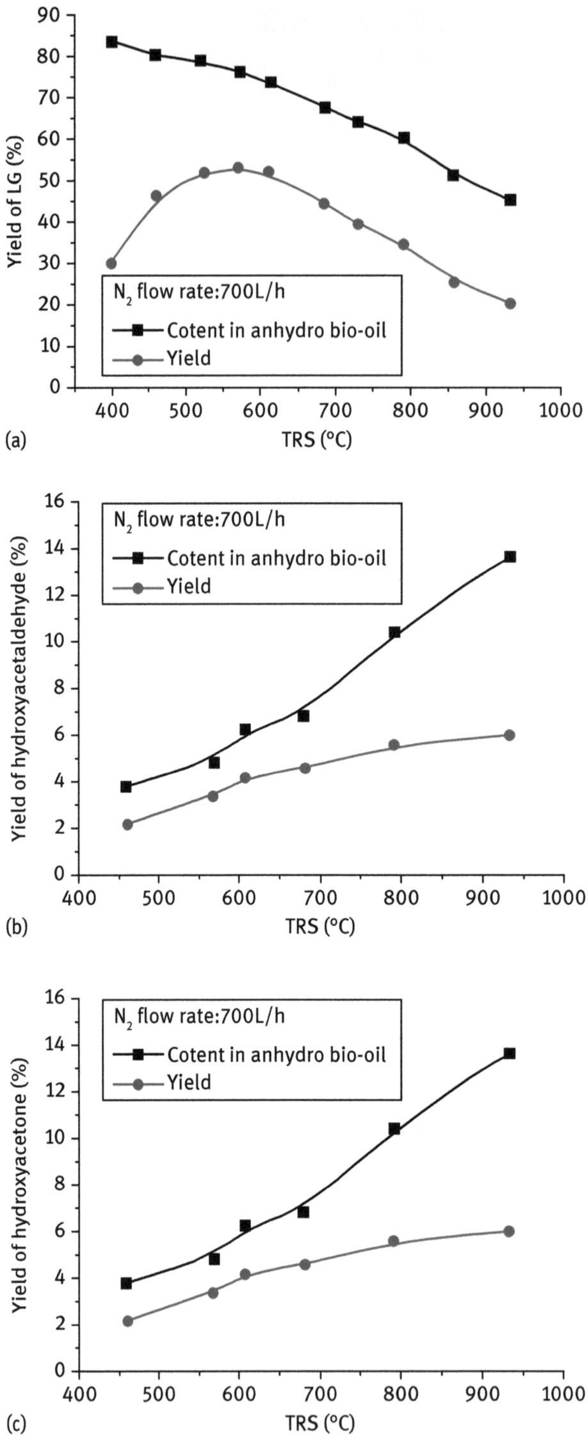

Fig. 2.7: Yields of main components in bio-oil at different TRS. Reprinted with permission from [11], © 2004 American Chemical Society.

clized to form aromatic hydrocarbons. Xin et al. [19] conducted cellulose pyrolysis in a fixed bed reactor and found that a large amount of levoglucosan was formed in the temperature range of 350 °C to 550 °C. When the temperature was up to 650 °C, the yield of levoglucosan dropped sharply while the yield of acetic acid and phenol showed a substantial increase, and the production of gases also improved markedly, suggesting that levoglucosan decomposed to other chemicals at higher temperature. The bicyclic structure of levoglucosan could be disrupted into light oxygenated products and radical fragments through the cleavage of C–O bonds at C_1 and C_2 positions, and the polymerization of radical fragments would result in the formation of phenol and aromatic compounds [20].

2.2.2 Effect of residence time

The residence time of biomass during pyrolysis mainly includes solid phase residence time and gas phase residence time, and generally it refers to the latter. Residence time is closely related to the flow rate of carrier gas, so the residence time can be adjusted by changing the carrier gas flow rate. Fig. 2.8 displays the effects of residence time on the product distribution of cellulose pyrolysis at 610 °C. Calculations show that the residence time of volatiles in the reaction zone was about 0.1 s when the carrier gas flow rate was 700 l/h. With an increased carrier gas flow, the residence time of pyrolytic volatiles decreases, the gas yield drops off a little while more bio-oil is produced. This can be ascribed to the secondary cracking of the volatiles. The higher the reaction temperature and the longer the residence time, the more severe secondary cracking occurs. Therefore, to maximize the yield of bio-oil, in addition to optimizing the reaction temperature, the residence time should also be shortened largely under the premise of complete pyrolysis [21].

It is very evident that the product distribution of cellulose pyrolysis changes with residence time. When the residence time increases, levoglucosan and its isomers will further decompose into small molecular products, such as hydroxyacetone and hydroxyacetaldehyde. The total yield of furans, including furfural and 5-hydroxymethylfurfural, decreases with increasing residence time. Long residence time favors the secondary decomposition of primary products, leading to a higher yield of CO. However, the production of other small molecular gases, such as CO_2, H_2O, CH_4, and C_2H_4 seems to be influenced little by changing residence time. In addition, the residence time shows only a small impact on the char yield. When the carrier gas flow rate increases, the carrier gas will absorb more heat and its temperature drops off, followed by the decrease in the temperature of cellulose particles, leading to an increase of primary char production. On the other hand, when the carrier gas flow rate decreases, though primary char production is suppressed, the formation of secondary char resulting from secondary cracking of volatiles increases under a longer gaseous residence time. Patwardhan et al. [22] carried out a detailed comparison of cellulose

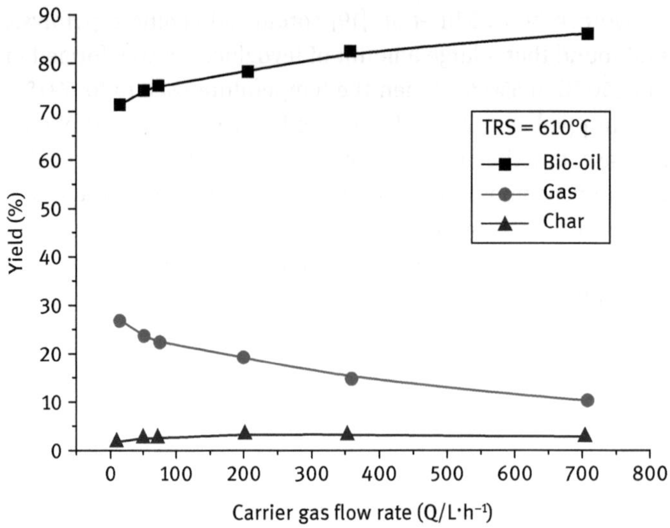

Fig. 2.8: Effect of carrier gas flow rate on product distribution in cellulose pyrolysis. Reprinted with permission from [11], © 2004 American Chemical Society.

pyrolysis behavior in a micro-pyrolyzer with a residence time of 15–20 ms and in a fluidized bed reactor with 1–2 s, respectively. They found that the secondary cracking reaction included the ring-opening of levoglucosan and the decomposition of furfural and 5-hydroxymethylfurfural to formic acid and propionic acid, which were further decomposed to produce CO. They also observed that the yield of CO increased sharply from 0.7 % to 4.3 %, and other gas yields remained almost unchanged. In addition, the char yield slightly declined.

2.2.3 Effect of acid pretreatment

Acid pretreatment of biomass feedstock is usually considered to be an effective means to remove the inorganic salts in it. Hague et al. [24] found that the presence of inorganic salts, especially alkali metal and alkaline earth metal salts, increased the yield of gases and char in cellulose pyrolysis, while they decreased the yield of liquid products, and the yield of levoglucosan was reduced significantly with a great improvement of hydroxyacetaldehyde production. Thus acid pretreatment of raw material may increase the yield of bio-oil. For pure cellulose, the effect of acid pretreatment has two aspects: on the one hand, the pretreatment may lead to some acid hydrolysis of cellulose and decrease the DP of cellulose macromolecules, which may further influence the activity of cellulose and the subsequent depolymerization. On the other hand, acids will enhance the intermolecular crosslinking reaction in the pyrolysis process and increase the yield of char. Moreover, acids also strongly catalyze dehydration, resulting in a significant decrease in tar yield and the apparent increase in water content. After soaking cellulose with dilute acids, it was found that acids have a significant impact on DP and crystallinity of cellulose. As shown in Fig. 2.9, the cellulose microstructure changes greatly after acid pretreatment. The microstructure of cellulose without acid pretreatment reveals that virgin cellulose is mainly composed of crystalline regions

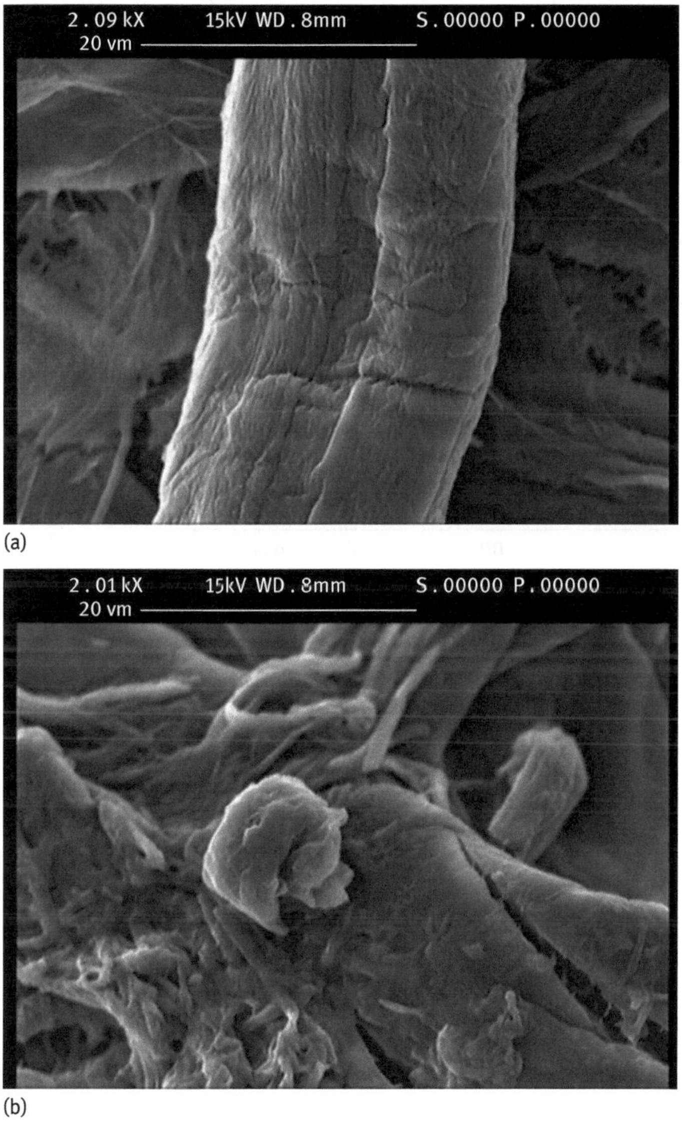

Fig. 2.9: SEM images of cellulose microstructure; (a) Before acid pretreatment, (b) after acid pretreatment. Reprinted with permission from [23].

with ordered and parallel molecular chains, and amorphous regions with relaxed and irregular molecular chains [25]. Crystalline regions, closely connected to each other with ordered arrangement, present poor chemical activity; while amorphous regions with loose structure participate easily in chemical reactions. The cellulose surface emerges greatly changed after pretreatment, and the ruptured structure visibly reflects the damage to the cellulose structure caused by acid. Compared with virgin cellulose, acid pretreated cellulose shows poor fragility, low tenacity and greater propensity to chemical reactions.

Using the viscosity method to measure the DP of cellulose before and after acid pretreatment, the results (Tab. 2.1) show that the DP of cellulose decreases with increasing concentration of hydrochloric acid. Celluloses pretreated by hydrochloric acid, sulfuric acid and phosphoric acid with the same concentration have similar DP, indicating the similar destructive effects of the acids on the cellulose macromolecular structure. Increasing the contact time between samples and acids leads to much severer damage of the structure, which may be ascribed to the extreme acid sensitivity of β-1,4-glycosidic bonds in cellulose macromolecules. With the action of hydrogen ions, glycosidic bonds will be hydrolyzed and broken, inducing the rapid decrease of DP, and the decrease will be more intense as the acid concentration, reaction temperature and pretreating time increase.

Tab. 2.1: DP of the cellulose pretreated under different acid conditions.

Pretreatment methods	DP	Decrease of DP (%)
Original cellulose	1145	/
3 % HCl soaking	553	52
5 % HCl soaking	456	60
7 % HCl soaking	406	65
7 % H_3PO_4 soaking	461	60
7 % H_2SO_4 soaking	426	63
7 % HCl soaking without further washing	156	86

After acid pretreatment, the product distribution of cellulose pyrolysis obviously changes, as presented in Fig. 2.10. After soaking with hydrochloric acid, the yield of bio-oil decreases, while the yield of gas increases slightly, and the yield of char remains almost the same. The changing trend is enhanced with increasing concentration of hydrochloric acid. When other dilute acids with the same concentration are used, similar results are obtained. Cellulose pretreated by sulfuric acid generates more char, small molecular gases and water, and less bio-oil. The effect of phosphoric acid is milder than that of sulfuric acid. Compared with hydrochloric acid, phosphoric acid shows slightly stronger effects on the formation of char and water. Cellulose after acid pretreatment can be more easily dehydrated. Furthermore, the short chain cellulose resulting from acid pretreatment, with damaged amorphous region and low DP, has higher reactivity and is prone to degradation during pyrolysis, which favors the improvement of gas yield. In addition, without further washing after acid soaking, the residual acid in cellulose will also have a recognizable influence on its pyrolytic products. In the dilute acid soaking process, water and acid will penetrate the surface of crystalline and amorphous regions of cellulose, which may weaken the hydrogen bond network between the molecular chains of cellulose. Thus, simple washing with water is not able to completely remove all the acids from the pore structure of the sample. The residual acids in the interior of the sample will participate in the reaction and affect the product distribution during the pyrolysis process. Pyrolysis of cellulose without further washing after acid pretreatment generates a higher yield of char, suggesting that the residual acids intensively catalyze the formation of char, which results from the intermolecular crosslinking reaction caused by acid catalysis at high temperature. Moreover, the dehydration promoted by acid catalysis also induces further cracking of volatiles to small molecular gas products. Compared to the other two acids, sulfuric acid shows the strongest catalytic effect on both the dehydration and the crosslinking reaction. The existence of acid evidently alters the composition distribution of bio-oil. Diluted acid soaking reduces the DP of cellulose, resulting in the partial increase of levoglucosan content in bio-oil. In addition, the acid, by catalyzing dehydration, causes an increase in the total content of anhydro-sugars in pyrolysis products. Gravitis et al. [26] found that acid pretreatment of biomass destroyed the whole hemicellulose fraction, and the pyrolysis of the residual cellulose generated a large amount of levoglucosan. Dobele et al. [27–30] also found that the DP of cellulose decreased after soaking with phosphoric acid, and its pyrolysis products contained more anhydro-sugar as a result of the severe dehydration reaction, and proposed that phosphoric acid promoted the transformation of levoglucosan to levoglucosenone.

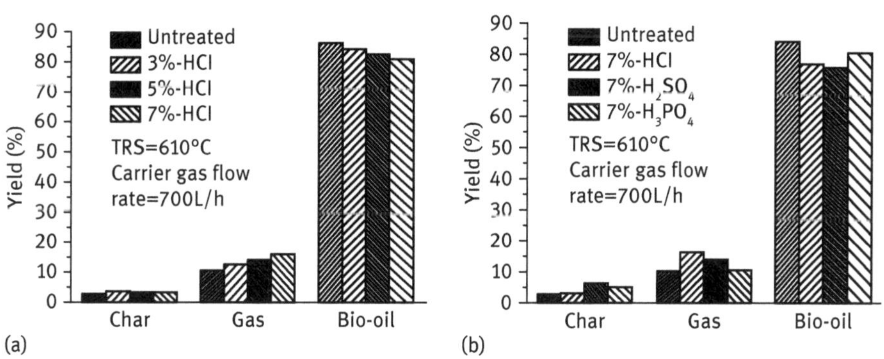

Fig. 2.10: Effects of acids pretreatment on the product distribution of cellulose pyrolysis. Reprinted with permission from [23].

2.2.4 Effect of other reaction factors

In the pyrolysis process, the selection of reaction temperature should also consider the effects of secondary cracking and other factors, whereas the adjustment of the gas phase residence time is restricted by the type of reactor and other specific reaction conditions. The particle size of feedstock has a certain effect on the pyrolysis behavior of biomass, which can affect the product composition by altering the residence time of volatiles in the reaction zone [31]. The pyrolysis process of particles with small size is mainly controlled by the intrinsic reaction kinetic rate, and the internal heat transfer resistance in particles can be ignored [32]. When particle size increases, internal heat transfer is suppressed, resulting in a difference in reaction temperature between the interior and exterior of particles, and the lower temperature in the interior of particles enhances the formation of char [33]. Moreover, the particles with larger size have higher mass transfer resistance against the release of volatiles from the interior of particles, which lengthens the residence time of the primary products in the high temperature zone and induces the formation of small molecular gases by further cracking. As observed in the experiments by infrared radiation reactor, bio-oil yield decreases slightly while the production of gases and char increase mildly with increasing cellulose sample thickness. Koufopanos et al. [34] also confirmed that larger sized particles of biomass feedstock enhanced the formation of small molecular gases and char in the pyrolysis process. Paulsen et al. [35] compared the pyrolysis products of thin-film and powder cellulose, and found that thin-film cellulose was heated more evenly, making the pyrolysis more efficient. The yield of levoglucosan from pyrolysis of thin-film cellulose was only 27 %, while that from power cellulose was 49 %.

Raw biomass always contains a substantial proportion of free water, which evaporates very slowly under natural conditions. Investigating the effect of free water content on biomass pyrolysis behavior can provide guidance for the pre-dehydration of raw feedstock by torrefaction. Artificially spraying quantitative water onto raw cellulose can yield cellulose materials with a specific water content. By comparing them with the pretreated cellulose after torrefaction, it has been found that adding water increases the formation of water in cellulose pyrolysis while it reduces the non-aqueous fraction content of bio-oil, indicating that water in feedstock is not simply evaporated at low temperature, but may participate in the pyrolysis process.

Reaction pressure also has some impact on the product distribution of cellulose pyrolysis. Low pressure will reduce the resistance of volatiles release, which can be released easily outside the sample surface and then flow into the reaction zone. Low pressure can suppress the occurrence of secondary cracking, which benefits a higher production of bio-oil [36]. Hoekstra et al. [37] performed cellulose pyrolysis under vacuum (< 30 Pa) in a metal mesh reactor and obtained a bio-oil yield of 95 %, and the gases were essentially composed of CO_2 and CO. The studies in the infrared radiation reactor also show that the yield of bio-oil increases by about 1.4 % when the reaction pressure drops from atmospheric condition to 91 kPa. Pindoria et al. [38] investigated the pyrolysis behavior of cellulose under pressure, from atmospheric pressure to 7 MPa, and found that rising pressure reduced the yield of bio-oil, especially in the range from atmospheric pressure to 1 MPa. While the yield remained almost unchanged after the pressure reached 4 MPa; the composition of bio-oil did not show a significant change with the increased pressure, indicating it was insensitive to the change of pressure.

Although cellulose pyrolysis is generally carried out under an inert atmosphere of nitrogen, some research has been performed under an atmosphere of steam [39, 40]. The change of reaction atmosphere affects the distribution of pyrolysis products. Giudicianni et al. [41] studied cellulose pyrolysis under a steam atmosphere and obtained 70.4 % bio-oil, 21 % char, and 8.6 % gases at 430 °C, and the composition of the gases varied obviously with the change of temperature. The majority of gases were CO_2 and CO when the temperature was below 430 °C, while the main gases were CH_4, C_2H_4, C_2H_6, and a small amount of H_2 when the temperature was above 430 °C. Sagehashi et al. [42] carried out cellulose pyrolysis under superheated steam and found that the main products were hydroxyacetaldehyde and 5-hydroxymethylfurfural at 250 °C, while they were hydroxyacetaldehyde, levoglucosan, 5-hydroxymethylfurfural and furfural above 325 °C.

2.3 Pyrolysis kinetic models for cellulose pyrolysis

Pyrolysis reaction kinetics includes mechanism reaction kinetics and apparent reaction kinetics. The former is to elucidate the mechanism of cellulose pyrolysis by analyzing cellulose pyrolysis kinetics from the perspective of chemical reactions. This research idea aroused a lot of interest in the early years, but due to the huge difficulty involved, researchers could only try to explain part of the reaction mechanism. In fact, more than one hundred species of cellulose pyrolysis products have been detected by Py-GC/MS and it is almost impossible to illuminate the mechanism of all the reactions involved. Therefore, research usually only focuses on the formation mechanism of several major products. Compared with mechanism reaction kinetics, the main task of apparent reaction kinetics is to seek appropriate apparent kinetic models that can accurately describe the global weight loss process without being concerned with the detailed reaction mechanism [43]. Biomass pyrolysis essentially belongs to solid state reactions driven by heat under inert atmosphere. The principle that has been widely applied in the conventional kinetics of solid decomposition is also suitable for biomass kinetic studies with small modifications. Thus biomass pyrolysis kinetics can be essentially described by equation (2.1), where the reaction rate $k(T)$ is expressed as equation (2.2) (Arrhenius theorem).

$$\frac{\mathrm{d}\alpha}{\mathrm{d}t} = k(T)\,f(\alpha) \tag{2.1}$$

$$k(T) = A \exp\left(\frac{-E}{RT}\right) \tag{2.2}$$

Generally, two methods are emphasized to carry out thermal kinetic analysis: the model-fitting methods (distributed activation energy theory is also included here, but due to its specificity, it is introduced separately) and the model-free methods, based on whether the reaction model $f(\alpha)$ is assumed prior to simulation. There are many different reaction models used in solid decomposition [44–46]. Among these models, the reaction-order model is commonly applied in biomass pyrolysis kinetic analyses based on model-fitting methods, which can be solved by using direct solution methods, for example the Freeman–Carroll method [47] and the Coats–Redfern method [48]. The application of model-free methods in biomass pyrolysis kinetics has recently become attractive. Without prior assumption of a reaction model, its foundational principle is that the reaction rate only depends on temperature at a constant conversion, as described in equation (2.3), and this method is also known as the isoconversion method. After the relationship between activation energy E and conversion rate α is determined, the reaction model can be obtained. The methods for solving the kinetics based on the model-free methods include the Friedman method [49], the Kissinger method [50], the Flynn–Wall–Ozawa (FWO) method [51, 52], modified Coats–Redfern method [53], etc. Thermal kinetic analysis methods also can be classified as isothermal methods and non-isothermal methods according to the experimental heating rate. The former means the experiments are carried out at a constant temperature, while the latter are usually executed at a constant heating rate or sometime at nonconstant heating rate, such as ladder heating method [54]. In addition, the heating rate can be dynamically adjusted so that the reaction rate is maintained at a certain value, which can reduce the effects of heat and mass transfer on the pyrolysis reaction, called constant rate thermal analysis (CRTA) method [55].

$$\left[\frac{\mathrm{d}\ln(\mathrm{d}\alpha/\mathrm{d}t)}{\mathrm{d}T^{-1}}\right]_{\alpha} = \frac{-E_{\alpha}}{R} \tag{2.3}$$

2.3.1 One-step global reaction model

The one-step global model has been widely used, as shown in Fig. 2.11. It assumes that cellulose generates only char and volatiles in the pyrolysis process and the process of volatiles release can be described by the Arrhenius equation. By performing TG experiments on microcrystalline cellulose at varied heating rates and then fitting the TG curves by differential or integral method, it has been found that the apparent

$$\text{Cellulose} \xrightarrow{\text{K}} \text{Char + Volatile}$$

Fig. 2.11: Single-step model for cellulose pyrolysis.

activation energy for cellulose pyrolysis is about 210 kJ/mol. Varhegyi et al. [56] obtained the activation energy of 234 kJ/mol for cellulose pyrolysis by TG analysis at a heating rate of 10 K/min. In a further report they found that the activation energy was 242 kJ/mol using filter paper as the sample [57]. Among the reports related to cellulose pyrolysis kinetics, the resulting apparent activation energies are quite dispersed with a range of 200–250 kJ/mol due to the difference in the experimental and data processing methods. With the premise of ignoring the gas-solid phase interactions and heat and mass transfer, TG curves of the cellulose pyrolysis process can be well described by a one-step and irreversible first-order reaction rate equation.

2.3.2 Two-step reaction model

Compared with the one-step global model, the two-step competitive reaction model and the two-step successive reaction model consider the secondary decomposition in cellulose pyrolysis and can reflect the actual process of cellulose pyrolysis more suitably. The most typical model is the B–S model proposed by Broido and Shafizadeh, which considers that cellulose pyrolysis undergoes two-step reactions with active cellulose as the intermediate in the pyrolysis process. In fact, studies of the cellulose pyrolysis mechanism originate from the research on cellulose combustion [57]. Broido conducted cellulose combustion tests and found that part of the cellulose was converted into anhydrocellulose via endothermic reaction under a low temperature heating condition. When the temperature was higher than 280 °C, cellulose depolymerized and formed a certain amount of volatiles in inert atmosphere and meanwhile the anhydrocellulose underwent further reaction to produce small molecular gas and char. Broido and Nelson [58] investigated the pyrolysis behavior of cellulose at 370 °C after the cellulose had been preheated at 230–275 °C for a long time. Their experimental results showed that char yield increased from 13.0 % to 27.6 % after pretreatment, while tar decreased greatly. They proposed that two parallel competing reactions occurred in cellulose pyrolysis. Based on previous research findings [59], they further inferred that one reaction mainly referred to the depolymerization for tar formation, while the other one to the formation of char and gas. Finally, the Broido and Nelson multistep first order model was proposed as in Fig. 2.12.

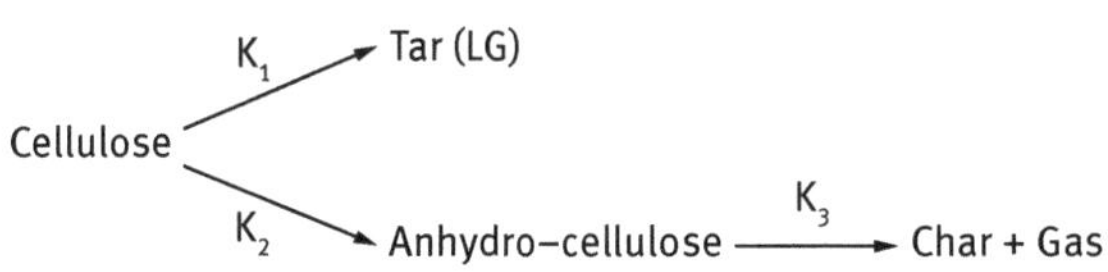

Fig. 2.12: Brodio and Nelson first order multistep model for cellulose pyrolysis.

Shafizadeh carried out isothermal pyrolysis of cellulose in the temperature range of 259–407 °C at low pressure and found there existed an acceleration process in the initial stage of weight loss. Then he proposed that there was a transition from a "non-activated state" to an "activate state" with a high activation energy in the initial cellulose pyrolysis process, and finally modified the Broido and Nelson model to the Broido–Shafizadeh model [1], as shown in Fig. 2.13. Cellulose first partially depolymerized to form active cellulose, followed by further dehydration and aromatization to generate char and small gases, or to produce tar with levoglucosan as the main component through dehydration of the edge groups and aldol condensation reaction. The calculation results based on the B–S model showed good agreement with the experimental results, and the activation energy for the formation of the intermediate active cellulose was about 242.8 kJ/mol [1].

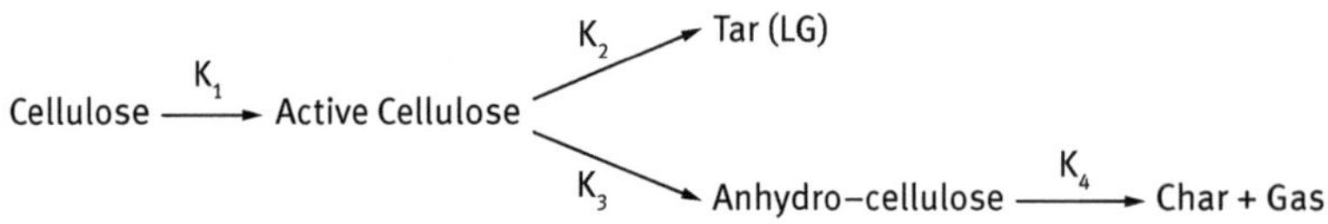

Fig. 2.13: Brodio–Shafizadeh model for cellulose pyrolysis.

Milosavljevic and Suuberg [60] analyzed a large amount of data on cellulose pyrolysis kinetics and found that the apparent kinetic parameters showed a transition at a temperature of about 330 °C. The activation energy was in the range of 140–155 kJ/mol when the temperature was higher than 330 °C, while it increased to about 218 kJ/mol at a temperature below 330 °C; which was interpreted as the formation of another kind of bio-oil at high temperature via a process with low activation energy. However, no specific proof was given in the report and it was just a possible hypothesis to illustrate the change in kinetic parameters. Antal et al. [61] put forward a different view that it was the increase of heat and mass transfer resistance between cellulose particles and reactor, resulting from the elevated temperature, which affected the apparent activation energy. By performing segmental simulation of the TG curves of cellulose, an obvious decrease of activation energy at the reaction temperature of about 330 °C occurs, indicating that the whole pyrolysis kinetics could be divided into two stages at this temperature. It should be noted that the small size of samples, low heating rate and a proper carrier gas flow rate in the TG experiments could effectively avoid the influence of heat and mass transfer. Actually, the B–S model can satisfactorily interpret this phenomenon. The cellulose weight loss process mainly takes place in the range of 280–400 °C, which can be denoted as a two-stage reaction: the formation of active cellulose and its subsequent consumption. When the temperature is below 330 °C, the formation rate is lower than the consumption rate due to high activation energy for active cellulose formation, and when the temperature is above 330 °C, the rate of active cellulose formation becomes much higher than its consumption. Thus from an apparent perspective, this reflects the kinetics for active cellulose formation when the temperature is below 330 °C, while presenting the kinetics for tar formation when the temperature is above 330 °C, in accordance with the results found by Shafizadeh [1].

After the B–S model was proposed, many researchers used various experiential methods and observed the existence of this reaction mechanism in which parallel and competitive reactions exist in the formation of char and tar. Later, the B–S model was widely accepted and has become the basic model for cellulose pyrolysis research [62]. Further modifications of the B–S model were also undertaken by some researchers in later studies. For example, Bradbury et al. [1] proposed a modified B–S model that included the formation of gases by the further decomposition of tar, wherein active cellulose was considered to be a kind of product with melting property. They also found the melting product formed in cellulose pyrolysis at low temperature was mainly composed of dextrans with carbon atom number of 2–7 [63].

Hydroxyacetaldehyde is a typical product from celluloses pyrolysis, whose formation shows a competitive relationship with that of char during cellulose pyrolysis [65]. Based on this observation, Piskorz et al. [66] modified the Bradbury model and proposed that char was the main product of cellulose slow pyrolysis at low temperature, accompanied by the formation of CO_2 and water; while in the process of cellulose fast pyrolysis at intermediate or high temperature, cellulose first depolymerized to form active cellulose and further decomposed to other products. One possible reaction path was that the pyran ring structure of active cellulose degraded completely, followed by decarbonylation and dehydration to generate small molecular chemicals, such as hydroxyacetaldehyde, acetic acid, formic acid, oxalaldehyde, pyroracemic aldehyde, and so on; the other possible reaction path was the cleavage of branched chains of active cellulose without ring-opening, leading to the formation of levoglucosan, cellobiose, glucose and fructose. A high temperature promoted the former reaction while a low temperature was conducive to the latter. Based on the collection and measurement of active cellulose, we also propose an improved B–S model, as shown in Fig. 2.14, which is similar to the mechanism model proposed by Piskorz et al. [17]. In

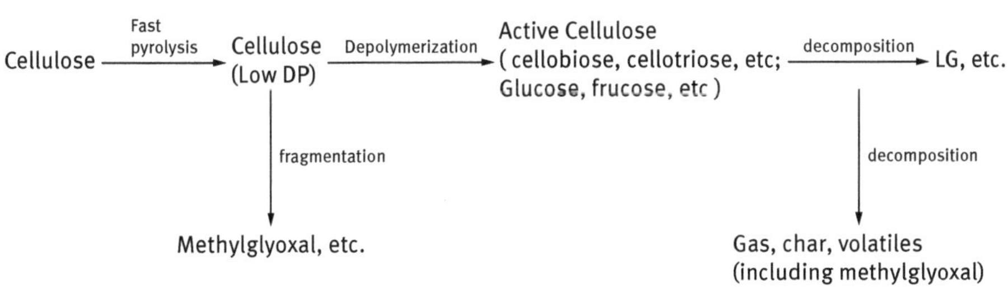

Fig. 2.14: Cellulose pyrolysis mechanism model based on the formation and consumption of active cellulose. Reprinted with permission from [64].

the pyrolysis process, the long-chain cellulose first depolymerizes to macromolecules with low DP, followed by two parallel competing reactions: one is to form methylglyoxal etc. via rupture of ring structures after decarbonylation and dehydration, which could be facilitated by high temperature [67]; the other one easily takes place at a lower temperature. Under the effect of transglycosylation, the macromolecules with low DP further depolymerize to form sugars such as cellobiose, cellotriose, glucose, fructose, etc., and the subsequent decomposition generates levoglucosan and other chemicals. Eventually the cleavage of glycosidic and some C–C bonds produces char, gases, and condensable volatiles.

2.3.3 Isoconversion methods

The application of isoconversion methods in cellulose pyrolysis kinetics analyses has gradually become hot in recent years. The average activation energy for cellulose pyrolysis obtained by the model-free kinetic model is about 192.3 kJ/mol, which is very close to the value obtained from the first-order reaction model based on the model-fitting method [68]. Capart et al. [69] analyzed the kinetics of microcrystalline cellulose pyrolysis by two isoconversion methods. The activation energy was 200 kJ/mol by the Kissinger method and results from the Friedman method showed that when the cellulose conversion rate was in the range of 0.1–0.7, the corresponding activation energy was about 200 kJ/mol. Furthermore, the activation energy sharply dropped to 136.2 kJ/mol at the conversion rate of 0.8, indicating that the Friedman method was no longer applicable at a higher conversion rate.

2.3.4 Distributed activation energy model

Distributed activation energy model (DAEM), an effective method developed in recent years to deal with the complex reaction kinetics, has been widely used in many areas, such as the pyrolysis of fossil fuel and the thermal regeneration of activated carbon etc. [70, 71]. DAEM is based on two assumptions: first, the reaction system is composed of numerous independent first-order reactions with different activation energies; second, based on the assumption of a distribution of activation energy, the activation energy of each reaction presents in the form of a continuous distribution function. The calculation of the kinetic parameters in DAEM can be performed with the aid of TG data by using the step approximation method, inflection tangent method, Miura differential method and Miura integral method etc. [58, 71–73]. The application of the first two methods only needs a single weight loss curve, while the second two methods need several TG curves obtained at various heating rates. DAEM is also suitable for the kinetic analysis of depolymerization, volatilization and crosslinking occurring in the formation of bio-oil and char. However, it cannot predict the release of gas products well [74]. Wang et al. [75] applied a DAEM model to analyze the kinetics of cellulose pyrolysis and obtained an activation energy of 142.6–167.7 kJ/mol at different weight loss rates. Zhang et al. [76] simulated the thermal behaviors of cellulose, hemicellulose and lignin using DAEM with a Gaussian function, and found that due to the regular and simple structure of cellulose, the application of one Gaussian function could achieve very high accuracy with the average activation energy of 240.2 kJ/mol.

2.4 Active cellulose

As mentioned above, all the widely accepted models for cellulose pyrolysis consider that the heated cellulose starts a series of complex initial reactions and generates an intermediate with relatively simple components and low DP, namely active cellulose. Active cellulose is recognized as the important intermediate in cellulose pyrolysis. Due to its extremely poor thermal stability, active cellulose is difficult to collect in a conventional pyrolysis reactor. Thus, whether active cellulose actually exists remained controversial for a long time [77]. Varhegyi et al. [56] believed that cellulose pyrolysis generated char, tar and gases directly, without the formation of active cellulose. However, in later studies, many scholars observed a "suspicious" intermediate material in the initial stage of cellulose pyrolysis. Boutin et al. [78, 79] collected a molten material with properties between those of cellulose and bio-oil by flash pyrolysis and rapid cooling, deducing that the molten material might be active cellulose. Piskorz et al. [63] also observed a similar phenomenon in their experiments, and further confirmed that the intermediate was not just confined to the surface layer, but the whole particle was in a molten state. Nowadays, more and more scholars have accepted the view that active cellulose does exist and plays an important role in the whole process of cellulose pyrolysis.

2.4.1 The collection and characterization of active cellulose

The collection of active cellulose is very difficult because it requires flash pyrolysis and rapid condensation. To achieve this, flash pyrolysis mechanism apparatus with a xenon lamp as high radiation source is set up, as shown in Fig. 2.15. A spherical high-pressure short arc xenon lamp with power ranging from 1.8 kW to 3 kW is used as radiation source, and by changing the optical path, it can achieve high radiation heating of the cellulose sample. The intensity of heat flux at the focal point is adjusted by changing the input power of the xenon lamp. The heating time of the sample is controlled by a steel rotary shutter with a response time of 0.01 s. In addition, a photocell with the sensitivity of 0.01 s is also placed at the focal point to determine the actual time of light passing through. Due to the extreme instability of active cellulose at high temperature, it needs to be rapidly cooled to room temperature at the moment heating

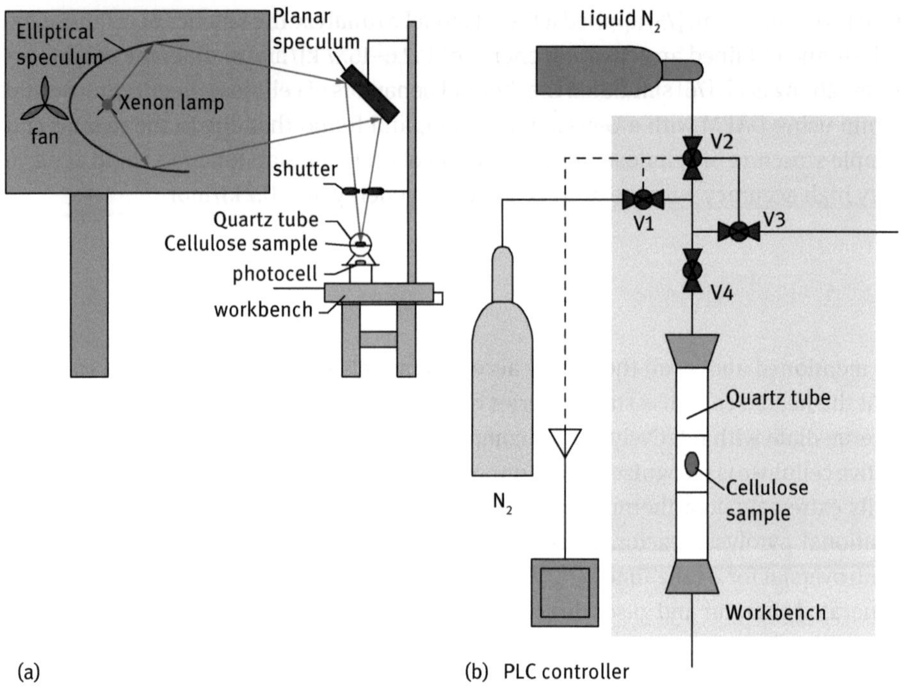

Fig. 2.15: Xenon lamp high radiation flash pyrolysis mechanism apparatus; (a) heating and optical system, (b) cooling and controlling system. Reprinted with permission from [64].

is completed. To achieve this, liquid nitrogen is used in the apparatus for its excellent efficiency of endothermic vaporization. As shown in Fig. 2.15 (b), three electronic heating rods with power of 1 kW are placed in the liquid nitrogen storage bottle. When heated, liquid nitrogen will vaporize to produce a certain pressure. After the valve is opened, liquid nitrogen will be sprayed for cooling.

Cellulose flash pyrolysis can be realized using the focused strong light from the xenon lamp. The experimental results show that a small amount of pale yellow product forms on the surface of the sample in a very short heating time. With increasing exposure time, the yield of pale yellow product increases, and the color gradually deepens to brown, finally, the sample turns into black char. After dissolving the heated sample in water followed by filtering and evaporating, some pale yellow gelatinous products are obtained. The images obtained from scanning electron micrograph of the product are shown in Fig. 2.16. Microcrystalline cellulose presents a granular shape with different sizes and the surface profiles are very clear. After being flash heated, the color of particles turns from white into pale yellow accompanied by the disappearance of surface profiles and appearing in molten state and cellulose particle shapes also change. Although each particle remains independent to some extent, they have begun to stick to each other and the molten particles start to agglomerate. Thus, it

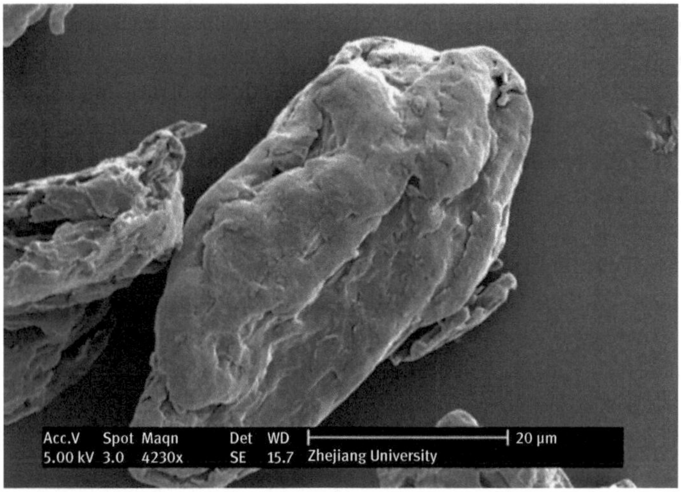

(a)

(b)

Fig. 2.16: SEM images of microcrystalline cellulose and the yellow gelatinous product.

can be inferred that during flash heating process, the microcrystalline cellulose goes through a series of chemical reactions and generates an intermediate that is unlike cellulose and tar.

Further structural analysis of the pale yellow gelatinous product shows that its FTIR spectrum is very similar to that of microcrystalline cellulose. The two substances have consistent characteristic absorption peaks with differences only in the intensity of some peaks, indicating that they may have similar molecular structure. The absorption peaks of C=O groups of the intermediate show stronger intensity, which may result from the formation of aldehyde groups on C_6 position or the ketone groups on C_2 and C_3 positions in the glucose units via dehydration and molecular rearrangement. The pale yellow gelatinous product was also analyzed by high performance liquid chromatography (HPLC), and the results are displayed in Fig. 2.17. Peak 1 corresponds to an unknown product with an average molecular weight of about 2,200. Considering that the molecular weight of β-D-glucose is 162, it can be inferred that peak 1 may correspond to the oligosaccharide with average DP of about 14. Peak 2 and peak 3 correspond to cellobiose and glucose, respectively, and both of them may be the decomposition product from the oligosaccharide mentioned above. Due to the randomness of glycosidic bond cleavage occurring during the depolymerization process, the products certainly contain cellotriose, cellotetraose, and other oligosaccharides. Methylglyoxal with a relatively high content is derived from the further decomposition of monosaccharides. Thus, the mixture of oligosaccharides and monosaccharides can stably exist at room temperature but easily decomposes at high temperature. Therefore, active cellulose can be referred to as the mixture of oligosaccharides and monosaccharides (the pale yellow gelatinous intermediate) produced from cellulose pyrolysis.

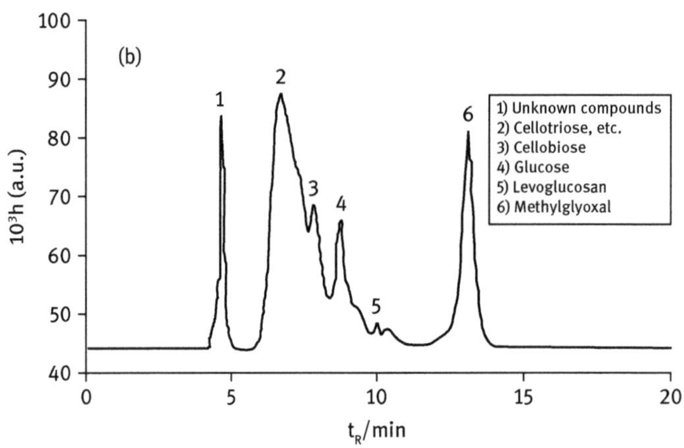

Fig. 2.17: HPLC analysis of the yellow intermediate product.

2.4.2 Effects of different factors on the characteristics of active cellulose

The experimental results show that heating time and intensity of heat flux influence the properties of the yellow product. Fig. 2.18 displays the effects of various heating times on the yield of yellow product with different heat flux. Under a high heat flux condition, the sample surface turns slightly yellow at a heating time of 0.09 s, and the corresponding yield of collected yellow product is only 0.001 g. With increased heating time, the extent of the reaction keeps deepening and the yield of yellow product increases obviously. At a heating time of 0.53 s, the yield increases to 0.014 g and then remains stable. When the heating time reaches 0.86 s, the yield increases again. Under a low heat flux, with increasing heating time, the yield of yellow product shows a similar changing trend, but it is relatively flat. The yield is only 0.006 g at the heating time of 1.15 s, which is much less than that at the same heating time under a high heat flux. When the heating time is longer than 1.64 s, the yield increases steadily again. Fig. 2.19 displays the yellow products obtained at a low heat flux with different heating times.

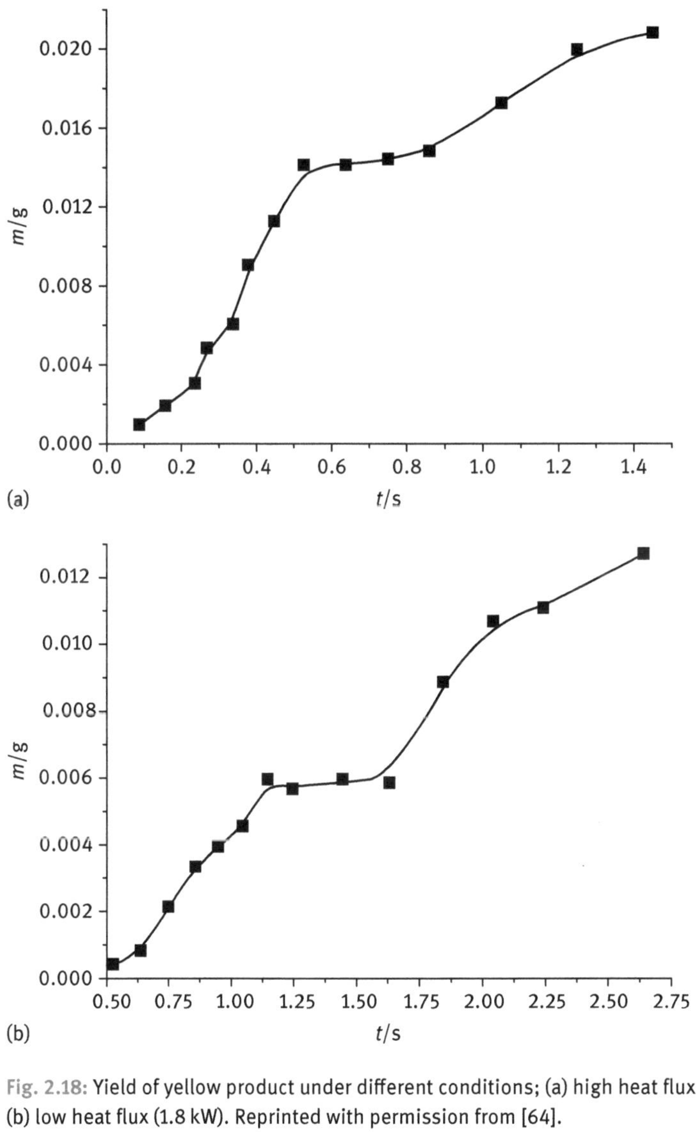

Fig. 2.18: Yield of yellow product under different conditions; (a) high heat flux (3 kW), (b) low heat flux (1.8 kW). Reprinted with permission from [64].

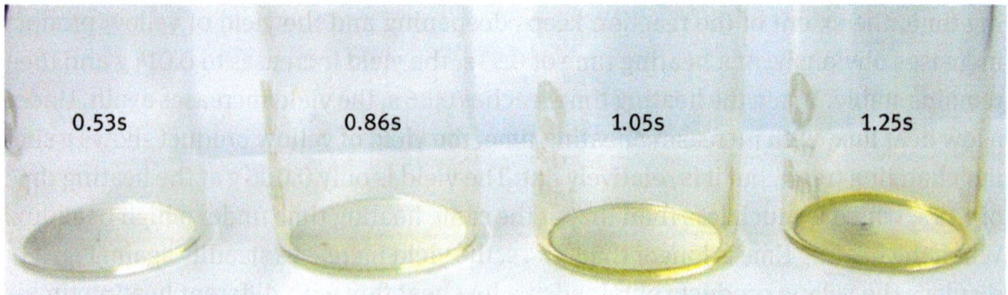

Fig. 2.19: Yellow solid product generated at lower heat flux with various flash times. Reprinted with permission from [64].

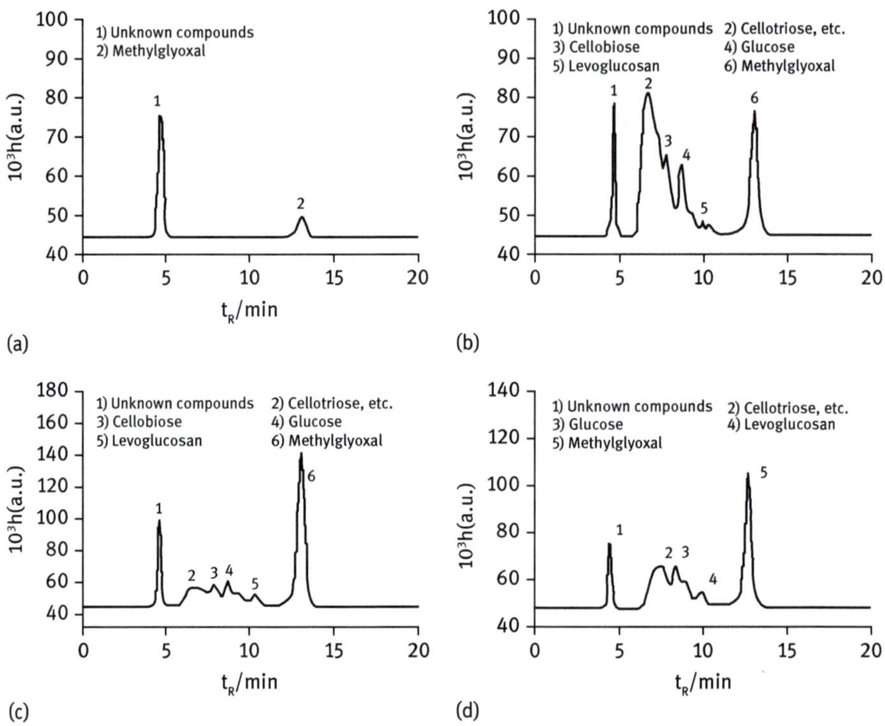

Fig. 2.20: Composition of the yellow products obtained under different reaction conditions; (a) high heat flux, 0.09 s, (b) high heat flux, 0.53 s, (c) high heat flux, 1.45 s, (d) low heat flux, 1.45 s. Reprinted with permission from [64].

The composition of the yellow products collected under typical conditions is shown in Fig. 2.20. At a high heat flux (3 kW), when heating time is 0.09 s, the products are mainly composed of an unknown product (peak 1) and a small amount of methylglyoxal. When the reaction time is extended to 0.53 s, a large amount of saccharides begin to be generated. The peaks 3, 4, and 5 correspond to cellobiose, glucose and levoglucosan, respectively. In addition, a small amount of fructose may exist as an isomer of glucose. Depending on the randomness of cleavage of glycosidic bonds, peak 2 corresponds to a mixture of oligosaccharides such as cellotriose, cellotetraose, and cellopentose. When heating time is 1.45 s, the types of product remain almost unchanged, but the intensities of the peaks corresponding to cellobiose, cellotriose, and glucose reduce obviously, indicating that these chemicals are instable at high temperature and decompose easily and rapidly. In contrast, the intensity of peak 6 corresponding to methylglyoxal is enhanced significantly, suggesting that the formation of oligosaccharides and methylglyoxal occurs simultaneously in competition or consecutively. Fig. 2.20 (d) shows the yellow product's composition obtained at a low heat flux (1.8 kW) when the heating time is 1.45 s. Here the broad peak 2 corresponds to a mixture of oligosaccharides such as cellobiose and cellotriose. The absence of a peak at the retention time of 6.5 min implies that the DP of the oligosaccharides generated at that time is relatively low.

By analyzing the soluble products obtained at different heat flux intensities, the production of the corresponding active cellulose can be compared, as shown in Fig. 2.21. At a high heat flux (3 kW), active cellulose starts to form at 0.09 s. Subsequently, the yield shows a rapid increase and its content in the soluble products is up to 68 %. The formation rate of active cellulose is much faster than its consumption rate in the initial stage, after that the consumption rate becomes dominant, leading to a continuous decrease in the active cellulose content. In addition, the monosaccharide content, such as glucose in the soluble products, is only about 10 %, and remains almost unchanged in the whole pyrolysis process. The formation and evolution of cellobiose and cellotriose, especially the dominant oligosaccharides with higher DP, can reflect the overall changing of active cellulose. At a low heat flux (1.8 kW), the yield of active cellulose also first increases and then decreases. After 1.8 s, the entire reaction tends to balance with a slightly decreasing trend, and the yield of active cellulose is moderate with a maximum of 57 %, which is much less than that at high heat flux. The different distribution of the saccharides indicates that the secondary pyrolysis of the oligosaccharides is suppressed, leading to a high yield of oligosaccharides. At low radiant heat flux, the reaction time is relatively long so that the generated oligosaccharides cannot be cooled immediately and will be further decomposed to the oligosaccharides with lower DP and even the small volatiles and char. Thus it confirms that active cellulose is unstable at high temperatures, as reflected by its formation and further evolution.

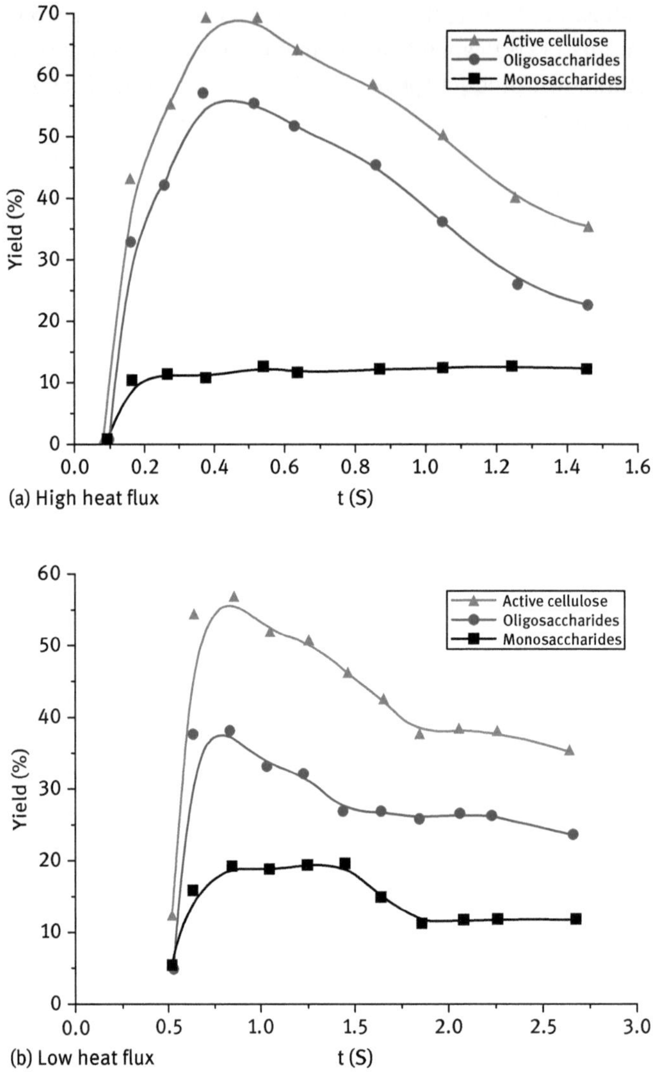

Fig. 2.21: Yield of active cellulose and its composition vs heating time. Reprinted with permission from [64].

2.5 Mechanism of cellulose pyrolysis based on the formation of products

An important approach to elucidating the mechanism of cellulose pyrolysis is research on the formation and evolution of pyrolysis products. Analyses of the composition of cellulose pyrolysis volatiles have found that the volatiles are mainly composed of pyrans such as levoglucosan and levoglucosenone, furans such as 5-hydroxymethylfurfural and furfural, linear small molecular chemicals such as hydroxyacetone and hydroxyacetaldehyde, and some small gases such as CO_2, CO, and CH_4 at medium or high reaction temperature.

Classifying the products according to their chemical structure will benefit mechanism research on cellulose pyrolysis. A possible scheme of cellulose pyrolysis is shown in Fig. 2.22. In the initial stage of cellulose pyrolysis, it first depolymerizes to oligosaccharides, followed by the further cleavage of the β-1,4-glycosidic bonds, leading to the formation of glucopyranose or other products. Typically there are three competing reaction paths that generate pyrans, furans and linear small molecular chemicals, respectively. The condensation reaction of the hydroxyl group at C_6 position and oxygen radical at C_1 position leads to the formation of levoglucosan, then the further dehydration reaction of the hydroxyl groups at C_3 and C_4 positions generates levoglucosenone. The formation of 1,4:3,6-Dianhydro-α-D-glucopyranose mainly results from the double dehydration reactions of the hydroxyl groups at C_1, C_4 and C_3, C_6 positions in the glucopyranose. In the process of pyran formation, the main reaction path is the dehydration of the generated levoglucosan to form levoglucosenone. The formation of acetaldehyde and hydroxyacetaldehyde competes with that of levoglucosan, and the yield of linear small molecular chemicals is enhanced by a reduction of the pyran yield. The furans, mainly furfural and 5-hydroxymethylfurfural, are more likely to be generated from the decomposition of glucopyranose units. The formation

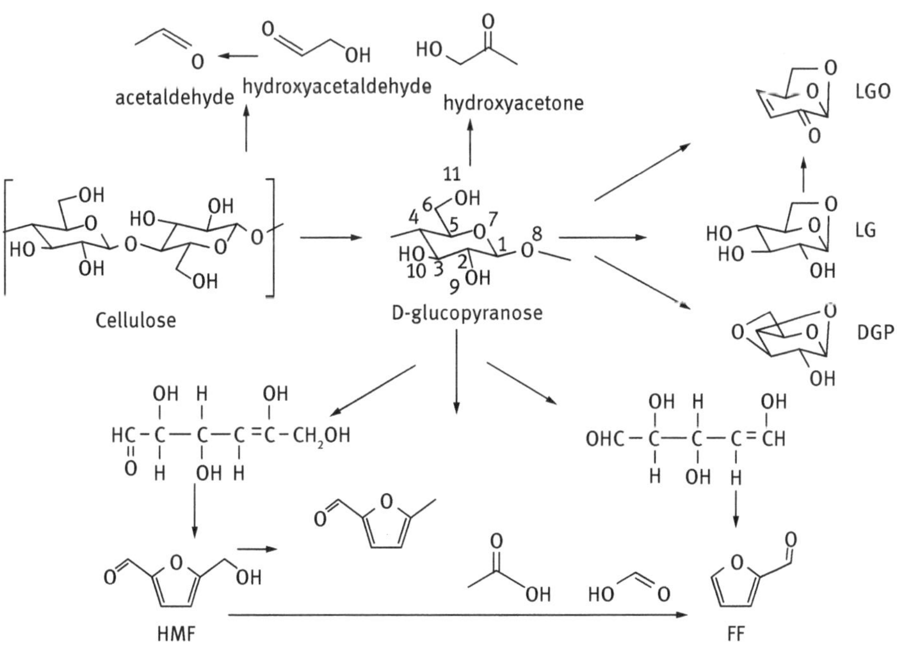

Fig. 2.22: Formation paths of the main products of cellulose pyrolysis.

of 5-hydroxymethylfurfural involves the following reactions: first, a ring-opening reaction through the C_1-O_7 bond of glucopyranose occurs, and subsequently intramolecular dehydration of the hydroxyl group generates the corresponding hexose with linear structure. The following dehydration of hydroxyl groups at C_2 and C_5 positions and final cyclization lead to the formation of 5-hydroxymethylfurfural. With respect to furfural, it can be generated from the removal of the hydroxymethyl of 5-hydroxymethylfurfural by secondary cracking.

In research on the mechanism of cellulose pyrolysis, cellobiose and glucose are usually selected as model compounds to represent cellulose. Cellobiose is the basic polymerized unit of cellulose, while glucose is the basic structural unit of cellulose. Compared to cellulose and cellobiose with polymeric structures, the thermal weight loss of glucose is less obvious, resulting from the absence of intermolecular glycosidic bonds, which leads to less active glucose structures and difficulty in cleaving chemical bonds [10, 80]. So a higher reaction temperature is required for glucose to get the same weight loss. However, the pyrolysis products of glucose are basically consistent with those of cellobiose and cellulose, only having some differences in the relative content. Among cellulose pyrolysis products, linear small molecular chemicals and pyrans like levoglucosan constitute the main content and furans also show a higher content in the products of cellobiose and glucose pyrolysis. Patwardhan et al. [9] further found that the formation of levoglucosan and other pyrans was directly correlated to the DP of sample: the higher the DP, the higher the yield of pyrans. Comparing the pyrolysis products of sucrose (a dimer of D-glucose and D-fructose) and cellulose, it was found that the furfural content of sucrose pyrolysis was up to 67.1 %, while that of cellulose pyrolysis was only 21 %; and furfural and 5-hydroxymethylfurfural were the main pyrolysis products of sucrose in the temperature range 350–850 °C [81]. By comparing the pyrolysis behaviors of cellobiose and glucose to that of cellulose, the difference in the products from the D-glucopyranose unit and its polymers can be investigated, which will be beneficial for deducing the products formation mechanism in cellulose pyrolysis.

2.5.1 Formation pathway of levoglucosan

Levoglucosan is a typical product of cellulose pyrolysis, and it is also an important intermediate to generate other volatile compounds. Its formation and evolution play a crucial role in cellulose pyrolysis. The maximum yield of levoglucosan reported in the literatures is up to 70 wt% [82]. In early studies, levoglucosan was considered to be generated from the further reaction of glucose units produced by the hydrolysis of cellulose. Ivanov et al. [83] carried out pyrolysis experiments with cellulose soaked in glucose solution and found that the yield of levoglucosan increased by 50 % compared to cellulose without pretreatment. However, further experiment revealed that glucose pyrolysis produced a much lower yield of levoglucosan compared with cellulose pyrolysis [9]. The results show that levoglucosan indeed can be generated from glucose pyrolysis, but the glucose unit is not the main precursor of levoglucosan in the cellulose pyrolysis process, and the formation of levoglucosan from cellulose pyrolysis is directly related to the cleavage of glycosidic bonds. With respect to the pattern of cleavage of glycosidic bonds, researchers have proposed various mechanism paths. Kislitsyn et al. [84] proposed a detailed reaction path of levoglucosan formation including hemolysis, rearrangement, and dehydration, which is based on a definite radical mechanism from the perspective of the moving and changing of atoms. Ponder et al. [85] put forward a two-step reaction model: cleavage of glycosidic bonds of cellulose occurred first and ion intermediates were generated, followed by the condensation of hydroxyl groups at C_1 and C_6 positions to form levoglucosan. Mamleev et al. [86] further deduced that in the cellulose pyrolysis process, the interaction between the hydroxyl group in hydroxymethyl group and glycosidic bond generated a transition state of a four-membered ring, then the concerted reaction broke the glycosidic bond and generated levoglucosan. In summary, the formation of levoglucosan mainly occurs in two steps: the cleavage of glycosidic bonds gives the glucopyranose free radicals, and the condensation of the hydroxyl group at C_6 position with the oxygen free radicals at C_1 position leads to the formation of levoglucosan.

It should be noted that the process of cellulose dehydration to form levoglucosan is different from that to form anhydrocellulose. The dehydration in the process of anhydrocellulose formation mainly refers to the cleavage of hydrogen bonds in the cellulose chain, which requires much lower energy, while the formation of levoglucosan involves the cleavage and arrangement of intramolecular chemical bonds, which needs plenty of heat. Thus levoglucosan is the product of cellulose pyrolysis at higher temperature [31]. Moreover, levoglucosan can be further dehydrated to levoglucosenone at pyrolysis temperature of 250–400 °C [87].

2.5.2 Formation pathway of 5-hydroxymethylfurfural

Furan compounds are another important product of cellulose pyrolysis, mainly composed of furfural and 5-hydroxymethylfurfural. Compared to pyrans, furans are more stable and difficult to decompose, so they can be recognized as the final products of cellulose pyrolysis [88]. 5-hydroxymethylfurfural is highly water soluble and is mainly derived from the pyrolysis of glucose units. As shown in Fig. 2.23, ring-opening of glucopyranose first occurs through the breakage of the C_1–O bond to form a glucose chain, followed by dehydration of hydroxyl groups at C_2 and C_5 positions and subsequent cyclization to form an intermediate, then double dehydration of the intermediate generates the final product 5-hydroxymethylfurfural. There are also other scholars who believe that in the formation of 5-hydroxymethylfurfural the glucose chain first undergoes double dehydration before cyclization to yield the final product [89]. It should be noted that both of the two paths are likely to occur and there is no definite conclusion at present. Recently, some researchers suggested that furans were directly generated from the decomposition of cellulose macromolecules rather than from the secondary cracking of the pyran rings resulting from the primary pyrolysis of cellulose [90]. Furfural is mainly generated from the removal of the hydroxymethyl from 5-hydroxymethylfurfural. Shin et al. [91] analyzed the chemical bond energy of 5-hydroxymethylfurfural and found that the two weakest bonds of 5-hydroxymethylfurfural are located on the hydroxymethyl group, and therefore the possibility of overall fracture of the hydroxymethyl group was much greater than that of the side aldehyde chain, leading to the predominance of furfural as the further decomposition product of 5-hydroxymethylfurfural. Furfural has very high chemical stability and is difficult to decompose generally.

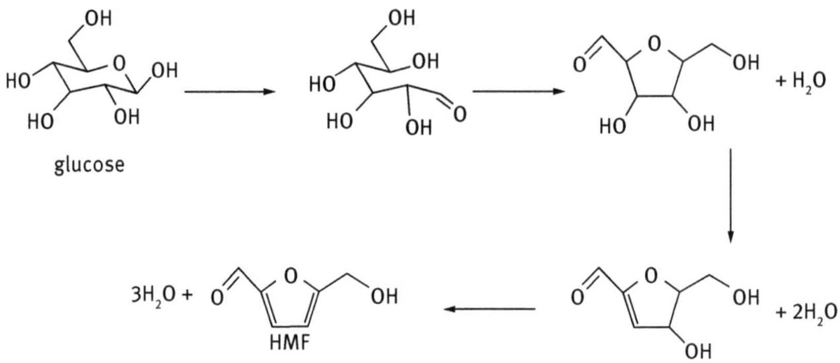

Fig. 2.23: Formation path of 5-hydroxymethylfurfural.

2.5.3 Formation pathway of hydroxyacetaldehyde and hydroxyacetone

Hydroxyacetaldehyde and hydroxyacetone belong to high yield linear small molecular products of cellulose pyrolysis. Due to the still undetermined competitive or successive relationship between levoglucosan and hydroxyacetaldehyde formation, research on the hydroxyacetaldehyde formation mechanism has been a focus in the field [92, 93]. At present, three paths of hydroxyacetaldehyde formation have been proposed, as shown in Fig. 2.24. In the first one, intramolecular dehydration of the glucose unit or levoglucosan occurs first and the subsequent cleavage of the C–C bond in the carbon skeleton and ring-opening of the hemiacetal group generate ethylene glycol, followed by arrangement to form hydroxyacetaldehyde; in addition, the resulting hydroxyacetaldehyde may further undergo dehydration of the hydroxyl group and the adjacent hydrogen atom to form ethanedial, which can decompose to CO and methanol. The second path proposes that the ring-opened glucose unit decomposes to formaldehyde and five-carbon molecular fragment, then bond cleavage and dehydration of the molecular fragment occur to generate hydroxyacetaldehyde. The third path

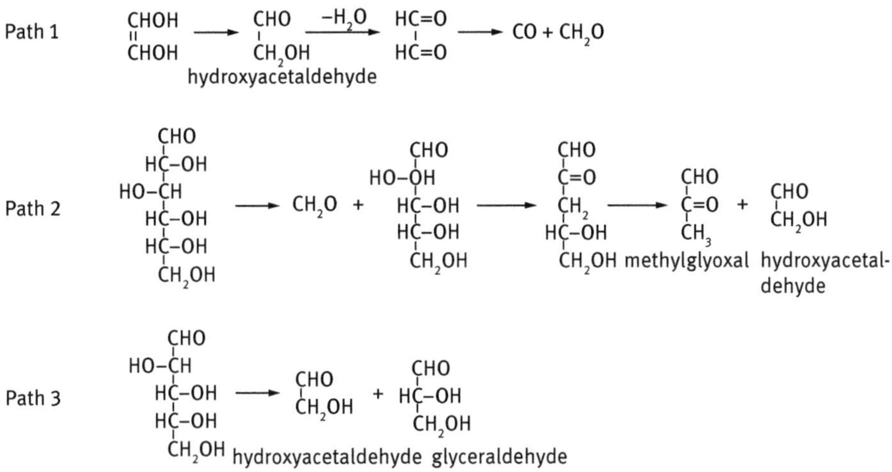

Fig. 2.24: Formation paths of hydroxyacetaldehyde.

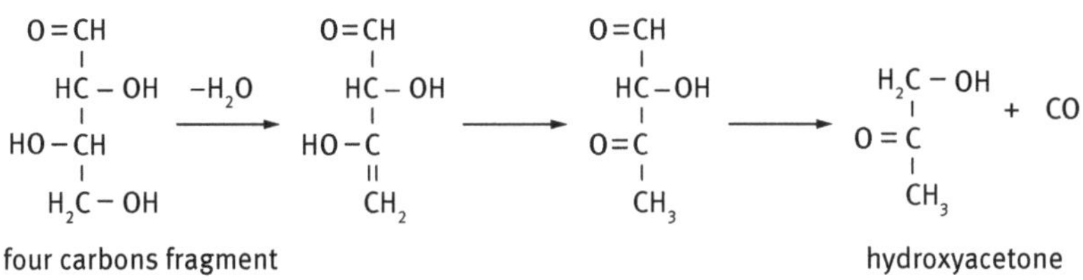

Fig. 2.25: Formation path of hydroxyacetone.

proposes that the five-carbon molecular fragment directly generates hydroxyacetaldehyde by cracking.

The formation mechanism of hydroxyacetone can be analyzed from the perspective of bond energy [17]. The lengths of C–C bonds in cellulose monomer D-glucopyranose structure are not all the same: the length of C_2–C_3 is the longest, and the hemiacetal C_1–O bond in cellulose monomer structure is more active than other C–O bonds. Thus these two bonds in the ring of the cellulose unit may be broken in the pyrolysis process, leading to the production of molecular fragments containing two carbons and four carbons. As shown in Fig. 2.25, the four carbon fragment undergoes successive dehydration and carbonyl transfer to form hydroxyacetone and with the strengthening of reaction conditions, hydroxyacetone is also prone to decompose into small molecular products such as formaldehyde and acetaldehyde. Paine et al. [94] performed pyrolysis of the isotope labeled D-glucose and found that the three carbon atoms in hydroxyacetone were derived from the C_4, C_5, and C_6 in glucose. They proposed that the formation process of hydroxyacetone mainly occurred through glucose dehydration and subsequent cleavage of the C_3–C_4 bond.

2.5.4 Formation pathway of small molecular gases

The small molecular gases of cellulose pyrolysis are mainly CO and CO_2. The existence of a large number of hydroxyl groups and the common occurrence of dehydration favors the generation of carbonyl group in the cellulose pyrolysis process. The unstable carbonyl groups are prone to being broken by arrangement and isomerization to form CO under the conditions of high temperature or long gas phase residence time. For example, hydroxyacetaldehyde will decompose to formaldehyde and CO at high temperature as shown in Fig. 2.24. As a permanent small molecular gas, the yield of CO is quite low in the initial pyrolysis process and increases with the deepening of the proceeding reaction, thus CO is always considered as the product of the secondary reaction of the primary volatiles.

Compared to CO, the formation of CO_2 mainly occurs at low temperature with a stable yield, while the secondary reaction that occurs at high temperature makes a small contribution to its formation [95]. It is generally recognized that CO_2 is generated from the secondary reaction of anhydrocellulose formed by cellulose carbonization at low temperature. In addition, the isomerization of the enone or enal structure of the primary volatiles in the secondary reaction process will generate carboxyl groups, which will further decompose to CO_2. However, formation of CO_2 from the latter path is much lower than that from the former. Usually, the formation of CO_2 is accompanied by the production of a variety of small molecular gases, such as propionaldehyde, acetone, 3-hydroxy-2-butanone, etc. Patwardhan et al. [22] studied the primary and secondary reactions of cellulose pyrolysis in a pyrolysis apparatus and fluidized bed reactor, and found that when only primary reactions occurred, the yield of CO and CO_2 were 0.7 wt% and 3.3 wt% respectively. They increased to 4.3 wt% and 3.9 wt%, respectively, when secondary reactions occurred, confirming that CO was mainly formed from the secondary reactions while CO_2 was mainly generated from the primary reactions.

In addition to CO and CO_2, there are also small quantities of small molecular gases, such as H_2, CH_4, and C_{2+} hydrocarbons in the products of cellulose pyrolysis. Both Yang et al. [96] and Qu et al. [7] observed the formation of the gases mentioned above, and their yields were enhanced by increasing temperature; H_2 was generally derived from the cracking and deformation of C=C and CH groups; and CH_4 was mainly formed from the cleavage of methoxy groups while C_{2+} hydrocarbons were generated from the secondary reactions at high temperature.

2.6 Mechanism of cellulose pyrolysis at molecular scale

Computational chemistry is an important branch of theoretical chemistry and mainly based on the theory of quantum chemistry. By simulation with special computer software, molecular structure and some properties relating to chemical reactions, such as binding energy, bond energy, molecular steady state, reaction energy barrier, vibration frequency, reactivity, etc., can be obtained and used to illustrate the relevant chemical phenomenon. With the development of computer hardware and software, a series of mature commercial computational chemistry software, such as Gaussian, VASP, Materials Studio, etc., are widely used and can achieve satisfactorily accurate simulation results at molecular level. After simplifying the reaction conditions and molecular structure, theoretical simulations of the pyrolysis process can be carried out based on previous experimental results. Such simulations are useful for judging the validity of the proposed pyrolysis mechanism from the perspective of reaction thermodynamics and kinetics. Detailed introductions to the computational theory are available in the reference [97]. The widely used density functional theory (DFT) is an accurate calculation method based on quantum chemistry. DFT method has developed very rapidly in recent years and now is successful in predicting molecular structure. DFT method takes the electron correlation effect of the system into consideration and can directly determine the ground state energy and electron density, without going through the complicated calculation of multi-electron wave function, thus DFT method can greatly reduce the complexity of calculating electron structure. Currently, the application of DFT method is being extended to studies involving excited state and time-dependent ground state.

The conversion paths of cellobiose or glucose monomer can be calculated using DFT method. Cellulose macromolecules containing thousands of atoms cannot be directly simulated by the current level of computing power due to the amazingly complex calculation even using DFT method. Therefore it is more reasonable to adopt molecular dynamics for simulating cellulose with a periodic crystal structure. Molecular dynamics is based on Newtonian mechanics without considering electron correlation effects, meaning that the structure of the molecular system is simulated by the force of atoms and molecules, which greatly reduces the calculation complexity. In addition, the thermodynamic parameters and other macroscopic properties of the system can be obtained by further calculations based on the resulting mechanical optimal structures.

The main role of computational chemistry in mechanism studies is to calculate the reaction activation energy (reaction energy barrier) to verify the proposed reaction paths and propose the most possible reaction path. The activation energy is defined as the energy difference between the transition state and the reactants. According to the complexity of the model and the different calculation methods, three kinds of saccharides, namely, glucose monomer, cellobiose/cellotriose and cellulose with periodic crystalline structure, are generally used in cellulose reaction mechanism simulation. The focus of simulation lies on two aspects, one is the ring-opening of the pyran ring and the other is the cleavage of the glycosidic bond.

2.6.1 Simulation of pyrolysis of cellulose monomer

As revealed in previous experiments, hydrolysis of cellulose with phosphoric acid in a microwave reactor could produce more than 90 % yield of glucose, indicating that the cleavage of glycosidic bonds in cellulose will preferentially occur to generate a large number of glucose monomers under certain hydrolysis conditions [98]. In the thermochemical conversion process, D-glucopyranose is also the most important intermediate in cellulose decomposition. Dobele et al. [30] hydrolyzed the residues of cellulose slow pyrolysis and the glucose yields obtained were used to judge the alteration of D-glucopyranose structure in cellulose with increasing temperature. The results showed that D-glucopyranose had good thermal stability and was well preserved in the cellulose structure when the temperature was below 280 °C. However, with increasing temperature, the glucopyranose yield decreased gradually and reached zero at about 400 °C. Cellulose pyrolysis mainly occurred in the temperature range of 300–400 °C, and the weight loss at the initial or terminal stage was very small [96]. Thus, the decomposition of glucose can be used to approximately represent the cellulose pyrolysis process. Compared to cellulose, the pyrolysis of D-glucopyranose generates more furans and fewer pyrans and small molecular gases. However, all the main products of cellulose pyrolysis can also be found in the products of D-glucopyranose pyrolysis [99]. Thus, it is an effective research method to select D-glucopyranose as a precursor to simulate the formation process of cellulose pyrolysis products. Huang et al. [100, 101] performed thermodynamic and kinetic simulation of the reaction process of β-D-glucopyranose pyrolysis by using DFT method with B3LYP/6-31++G(d,p) basis set with the help of the Gaussian 03 suite of programs.

Based on the results of Py-GC/MS experiments, the product formation paths of D-glucopyranose pyrolysis have been deduced and are shown in Fig. 2.26. Path 1 is the formation of levoglucosan (P1) by dehydration of the hydroxyl groups at C_6 and C_1 positions of the D-glucopyranose monomer; path 2 is the formation of 3,4-anhydro-altrose (P2) by the dehydration of hydroxyl groups at C_3 and C_4 and the subsequent isomerization; path 3 is related to the formation of 5-hydroxymethylfurfural, which involves the cleavage of the O_4–C_1 bond to form a carbon-carbon linear chain and further cyclization to give a furan ring unit, followed by double dehydration to form 5-hydroxymethylfurfural (P3); path 4 is the formation of furfural by the direct removal of the hydroxymethyl group of 5-hydroxymethylfurfural, simultaneously accompanied by the formation of formaldehyde (P5). In Fig. 2.26, IM is intermediate, TS is transition state. All the calculations are performed using the Gaussian 03 suite of programs with the B3LYP/6-31++G(d,p) basis set.

Firstly, the bond strength of D-glucopyranose is assessed to determine the bonds where a ring-opening reaction most easily occurs and Mulliken populations of the bonds are selected as the judgment [102]. The larger the Mulliken population, the more stable the bond. As shown in Tab. 2.2, the Mulliken populations of C–H bonds have the maximum valves with a range of 0.34–0.36, indicating their high stability;

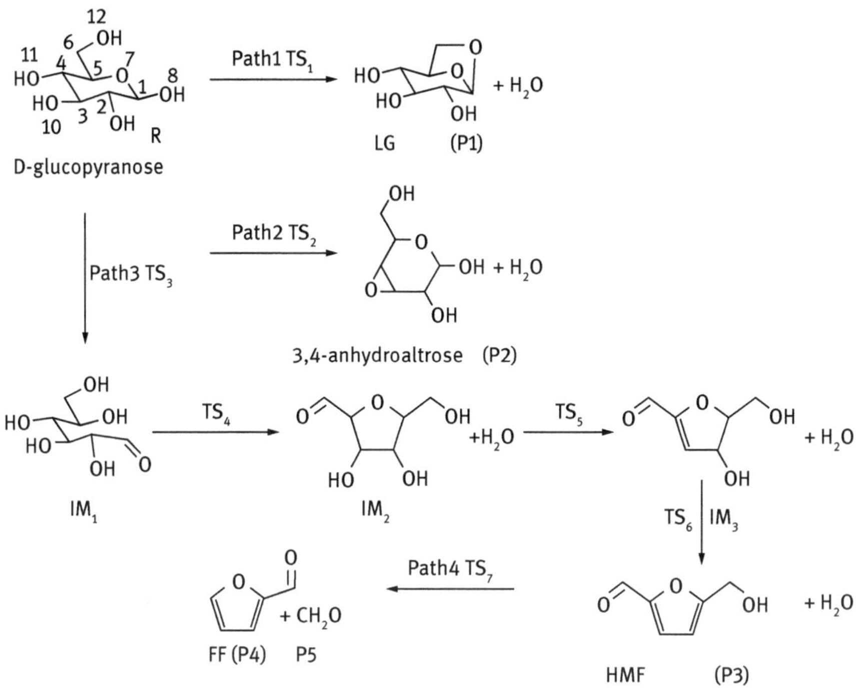

Fig. 2.26: Reaction paths of D-glucopyranose monomer pyrolysis.

Tab. 2.2: Mulliken populations of each bond in glucopyranose structure.

Bond	Mulliken population	Bond	Mulliken population	Bond	Mulliken population
C_1–C_2	0.17	C_2–O_9	0.18	C_5–H_{17}	0.35
C_2–C_3	0.20	C_3–O_{10}	0.21	C_6–H_{18}	0.34
C_3–C_4	0.17	C_4–O_{11}	0.18	C_6–H_{19}	0.36
C_4–C_5	0.17	C_6–O_{12}	0.22	O_8–H_{20}	0.24
C_5–C_6	0.26	C_1–H_{13}	0.36	O_9–H_{21}	0.24
C_5–O_7	0.17	C_2–H_{14}	0.34	O_{10}–H_{22}	0.24
C_1–O_7	0.15	C_3–H_{15}	0.35	O_{11}–H_{23}	0.24
C_1–O_8	0.21	C_4–H_{16}	0.35	O_{12}–H_{24}	0.23

followed by the Mulliken populations of O–H bonds with a range of 0.23–0.24. The Mulliken populations of both the C–C and the C–O bonds are distributed over quite a wide range, among these bonds, the C_5–C_6 bond has the largest Mulliken population and is thus difficult to break, and the C_1–O_7 bond has the smallest Mulliken population, indicating that C_1–O_7 is easily broken, involving the ring-opening of the pyran ring. The hydroxyl groups at C_2 and C_4 positions are more active due to their lower Mulliken populations compared to other hydroxyl groups.

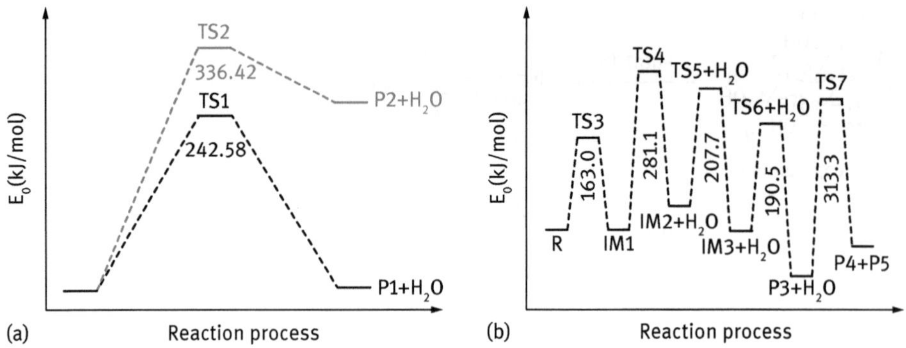

Fig. 2.27: Potential energy profiles of reaction paths of glucose pyrolysis.

By performing further calculations on the four proposed reaction paths mentioned above, the resulting energy profiles of the reaction paths were obtained and are shown in Fig. 2.27. Comparing path 1 with path 2, it can be seen that the energy barrier of 3,4-anhydro-altrose formation (336 kJ/mol) is much higher than that of levoglucosan formation (243 kJ/mol) through dehydration. Therefore, it is easier to generate levoglucosan from the initial dehydration of D-glucopyranose, which is in line with observations in experiments. Lu et al. [103] used Py-GC/MS to analyze the products of cellulose pyrolysis, and found that the levoglucosan content was greatly higher than that of 3,4-anhydro-altrose, and actually the latter content generally was very low. In path 3, D-glucopyranose undergoes four transition states with the formation of three intermediates to finally generate 5-hydroxymethylfurfural. Among the four transition states, TS4 requires the highest energy, corresponding to cyclization, while the energy required for TS3 is the lowest, indicating that the ring-opening of the glucose is the easiest reaction. In path 4, 5-hydroxymethylfurfural is directly converted to furfural and formaldehyde via TS7 by the removing of the hydroxymethyl group with the energy barrier of 313 kJ/mol. All of the Mulliken populations of the C–H bond in 5-hydroxymethylfurfural are quite large, while the Mulliken populations of the two bonds of the aldehyde group are smaller, indicating the easy removal of the hydroxymethyl group to form furfural. Due to the largest Mulliken population of the aldehyde group connected to C_1, 5-hydroxymethylfurfural is prone to form furfural by removal of the hydroxymethyl group in the pyrolysis process. In addition to glucose, anhydroglucose is also used as model compound of cellulose monomer to simulate pyrolysis. Zhang et al. [104] discussed the formation paths of levoglucosan and formaldehyde and concluded that the formation of levoglucosan was easier than that of formaldehyde after comparing the energy barriers. Next, Zhang et al. [105] also simulated the pyrolysis of levoglucosan. The results showed that levoglucosan could decompose to 1-pentene-3,4-dione, acetaldehyde, 2,3-dihydroxypropanal, and propanedialdehyde by the cleavage of C–O bonds, and to 1,2-dihydroxyethene and hydroxyacetic acid vinyl ester by the cleavage of C–C bonds, and the paths involving the cleavage of C–O bonds had a lower energy barrier.

2.6.2 Simulation of pyrolysis of cellobiose and cellotriose

Cellobiose is the disaccharide formed from the combination of two molecules of β-D-glucose via a β(1→4) glycosidic bond, and is the basic repeating unit of cellulose. Compared to glucopyranose monomer, a cellobiose molecule contains a glycosidic bond, so carrying out simulation of cellobiose pyrolysis can offer additional information about the breaking of intermolecular bonds. Huang et al. [106] used DFT method with UB3LYP/6-31G(d) basis set to simulate cellobiose pyrolysis, and found that the Mulliken population of the glycosidic bond between the two glucose units was the lowest, indicating the most likely occurrence of a cleavage of the glycosidic bond; and further kinetics analyses of three proposed reaction paths revealed that one cellobiose molecule most likely decomposed to one levoglucosan molecule and one glucopyranose molecule by synergistic reaction via the transition state of four-membered ring structure, and the energy barrier of this reaction was 378 kJ/mol.

In a natural cellulose structure, cellobiose is the basic polymerized unit. However, there are also other molecular groups connected to the terminal of the cellulose molecule, and the terminal position shows very strong activity and very easily forms hydrogen bonds. Therefore, methyl cellobiose is often used as a substitute for cellobiose for pyrolysis simulation [107]. In order to study the cleavage of glycosidic bonds between glucose units, in recent years, many scholars performed simulations with cellotriose or the oligosaccharides with higher DP. Zhang et al. [108] used DFT method to simulate the pyrolysis process of cellotriose, and found that the dehydration behavior of the hydroxyl groups at different positions differed significantly, wherein the dehydration of the hydroxyl group at C_2 position occurred more easily.

2.6.3 Simulation of pyrolysis of cellulose crystal with periodically repeated structure

Atalla and Vanderhart et al. [109] analyzed the structural characteristics of cellulose crystal by ^{13}C CP/MAS NMR and found that cellulose in nature exists in two different crystalline phases, namely, I_α and I_β, corresponding to a triclinic unit cell with one chain and a monoclinic unit cell with two chains, respectively [110]. Valonia and bacterial celluloses are rich in I_α crystal, while the celluloses of angiosperms are mainly composed of I_β crystal. Recently, with the development of quantum chemistry theory and computer hardware and software, simulation of the reaction system with huge numbers of atoms and periodic structure can be carried out and the results are acceptable. Compared to monomers or oligosaccharide with limited units, the cellulose macromolecule will be more representative.

Recent research shows that using the CPMD (Car–Parrinello molecular dynamics) method to simulate cellulose crystal pyrolysis can achieve good results [111]. Mettler et al. [90] employed CPMD software to simulate the pyrolysis of α-cyclodextrin, one kind of cyclic oligosaccharide with two-dimensional periodic structure formed from the polymerization of glucopyranose that can be approximately considered a cellulose molecule with sheet-like structure. They concluded that furan products were formed from the direct decomposition of the whole cellulose, rather than from the further reactions of glucopyranose or anhydro-sugar resulting from cellulose pyrolysis and the combined effect of homolytic cleavage of glycosidic bonds and internal dehydration of pyran units was the key to the whole pyrolysis process. Agarwal et al. [112] selected the cellulose crystal with three-dimensional periodic structure as the representative of cellulose and also used the CPMD method to study the reactions that occurred at the initial stage of cellulose pyrolysis, including depolymerization, fragmentation, ring-opening and cyclic condensation. The results showed that the cyclic condensation of cellulose had the lowest energy barrier of 84 kJ/mol at the temperature of 327 °C, indicating that the pyran ring could easily transform into a furan ring by cyclic condensation, and further be converted to furans via cleavage of molecular chains. At a temperature of 600 °C, the configuration of the expanded cellulose crystal had an important impact on the formation of pre-levoglucosan (the precursor that can be further converted into levoglucosan).

References

[1] Bradbury AG, Sakai Y, Shafizadeh F. A kinetic model for pyrolysis of cellulose. Journal of Applied Polymer Science. 1979;23(11):3271–3280.

[2] Li Y. Thermal analysis. Tsinghua University Press, Beijing; 1987.

[3] Maciel AV, Job AE, Da Nova Mussel W, et al. Bio-hydrogen production based on catalytic reforming of volatiles generated by cellulose pyrolysis: An integrated process for zno reduction and zinc nanostructures fabrication. Biomass and Bioenergy. 2011;35(3):1121–1129.

[4] Biagini E, Barontini F, Tognotti L. Devolatilization of biomass fuels and biomass components studied by TG/FTIR technique. Industrial & Engineering Chemistry Research. 2006;45(13):4486–4493.

[5] Lu Q, Xiong W, Li W, et al. Catalytic pyrolysis of cellulose with sulfated metal oxides: a promising method for obtaining high yield of light furan compounds. Bioresource Technology. 2009;100(20):4871–4876.

[6] Wu S, Shen D, Hu J, et al. TG-FTIR and Py-GC-MS analysis of a model compound of cellulose-glyceraldehyde. Journal of Analytical and Applied Pyrolysis. 2013;101:79–85.

[7] Qu TT, Guo WJ, Shen LH, et al. Experimental study of biomass pyrolysis based on three major components: hemicellulose, cellulose, and lignin. Industrial& Engineering Chemistry Research. 2011;50(18):10424–10433.

[8] Zhou XW, Nolte MW, Mayes HB, et al. Experimental and mechanistic modeling of fast pyrolysis of neat glucose-based carbohydrates. 1. Experiments and development of a detailed mechanistic model. Industrial & Engineering Chemistry Research. 2014;53(34):13274–13289.

[9] Patwardhan PR, Satrio JA, Brown RC, et al. Product distribution from fast pyrolysis of glucose-based carbohydrates. Journal of Analytical and Applied Pyrolysis. 2009;86(2):323–330.

[10] Mettler MS, Paulsen AD, Vlachos DG, et al. The chain length effect in pyrolysis: bridging the gap between glucose and cellulose. Green Chemistry. 2012;14(5):1284–1288.

[11] Luo Z, Wang S, Yanfen Liao, et al. Mechanism Study of Cellulose Rapid Pyrolysis. Industrial & Engineering Chemistry Research. 2004;43(18):5605–5610.

[12] Richards GN, Zheng G. Influence of metal ions and of salts on products from pyrolysis of wood: applications to thermochemical processing of newsprint and biomass. Journal of Analytical and Applied Pyrolysis. 1991;21(1):133–146.

[13] Mok WS, Antal MJ. Effects of pressure on biomass pyrolysis. II. Heats of reaction of cellulose pyrolysis. Thermochimica Acta. 1983;68(2):165–186.

[14] Luo Y, Yu FW, Nie Y, et al. Study on the pyrolysis of cellulose. Kezaisheng Nengyuan/ Renewable Energy Resources. 2010;28(1):40–43.

[15] Kojima E, Miao Y, Yoshizaki S. Pyrolysis of cellulose particles in a fluidized bed. Journal of Chemical Engineering of Japan. 1991;24(1):8–14.

[16] Rutkowski P. Pyrolytic behavior of cellulose in presence of montmorillonite K10 as catalyst. Journal of Analytical and Applied Pyrolysis. 2012;98:115–122.

[17] Piskorz J, Radlein D, Scott D S. On the mechanism of the rapid pyrolysis of cellulose. Journal of Analytical and Applied Pyrolysis. 1986;9(2):121–137.

[18] Norinaga K, Shoji T, Kudo S, et al. Detailed chemical kinetic modelling of vapour-phase cracking of multi-component molecular mixtures derived from the fast pyrolysis of cellulose. Fuel. 2013;103:141–150.

[19] Xin S, Yang H, Chen Y, et al. Assessment of pyrolysis polygeneration of biomass based on major components: Product characterization and elucidation of degradation pathways. Fuel. 2013;113:266–273.

[20] Lanza R, Dalle Nogare D, Canu P. Gas phase chemistry in cellulose fast pyrolysis. Industrial & Engineering Chemistry Research. 2008;48(3):1391–1399.

[21] Drummond AF, Drummond IW. Pyrolysis of sugar cane bagasse in a wire-mesh reactor. Industrial & Engineering Chemistry Research. 1996;35(4):1263–1268.

[22] Patwardhan PR, Dalluge DL, Shanks BH, et al. Distinguishing primary and secondary reactions of cellulose pyrolysis. Bioresource Technology. 2011;102(8):5265–5269.

[23] Wang S, Liao Y, Liu Q, et al. Experimental study of the influence of acid wash on cellulose pyrolysis. Journal of Fuel Chemistry and Technology. 2006;34(2):179–183.

[24] Hague RA. Pre-treatment and pyrolysis of biomass for the production of liquids for fuels and speciality chemicals. Aston University, 1998.

[25] Zhang YZ, Liu J, Gao PJ. Study on the ultrastructure of microcrystalline cellulose with scanning tunneling microscope. Journal of Chinese Electron Microscopy Society. 1997;6:48–50.

[26] Gravitis J, Vedernikov N, Zandersons J, et al. Furfural and levoglucosan production from deciduous wood and agricultural wastes: ACS Symposium Series. ACS Publications, Washington, DC; 2001.

[27] Dobele G, Meier D, Faix O, et al. Volatile products of catalytic flash pyrolysis of celluloses. Journal of Analytical and Applied Pyrolysis. 2001;58:453–463.

[28] Dobele G, Dizhbite T, Rossinskaja G, et al. Pre-treatment of biomass with phosphoric acid prior to fast pyrolysis: a promising method for obtaining 1,6-anhydrosaccharides in high yields. Journal of Analytical and Applied Pyrolysis. 2003;68:197–211.

[29] Dobele G, Rossinskaja G, Dizhbite T, et al. Application of catalysts for obtaining 1,6-anhydrosaccharides from cellulose and wood by fast pyrolysis. Journal of Analytical and Applied Pyrolysis. 2005;74(1):401–405.

[30] Dobele G, Rossinskaja G, Telysheva G, et al. Cellulose dehydration and depolymerization reactions during pyrolysis in the presence of phosphoric acid. Journal of Analytical and Applied Pyrolysis. 1999;49(1–2):307–317.

[31] Stiles HN, Kandiyoti R. Secondary reactions of flash pyrolysis tars measured in a fluidized bed pyrolysis reactor with some novel design features. Fuel. 1989;68(3):275–282.

[32] Lanzetta M, Di Blasi C, Buonanno F. An experimental investigation of heat-transfer limitations in the flash pyrolysis of cellulose. Industrial & Engineering Chemistry Research. 1997;36(3):542–552.

[33] Chan WCR, Kelbon M, Krieger-Brockett B. Single-particle biomass pyrolysis: correlations of reaction products with process conditions. Industrial & Engineering Chemistry Research. 1988;27(12):2261–2275.

[34] Koufopanos CA, Lucchesi A, Maschio G. Kinetic modelling of the pyrolysis of biomass and biomass components. The Canadian Journal of Chemical Engineering. 1989;67(1):75–84.

[35] Paulsen AD, Mettler MS, Dauenhauer PJ. The role of sample dimension and temperature in cellulose pyrolysis. Energy & Fuels. 2013;27(4):2126–2134.

[36] Howard JB. Fundamentals of coal pyrolysis and hydropyrolysis. Second supplementary volume. Wiley, New York; 1981.

[37] Hoekstra E, Van Swaaij WP, Kersten SR, et al. Fast pyrolysis in a novel wire-mesh reactor: Decomposition of pine wood and model compounds. Chemical Engineering Journal. 2012;187:172–184.

[38] Pindoria RV, Chatzakis IN, Lim J, et al. Hydropyrolysis of sugar cane bagasse: effect of sample configuration on bio-oil yields and structures from two bench-scale reactors. Fuel. 1999;78(1):55–63.

[39] Fushimi C, Katayama S, Tasaka K, et al. Elucidation of the interaction among cellulose, xylan, and lignin in steam gasification of woody biomass. Aiche Journal. 2009;55(2):529–537.

[40] Fushimi C, Katayama S, Tsutsumi A. Elucidation of interaction among cellulose, lignin and xylan during tar and gas evolution in steam gasification. Journal of Analytical and Applied Pyrolysis. 2009;86(1):82–89.

[41] Giudicianni P, Cardone G, Ragucci R. Cellulose, hemicellulose and lignin slow steam pyrolysis: Thermal decomposition of biomass components mixtures. Journal of Analytical and Applied Pyrolysis. 2013;100:213–222.

[42] Sagehashi M, Miyasaka N, Shishido H, et al. Superheated steam pyrolysis of biomass elemental components and Sugi (Japanese cedar) for fuels and chemicals. Bioresource Technology. 2006;97(11):1272–1283.

[43] Liu N. Kinetics of thermal decomposition for biomass and relevant kinetic analysis method. University of Science and Technology of China, Heifei; 2000.

[44] Vyazovkin S, Wight CA. Isothermal and non-isothermal kinetics of thermally stimulated reactions of solids. International Reviews in Physical Chemistry. 1998;17(3):407-433.

[45] Vyazovkin S, Burnham AK, Criado JM, et al. ICTAC Kinetics committee recommendations for performing kinetic computations on thermal analysis data. Thermochimica Acta. 2011;520(1):1–19.

[46] Dahiya JB, Kumar K, Muller-Hagedorn M, et al. Kinetics of isothermal and non-isothermal degradation of cellulose: model-based and model-free methods. Polymer International. 2008;57(5):722, 729.

[47] Freeman ES, Carroll B. The application of thermoanalytical techniques to reaction kinetics: the thermogravimetric evaluation of the kinetics of the decomposition of calcium oxalate monohydrate. The Journal of Physical Chemistry. 1958;62(4):394–397.

[48] Coats AW, Redfern JP. Kinetic Parameters from Thermogravimetric Data. Nature. 1964;201(491):68–69.

[49] Friedman HL. Kinetics of thermal degradation of char-forming plastics from thermogravimetry. Application to a phenolic plastic: Journal of Polymer Science Part C: Polymer Symposia. 1964; Wiley Online Library.

[50] Kissinger HE, Chem. A. Reaction kinetics in differential thermal analysis. Analytical Chemistry. 2002;29(11):1702–1706.

[51] Flynn JH, Wall LA. General treatment of the thermogravimetry of polymers. Journal of Research of the National Bureau of Standards–a Physics & Chemistrya. 1966;70(6):487–523.
[52] Ozawa T. A new method of analyzing thermogravimetric data. Bulletin of the Chemical Society of Japan. 1965;38(11):1881–1886.
[53] Burnham AK, Braun RL. Global kinetic analysis of complex materials. Energy & Fuels. 1999;13(1):1–22.
[54] Becidan M, Varhegyi G, Hustad JE, et al. Thermal decomposition of biomass wastes. A kinetic study. Industrial & Engineering Chemistry Research. 2007;46(8):2428–2437.
[55] Koga N, Criado J. The influence of mass transfer phenomena on the kinetic analysis for the thermal decomposition of calcium carbonate by constant rate thermal analysis (CRTA) under vacuum. International Journal of Chemical Kinetics. 1998;10(30):737–744.
[56] Varhegyi G, Antal Jr MJ, Szekely T, et al. Kinetics of the thermal decomposition of cellulose, hemicellulose, and sugarcane bagasse. Energy & Fuels. 1989;3(3):329–335.
[57] Antal M JJ, Varhegyi G. Cellulose pyrolysis kinetics: the current state of knowledge. Industrial & Engineering Chemistry Research. 1995;34(3):703–717.
[58] Broido A, Nelson MA. Char yield on pyrolysis of cellulose. Combustion and Flame. 1975;24:263–268.
[59] Broido A, Weinstein M. Low temperature isothermal pyrolysis of cellulose. Thermal Analysis. Springer/Birkhäuser, Basel; 1972. pp. 285–296.
[60] Milosavljevic I, Suuberg EM. Cellulose thermal decomposition kinetics: global mass loss kinetics. Industrial & Engineering Chemistry Research. 1995;34(4):1081–1091.
[61] Antal Jr MJ, Friedman HL, Rogers FE. Kinetics of cellulose pyrolysis in nitrogen and steam. Combustion Science and Technology. 1980;21(3–4):141–152.
[62] Mok WS, Antal MJ. Effects of pressure on biomass pyrolysis. I. Cellulose pyrolysis products. Thermochimica Acta. 1983;68(2):155–164.
[63] Piskorz J, Majerski P, Radlein D, et al. Flash pyrolysis of cellulose for production of anhydro-oligomers. Journal of Analytical and Applied Pyrolysis. 2000;56(2):145–166.
[64] Liu Q, Wang S, Wang K, et al. Mechanism of formation and consequent evolution of active cellulose during cellulose pyrolysis. Acta Physico-Chimica Sinica. 2008;24(11):1957–1963.
[65] Di Blasi C. Modeling chemical and physical processes of wood and biomass pyrolysis. Progress in Energy and Combustion Science. 2008;34(1):47–90.
[66] Piskorz J, Radlein DSA, Scott DS, et al. Liquid products from the fast pyrolysis of wood and cellulose. Research in Thermochemical Biomass Conversion. Springer, Netherlands; 1988. pp. 557–571.
[67] Scott G. Atmospheric oxidation and antioxidants. Elsevier, Amsterdam; 1965.
[68] Şerbănescu C. Kinetic analysis of cellulose pyrolysis: a short review. Chemical Papers. 2014;68(7):847–860.
[69] Capart R, Khezami L, Burnham AK. Assessment of various kinetic models for the pyrolysis of a microgranular cellulose. Thermochimica Acta. 2004;417(1):79–89.
[70] Hashimoto K, Miura K, Watanabe T. Kinetics of thermal regeneration reaction of activated carbons used in waste water treatment. Aiche Journal. 1982;28(5):737–746.
[71] Liu X, Li B, Miura K. Analysis of pyrolysis and gasification reactions of hydrothermally and supercritically upgraded low-rank coal by using a new distributed activation energy model. Fuel Processing Technology. 2001;69(1):1–12.
[72] Miura K. A new and simple method to estimate f (E) and k0 (E) in the distributed activation energy model from three sets of experimental data. Energy & Fuels. 1995;9(2):302–307.
[73] Miura K, Maki T. A simple method for estimating f (E) and k 0 (E) in the distributed activation energy model. Energy & Fuels. 1998;12(5):864–869.

[74] Solomon PR, Hamblen DG, Carangelo RM, et al. General model of coal devolatilization. Energy & Fuels. 1988;2(4):405–422.
[75] Wang G, Li W, Li BQ, et al. TG study on pyrolysis of biomass and its three components under syngas. Fuel. 2008,87(4–5):552–558.
[76] Zhang J, Chen T, Wu J, et al. A novel Gaussian-DAEM-reaction model for the pyrolysis of cellulose, hemicellulose and lignin. RSC Advances. 2014;4(34):17513–17520.
[77] Demirbaş A. Mechanisms of liquefaction and pyrolysis reactions of biomass. Energy Conversion and Management. 2000;41(6):633–646.
[78] Boutin O, Ferrer M, Lédé J. Radiant flash pyrolysis of cellulose—Evidence for the formation of short life time intermediate liquid species. Journal of Analytical and Applied Pyrolysis. 1998;47(1):13–31.
[79] Boutin O, Ferrer M, Lédé J. Flash pyrolysis of cellulose pellets submitted to a concentrated radiation: experiments and modelling. Chemical Engineering Science. 2002;57(1):15–25.
[80] Fahey P J. Insights from glucose pyrolysis for cellulose pyrolysis: 2012 AIChE Annual Meeting, Pittsburgh; 2012.
[81] Zhu YP, Zajicek J, Serianni AS. Acyclic forms of [1-13C] aldohexoses in aqueous solution: Quantitation by 13C NMR and deuterium isotope effects on tautomeric equilibria. The Journal of Organic Chemistry. 2001;66(19):6244–6251.
[82] Kwon GJ, Kim DY, Kimura S, et al. Rapid-cooling, continuous-feed pyrolyzer for biomass processing: Preparation of levoglucosan from cellulose and starch. Journal of Analytical & Applied Pyrolysis. 2007;80(1):1–5.
[83] Ivanov VI, Golova OP, Pakhomov AM. Main direction of reaction in the thermal decomposition of cellulose in a vacuum. Russian Chemical Bulletin. 1956;5(10):1295–1296.
[84] Kislitsyn AN, Rodionova ZM, Savinykh VI. Thermal decomposition of monophenylglycol ether. Khim Drev. 1971;9:131–136.
[85] Ponder GR, Richards GN, Stevenson TT. Influence of linkage position and orientation in pyrolysis of polysaccharides: A study of several glucans. Journal of Analytical and Applied Pyrolysis. 1992;22(3):217–229.
[86] Mamleev V, Bourbigot S, Le Bras M, et al. The facts and hypotheses relating to the phenomenological model of cellulose pyrolysis: Interdependence of the steps. Journal of Analytical and Applied Pyrolysis. 2009;84(1):1–17.
[87] Kawamoto H, Murayama M, Saka S. Pyrolysis behavior of levoglucosan as an intermediate in cellulose pyrolysis: polymerization into polysaccharide as a key reaction to carbonized product formation. Journal of Wood Science. 2003;49(5):469–473.
[88] Kato K. Pyrolysis of cellulose part III. Comparative studies of the volatile compounds from pyrolysates of cellulose and its related compounds. Agr. Biol. Chem. 1967;31(6):657–663.
[89] Shen DK, Gu S. The mechanism for thermal decomposition of cellulose and its main products. Bioresource Technology. 2009;100(24):6496–6504.
[90] Mettler MS, Mushrif SH, Paulsen AD, et al. Revealing pyrolysis chemistry for biofuels production: Conversion of cellulose to furans and small oxygenates. Energy & Environmental Science. 2012;5(1):5414–5424.
[91] Shin E, Nimlos MR, Evans RJ. Kinetic analysis of the gas-phase pyrolysis of carbohydrates. Fuel. 2001;80(12):1697–1709.
[92] Richards GN. Glycolaldehyde from pyrolysis of cellulose. Journal of Analytical and Applied Pyrolysis. 1987;10(3):251–255.
[93] Shafizadeh F, Lai YZ. Thermal Degradation of 1,6-Anhydro-β-D-glucopyranose. Journal of Organic Chemistry. 1972;37(2):278–284.
[94] Paine III JB, Pithawalla YB, Naworal JD. Carbohydrate pyrolysis mechanisms from isotopic labeling: Part 3. The Pyrolysis of d-glucose: Formation of C_3 and C_4 carbonyl compounds and

a cyclopentenedione isomer by electrocyclic fragmentation mechanisms. Journal of Analytical and Applied Pyrolysis. 2008;82(1):42–69.
[95] Banyasz JL, Li S, Lyons-Hart J L, et al. Cellulose pyrolysis: the kinetics of hydroxyacetaldehyde evolution. Journal of Analytical and Applied Pyrolysis. 2001;57(2):223–248.
[96] Yang H, Yan R, Chen H, et al. Characteristics of hemicellulose, cellulose and lignin pyrolysis. Fuel. 2007;86(12):1781–1788.
[97] G X, LL. Quantum chemistry. Beijing: Science Press, 2008.
[98] Orozco A, Ahmad M, Rooney D, et al. Dilute acid hydrolysis of cellulose and cellulosic bio-waste using a microwave reactor system. Process Safety and Environmental Protection. 2007;85(5):446–449.
[99] Evans RJ, Milne TA, Soltys MN, et al. Mass spectrometric behavior of levoglucosan under different ionization conditions and implications for studies of cellulose pyrolysis. Journal of Analytical and Applied Pyrolysis. 1984;6(3):273–283.
[100] Huang J, Liu C, Wei S, et al. Density functional theory studies on pyrolysis mechanism of β-d-glucopyranose. Journal of Molecular Structure: Theochem. 2010;958(1):64–70.
[101] Huang J, Liu C, Wei S. Thermodynamic Studies of Pyrolysis Mechanism of Cellulose Monomer. Acta Chimica Sinica. 2009;67(18):2081–2086.
[102] Mulliken RS. Electronic population analysis on Lcao [single bond] MO molecular wave functions. I. The Journal of Chemical Physics. 1955;23(10):1833.
[103] Lu Q, Yang X, Dong C, et al. Influence of pyrolysis temperature and time on the cellulose fast pyrolysis products: Analytical Py-GC/MS study. Journal of Analytical and Applied Pyrolysis. 2011;92(2):430–438.
[104] Zhang X, Li J, Yang W, et al. Formation mechanism of levoglucosan and formaldehyde during cellulose pyrolysis. Energy & Fuels. 2011;25(8):3739–3746.
[105] Zhang X, Yang W, Blasiak W. Thermal decomposition mechanism of levoglucosan during cellulose pyrolysis. Journal of Analytical and Applied Pyrolysis. 2012;96:110–119.
[106] Huang J, Liu C, Wei S, et al. A theoretical study on the mechanism of levoglucosan formation in cellulose pyrolysis. Jounral of Fuel Chemistry and Technology. 2011;39(8):590–594.
[107] Mayes HB, Broadbelt LJ. Unraveling the reactions that unravel cellulose. The Journal of Physical Chemistry A. 2012;116(26):7098–7106.
[108] Zhang M, Geng Z, Yu Y. Density functional theory (DFT) study on the dehydration of cellulose. Energy & Fuels. 2011;25(6):2664–2670.
[109] Atalla RH, Vanderhart DL. Native cellulose: a composite of two distinct crystalline forms. Science. 1984;223(4633):283–285.
[110] Zhang KC. Modern crystallography. Science Press, Beijing; 2011.
[111] Car R, Parrinello M. Unified approach for molecular dynamics and density-functional theory. Physical Review Letters. 1985;55(22):2471–2474.
[112] Agarwal V, Dauenhauer PJ, Huber GW, et al. Ab initio dynamics of cellulose pyrolysis: nascent decomposition pathways at 327 and 600 °C. Journal of the American Chemical Society. 2012;134(36):14958–14972.

3 Pyrolysis of hemicellulose

3.1 Fundamental process of hemicellulose pyrolysis

Hemicellulose, whose structure is amorphous, is composed of heteropolysaccharides with a high degree of branching. Hemicellulose from various biomass differs greatly in its chemical structures. Among the three components of lignocellulosic biomass, hemicellulose has a low degree of polymerization and starts to decompose at a lower temperature. Generally, the decomposition of hemicellulose starts at about 200 °C, and volatiles are released in a narrow temperature range of 200–400 °C. Compared with cellulose pyrolysis, hemicellulose pyrolysis yields more gaseous products and fewer liquid products. The main pyrolytic products include hydroxyacetone, acetic acid, furfural, formic acid, propionic acid, 5-hydroxymethylfurfural, cyclopentenone, CO_2 and CO, of which furfural and acetic acid are considered the two typical products from hemicellulose pyrolysis. Although hemicellulose is usually the abundant component in biomass, isolating it without damaging its structure is very difficult, hence many model compounds with simplified structure are popularly used as representatives of hemicellulose.

3.1.1 Pyrolysis of hemicellulose-based monosaccharides

Xylose, mannose, galactose, and arabinose are the four basic monosaccharides constituting hemicellulose. Xylose and arabinose are pentoses, while mannose and galactose are hexoses. Xylan, mainly composed of xylose units, is another popular model compound used in pyrolysis mechanism studies of hemicellulose. The FTIR spectra for xylan, mannose, galactose and arabinose are shown in Fig. 3.1. The strong peak between 3,600–3,200 cm^{-1} corresponds to O–H stretching; the band at 1,320–1,210 cm^{-1} is assigned to C–O bonds; and the peaks at 1,725–1,705 cm^{-1} and 1,150–1,070 cm^{-1} originate from the stretching of C=O and C–O–C, respectively.

Thermogravimetric (TG) analysis shows that the devolatilization of all the hemicellulose-based monosaccharides is very similar. Fig. 3.2 shows the TG/DTG curves for the pyrolysis of mannose. It indicates that mannose mainly decomposes at 171–525 °C. Below this temperature range the evaporation of hydrate water is the main contributor to weight loss. The shoulder peak at 230 °C results from intermolecular dehydration between monosaccharides. The maximum weight loss is achieved at 300 °C. Mannose degrades extensively around this temperature, and most of the C–O and C–C bonds break to form volatiles [1]. The weight loss declines above 500 °C, and the final yield of solid residue for mannose pyrolysis was about 20 %.

DOI 10.1515/9783110369632-005

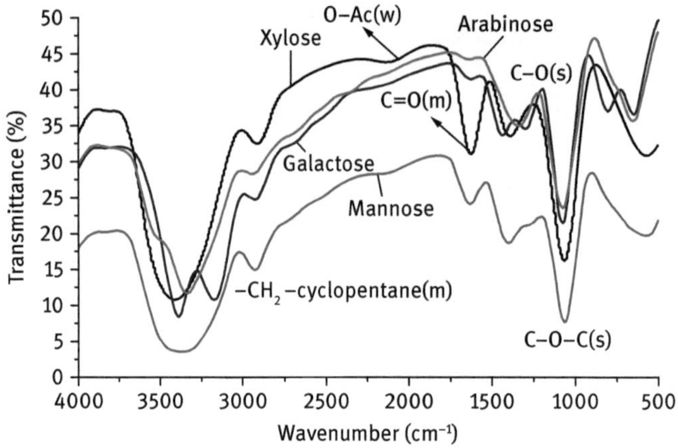

Fig. 3.1: FTIR spectra for xylan, mannose, galactose and arabinose.

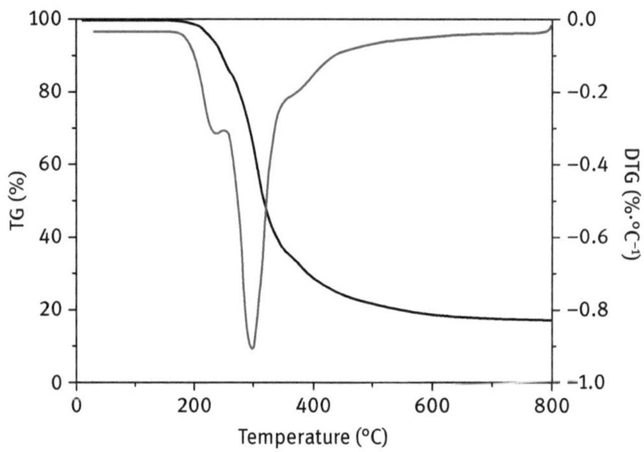

Fig. 3.2: TG/DTG curves for mannose pyrolysis. Reprinted with permission from [2], © 2013 Elsevier.

Based on Py-GC/MS analysis, the product distributions for pyrolysis of mannose, galactose and arabinose are similar, mainly including acids, furans, ketones, alcohols, anhydro-sugars, etc. The typical acid products are acetic acid, formic acid, and propionic acid. Furans mainly include furfural (for pentose) and 5-hydroxymethylfurfural (for hexose). The distribution of ketones is very complex, and the most abundant ketones are hydroxyacetone, furanones, pyranones, cyclopentanones, and cyclopentenones. Anhydro-sugars, such as levoglucosan, are mainly detected in hexose pyrolysis. Among these products, furfural and 5-hydroxymethylfurfural are the most abundant products from the pyrolysis of pentose and hexose, respectively. Shafizadeh et al. [3] investigated the pyrolysis behaviors of an oligomer of xylose and found that the glycosidic bond decomposes first when the temperature rises. Räisänen et al. [4] studied the product distributions for pyrolysis of xylose, arabinose, mannose and galactose. The results showed that the main pyrolytic products include furans, ketones and anhydro-sugars; the products from mannose pyrolysis are the most abundant, while the pyrolysis of xylose and arabinose yields many small molecular compounds. Gardiner et al. [5] performed pyrolysis of mannose and galactose in a vacuum environment, and found that the main products from mannose pyrolysis are 1,6-anhydro-β-D-mannopyranose and 1,4:3,6-dianhydro-D-mannopyranose, while those from galactose pyrolysis are 1,6-anhydro-β-D-galactopyranose and 5-hydroxymethylfurfural.

3.1.2 Pyrolysis of xylan and glucomannan

The weight loss behavior of xylan during pyrolysis is shown in Fig. 3.3. Compared with cellulose, xylan decomposes in a lower and narrower temperature range. The DTG curve shows a small weight loss peak corresponding to free water evaporation below 150 °C and a single main peak corresponding to the primary pyrolysis stage. The main products released from xylan pyrolysis include acids, furans, ketones, cyclopentanones, cyclopentenones, H_2O, CO_2, and CO, which are similar to those from the pyrolysis of hemicellulose-based monosaccharides. Yang et al. [6, 7] also found similar phenomenon during xylan pyrolysis. The main degradation stage was between 220–315 °C, exhibiting a single weight loss peak, and the final char yield was about 20 %. Biagini et al. [8] studied the pyrolytic devolatilization behavior of xylan by using TG-FTIR. The results show that xylan mainly decomposes at 253–308 °C, and the amount of volatiles released during this range accounts for 76.3 % of all volatiles.

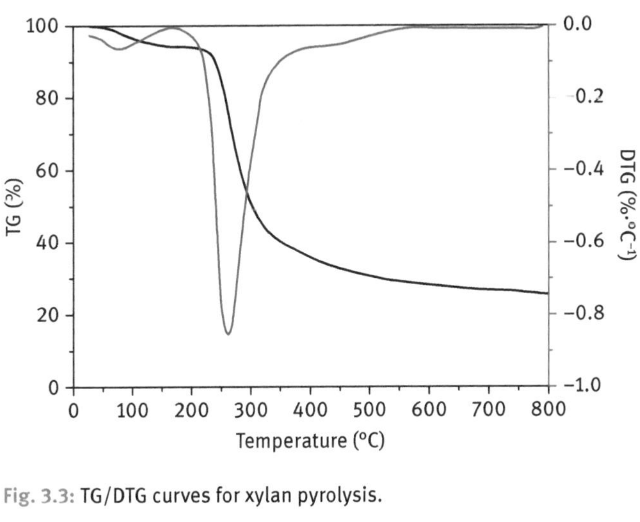

Fig. 3.3: TG/DTG curves for xylan pyrolysis.

However, Šimkovic et al. [9] and Jensen et al. [10] observed two overlapped weight loss peaks in the main pyrolysis stage of xylan, which might be due to the different structures of the xylan used in the studies. Šimkovic et al. [9] detected the release of methanol, H_2O, CO_2, formyl groups, and furfural in the first DTG curve peak and in the second peak acetone, formic acid, formaldehyde, acrolein, acetic acid, and 3-hydroxy-2-penteno-1,5-lactone. They proposed that dehydration, decarboxylation and decarbonylation occur in both peaks and the second peak mainly results from the further degradation of the produced intermediates. Beaumont et al. [11] also found a similar product distribution from xylan pyrolysis, which mainly included alcohols, acids, furans, and ketones.

Based on the results of Py-GC/MS analysis, the specific products from xylan pyrolysis are hydroxyacetone, acetic acid, furfural, formic acid, propionic acid and cyclopentenone etc. Furfural and acetic acid are considered the two typical products from xylan pyrolysis since both of them usually have high contents in the final product distribution. Nowakowski et al. [12, 13] investigated the pyrolysis behavior of xylan by using Py-GC/MS and found that the main products are hydroxyacetone, acetic acid, furfural, 3-methyl-2-2-hydroxy-cyclopenten-1-one, of which the acetic acid and furfural content are high. Acids mainly include formic acid, acetic acid and propionic acid, accounting for 29.87 % of the total products; furans include furfural, furfuryl alcohol and 5-hydroxymethylfurfural, together sharing a content of 27.74 %; hydroxyacetone, furanones and cyclopentenones are the main ketone products, showing yields of 4.91 %, 4.98 % and 3.04 %, respectively; in addition, some pyranones, such as levoglucosenone and 2,3-2H-3,5-dihydroxyl-6-pyranone, have a total content of 6.20 %.

Hosoya et al. [14] isolated glucomannan from Japanese cedar; the neutral sugar analysis showed that it was composed of 16.5 % glucose, 68.1 % mannose, 4.7 % galactose, 6.4 % xylose and 1.9 % arabinose. In the following pyrolysis experiment, 41.3 % water-free bio-oil was obtained. The main composition included C_2–C_3 carbonyls (glycolaldehyde and hydroxyacetone), carboxylic acids (formic acid, acetic acid and glycolic acid), anhydro-sugars (levoglucosan, 1,6-anhydro-β-D-glucofuranose and levomannosan) and furans (furfural and 5-hydroxymethylfurfural).

Aho et al. [15] performed the pyrolysis of galactoglucomannan in a fluidized bed reactor and achieved 12.1 % water-free bio-oil and 31.2 % char yields. The main components of the bio-oil were acetic acid, furfural, 2-furanmethanol, levoglucosan and hydroxyacetone. Acetic acid had the highest yield and was mainly formed from the dissociation of O-acetyl in mannose residue. CO_2 and CO were the two major gaseous products from galactoglucomannan pyrolysis. CO_2 was formed at 200 °C, and achieved maximum yield at 320 °C, while CO started to be released at 250 °C and reached the highest yield at 300 °C.

3.1.3 Pyrolysis of the isolated hemicellulose

Hemicellulose can be extracted from delignified holocellulose. Xiao et al. [16] isolated three hemicelluloses from maize stems, rye straw and rice straw. Their structural analysis showed that both hemicelluloses isolated from maize stems and rye straw were composed of glucuronoarabinoxylans, and the rice straw hemicellulose was dominant in α-glucan and L-arabino-(4-O-methyl-D-glucurono)-D-xylan. The TG analysis indicated that all of the three hemicelluloses decomposed in the temperature range of 200–300 °C. Yu et al. [17] compared the pyrolysis behaviors of water-soluble hemicellulose and alkali-soluble hemicellulose isolated from bamboo stem. Alkali-soluble hemicellulose showed stronger thermal stability due to the higher lignin fragment content and it also produced a higher char yield, indicating that the isolation process had a significant influence on the resulting pyrolysis behavior. Peng et al. [18] investigated the neutral sugar composition of hemicellulose isolated from sugarcane bagasse, and found that xylose was the dominant monosaccharide, followed by arabinose and glucose, and a little galactose, glucuronic acid and galacturonic acid. The devolatilization behavior of sugarcane bagasse hemicellulose was similar to that of xylan, which could be divided into four stages, namely water evaporation, initial pyrolysis stage, primary pyrolysis stage and char formation stage. As the temperature rises, acetic acid, hydroxyacetone, 1-hydroxyl-2-butanone, furfural and cyclopentenones are formed and released and can be further decomposed into H_2O, CO, CO_2, and CH_4. At a low temperature range, CO and CO_2 are the major gaseous products, while at a high temperature range, the H_2 and CH_4 contents increase [19]. The char formation stage occurred above 400 °C, where weight loss tended to decline and the final char yield was about 20 % [20]. The pyrolysis of hemicellulose isolated from corn straw also presented similar results, in addition to H_2, CO, CO_2, and CH_4; C_2H_4 and C_2H_6 were also detected in the gaseous products, and the liquid products mainly included ketones, furans, acids and alcohols [21].

3.1.4 Comparison of the pyrolysis behaviors of hemicellulose-based monosaccharides and xylan

Mannose, galactose, arabinose and xylan were selected as model compounds to study the differences in the pyrolysis behaviors of hemicellulose-based monosaccharides and xylan, especially for the dissociation of O-acetyl groups and the different degradation behaviors between pentose and hexose. As shown in Fig. 3.4, all four curves exhibit similar tendencies: weight loss starts at about 200 °C and achieves the maximum rate at 250–300 °C, above this temperature weight loss slows down and almost terminates after 500 °C. Xylan pyrolysis led to the highest char yield of 26.5 %, which might be due to the macromolecular radicals produced from the cleavage of side branches in xylan during pyrolysis, which can repolymerize with each other, leading to char

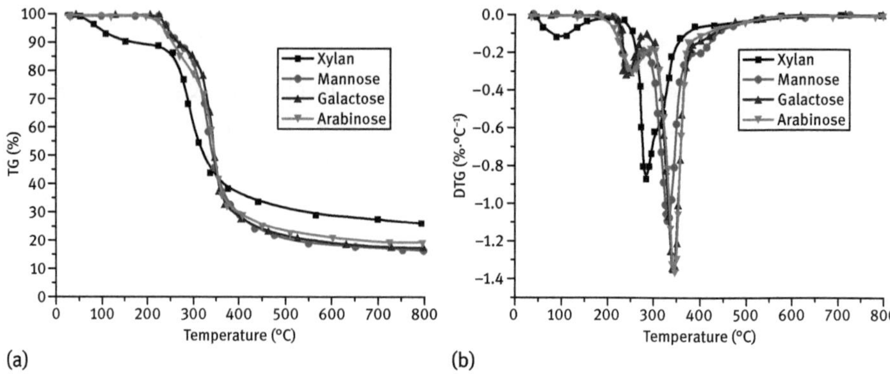

Fig. 3.4: TG/DTG analysis for the different hemicellulose model compounds.

formation. Furthermore, the char yield from pentose pyrolysis (19.2 % for arabinose) is higher than that from hexose pyrolysis (17.3 % and 17.8 for mannose and galactose, respectively). This might be due to the additional hydroxymethyl in hexose structure, which can dissociate as formaldehyde and then form small molecular radicals that could stabilize the molecular radicals to avoid their polymerization with each other to form char [22].

With regards to the evolution behaviors of some typical products (see Fig. 3.5), the release of water occurs at 200–450 °C; the water released below 100 °C results from the evaporation of physical water. During the pyrolysis of three monosaccharides, the water produced at 250 °C mainly originates from intermolecular dehydration during polymerization, while the water released at the primary weight loss stage is from the intramolecular dehydration of hydroxyls in monosaccharides. The weak absorbance signal of water shown above 500 °C is ascribed to condensation during char formation.

The initial release temperature of CO_2 from xylan pyrolysis is at about 200 °C, while that from pyrolysis of monosaccharides is at 300 °C. Generally, the abundant uronic acid moieties in xylan structure are the primary source of CO_2 [22]. CO_2 from pyrolysis of monosaccharides is mainly produced from the further decomposition of ring-opened intermediates. The evolution of CO from the pyrolysis of three monosaccharides is similar, mainly resulting from the decarbonylation of the ring-opened products with C=O groups. Xylan pyrolysis exhibits two absorbance peaks in CO release, the first peak might be related to the decarbonylation of ring-opened products and the second one originates from the further degradation of aldehydes [23].

Acids formed in xylan pyrolysis are mainly from the dissociation of O-acetyl groups substituted at xylan backbone, which occurs easily at low temperatures due to its poor thermal stability. Acid formation during pyrolysis of monosaccharides exhibits some complex pathways and they are released at a high temperature. With regard to furan products, these were mainly furfural for pentose pyrolysis and 5-hydroxymethylfurfural for hexose pyrolysis, respectively [24]. It has been found

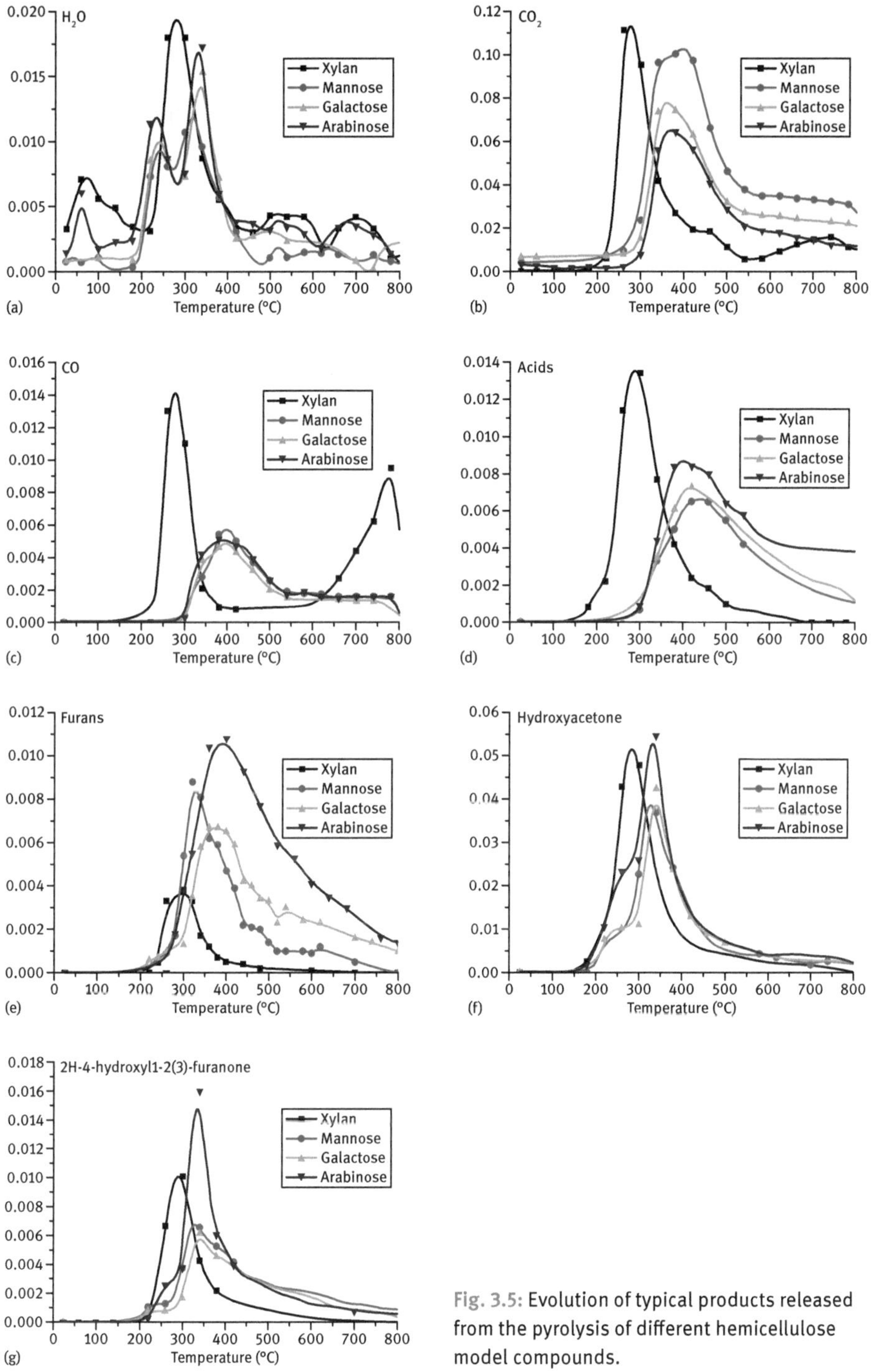

Fig. 3.5: Evolution of typical products released from the pyrolysis of different hemicellulose model compounds.

that the furans produced from pyrolysis of monosaccharides show higher release intensities, indicating that furans are mainly produced from the decomposition of monosaccharide units.

With regard to the two typical ketones produced from hemicellulose pyrolysis, hydroxyacetone and 2H-4-hydroxyl-2(3H)-furanone, their evolution behaviors for the two hexoses are similar and the release intensities are stronger than those for xylose and xylan. This might be due to the dissociation of additional hydroxymethyl in hexose structure, which has a higher energy barrier.

3.2 Effect of reaction parameters on the pyrolysis behavior of hemicellulose

3.2.1 Effect of reaction temperature

Similar to cellulose pyrolysis, temperature is the key factor influencing the pyrolysis behavior of hemicellulose. Fig. 3.6 shows the product distribution from xylan pyrolysis in an infrared radiation heating reactor. As temperature increases, the tar yield increases and achieves the maximum value of 44 %, then decreases due to the further decomposition of volatiles to form small molecular gas products at higher temperature. The char yield decreases with rising temperature, especially within the range of 600–735 °C, and then remains at about 22 % at higher temperatures. This is in agreement with the results from TG and Py-GC/MS analysis. The analysis of obtained tar indicates that the main components are acids, furans, aldehydes, ketones, and pyrans

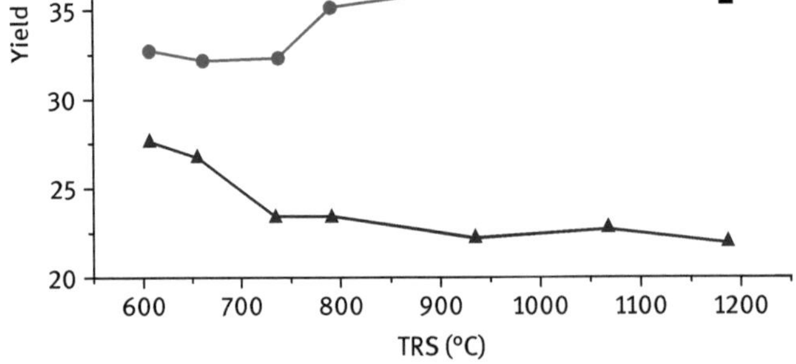

Fig. 3.6: The effect of temperature on the product distribution from xylan pyrolysis.

and that the aldehydes and ketones content changed greatly with temperature. CO and CO_2 accounted for more than 90 % of the gaseous products. Furthermore, the CO_2 content was always higher than that of CO over the whole stage. In general, the yield of small molecular products is enhanced as temperature increases, while the char yield shows an opposite tendency [25].

Liu et al. [26] studied the effect of temperature on xylan pyrolysis by using CP-GC/MS and found that the pyrolysis could be divided into two stages. When the temperature was below 300 °C, furfural and saccharide compounds were formed by the cleavage of the glycosidic bond. When the temperature exceeded 300 °C, various small molecular aldehydes and ketones were formed with increasing yields, while the production of furfural and saccharide compounds decreased.

Qu et al. [27] found that acids formed the majority of the products from xylan pyrolysis in a fixed bed reactor and that their yield first increased and then decreased with increasing temperature. The gaseous products included CO_2, CO, H_2, CH_4, and a small amount of C_2H_6 and C_3H_8, of which CO_2 showed the highest content. The CO_2 content was sensitive to temperature, and decreased from 60 % to 40 % when the temperature increased from 500 °C to 550 °C whereas the CO content remained about 30 % at 350–650 °C. The yields of H_2 and CH_4 increased sharply at temperatures above 500 °C. Zhao et al. [28] also observed similar phenomena when performing pyrolysis of xylan in a tube furnace.

Branca et al. [29] conducted pyrolysis of glucomannan on a fluidized bed reactor and found that when temperature increased, the yield of gaseous products improved slowly, while char production decreased. The liquid products mainly included acetic acid, furfural, hydroxyacetone and glycolaldehyde, with the highest yield of 18 % at 370 °C. CO_2 was the main gaseous product, having a high production of 10 % at about 330 °C.

Alén et al. [30] studied the effect of temperature on the pyrolysis behaviors of xylan and glucomannan. The results showed that for the pyrolysis of glucomannan, the yield of small moleculars, such as formic acid, acetic acid and glycolaldehyde, was enhanced as temperature increased whereas the production of large molecular compounds, such as anhydro-sugars and furans, declined. Levoglucosenone was the dominant pyrolytic product at 400 °C; the yield of anhydro-sugars increased above 400 °C, while the production of levoglucosenone and furans decreased. At high temperatures of 800–1,000 °C, the formation of aliphatic hydrocarbons and aromatics was observed. Compared with glucomannan, xylan showed weaker thermal stability. As the temperature increased, xylan pyrolysis yielded more small molecular volatiles, while the yield of lactones and furans decreased. Aromatics were formed at a high temperature range, including naphthalene, phenanthrene and acenaphthylene.

3.2.2 Effect of residence time

The volatiles produced from hemicellulose pyrolysis can further decompose into smaller molecular compounds. Longer residence time can favor the further decomposition of volatiles, and enhance the yield of gaseous products. Residence time largely influences the product distribution of tar. Our study of xylan pyrolysis with residence time in the range of 2–20 s using Py-GC/MS found that as residence time increased, the yield of acids increased while that of furans decreased.

3.2.3 Effect of other reaction factors

Raw biomass usually has a high water content that can be removed by torrefaction. This pretreatment inevitably changes the chemical structure of biomass, especially the amorphous structure of hemicellulose. It was found that xylose, the basic unit in xylan, started to decompose at 210 °C during torrefaction. The weight loss of xylan after torrefaction at 230 °C and 260 °C was 14.16 % and 17.10 %, respectively. The temperature corresponding to the maximum weight loss peak of DTG curves moved to a high temperature range when the torrefaction temperature increased [18]. Compared with cellulose and lignin, hemicellulose was more sensitive to torrefaction, and in order to avoid its influence, the torrefaction temperature should be lower than 230 °C [31].

The reaction atmosphere was another important factor influencing product distribution. Usually pyrolysis was performed under inert atmosphere, however some experimental research has been conducted under a different atmosphere. Wang et al. [32] performed TG analysis of xylan under synthesis gas atmosphere. It was found that xylan mainly decomposed at 196–340 °C. Compared with N_2, synthesis gas lowered and widened the decomposition temperature of xylan. Sagehashi et al. [33] studied the degradation behavior of xylan under superheated steam atmosphere. The results showed that large amounts of condensable volatiles, mainly glycolaldehyde, acetic acid and furfural, are produced between 200–250 °C and are little influenced by temperature. Giudicianni et al. [34] studied the pyrolysis of xylan at 430 °C and 600 °C under steam atmosphere and also found that as temperature increased, the yield of liquid products changed little, while the production of gaseous products increased sharply. In addition to CO_2, the dominant gaseous compounds including CO, H_2, CH_4, and C_{2+} hydrocarbons were also detected. Fushimi et al. [35, 36] also performed xylan pyrolysis under steam atmosphere and found that as the residence time increased, the tar yield was significantly lowered. They also indicated that two decomposition stages might exist in xylan degradation under steam atmosphere.

3.3 Mechanism of hemicellulose pyrolysis

3.3.1 Pyrolysis kinetic model for hemicellulose pyrolysis

3.3.1.1 One-step global reaction model

The widely used kinetic models for xylan pyrolysis are the one-step global reaction model and the multi-step reaction model. The one-step global reaction model is the simplest one, it assumes that hemicellulose decomposes through a single step to form the final products. The kinetic parameters of the one-step global reaction model calculated by some researchers are listed in Tab. 3.1. The isolated hemicelluloses differed greatly, especially with regard to the pre-exponential factor, because that they were isolated from different biomass species. With regard to xylan, it was found that the data obtained by Bilbao et al. [37] were lower than those obtained by Williams et al. [38] and as the heating rate increased, Bilbao's parameters increased while Williams' decreased. This difference might be due to their different experimental conditions and data processing methods used in their kinetic studies. Although the one-step global reaction model can effectively describe the pyrolysis rate of hemicellulose, it cannot be used to predict the product distribution.

Tab. 3.1: One-step global reaction model for hemicellulose (or xylan) pyrolysis.

Sample	Heating rate (°C/min)	Temperature (°C)	Kinetic parameters		Reference
			E (kj/mol)	*A* (s^{-1})	
Xylan	Isothermal	215–250	117–134	2.2×10^9	Ramiah [39]
Xylan	1.5	200–400	43	9.8	Bilbao et al. [37]
	20		72	2.0×10^4	
	80		86	7.7×10^5	
Xylan	5	RT–720	259	1.9×10^{22}	Williams and Besler [38]
	20		257	2.7×10^{22}	
	40		194	2.9×10^{15}	
	80		125	1.6×10^9	
Isolated hemicellulose[a]	Isothermal	110–220	112	3.6×10^{10}	Stamm [40]
Isolated hemicellulose[b]	30	RT–560	124	1.5×10^9	Min [41]

a Isolated from Douglas fir.
b Isolated from fir and birch, the kinetic parameters were the fitted values from the two hemicelluloses.

Tab. 3.2: Multi-step reaction model for hemicellulose (or xylan) pyrolysis.

Sample	Heating rate (°C/min)	Temperature (°C)	Kinetic parameters		Reference
			E (kj/mol)	*A* (s^{-1})	
Xylan	4	195–225	54–71	/	Ramiah [39]
		225–265	105–113	/	
Xylan	10	200–350	193	7.9×10^{16}	Varhegyi et al. [42]
	80		195	7.9×10^{16}	
Isolated hemicellulose from oak	Isothermal	300	73	3.3×10^{6}	Koufopanos et al. [43]
		350	174	1.1×10^{14}	
		400	172	2.5×10^{13}	
Isolated hemicellulose from cerasus	1.7	RT–400	187	2.1×10^{16}	Ward and Braslaw [44]
	5.1		216	4.95×10^{17}	
	9.1		251	2.0×10^{17}	

3.3.1.2 Multi-step reaction model

The most popular multi-step reaction model is the two-step model, which assumes that hemicellulose first forms an intermediate during pyrolysis and then further decomposes into the final products. Some other researchers believe that hemicellulose experiences more than two steps during its pyrolysis. As shown in Tab. 3.2, in the same temperature range, the activation energies obtained by Varhegyi et al. [42] at heating rates of 10 °C/min and 80 °C/min were higher than those obtained by Ramiah [39] at a heating rate of 4 °C/min. There were also certain differences in the kinetic parameters of hemicellulose isolated from different biomass. In addition, both the one-step global reaction model and the multi-step reaction model could be selected to study the kinetics of hemicellulose pyrolysis according to the reaction conditions. Varhegyi et al. [42] employed many kinetic models to simulate the pyrolysis of xylan, and found that the two-step reaction model had a high accuracy and was closer to the actual pyrolysis process.

During xylan pyrolysis, there are two main weight loss stages during devolatilization, within the temperature range of 215–300 °C and 300–500 °C, corresponding to the formation of intermediates and condensation to form char. The kinetic parameters for these two stages are listed in Tab. 3.3. As the heating rate increases, the activation energy for the first stage changes a little between 117–119 kJ/mol whereas that for the second stage changes greatly (31–71 kJ/mol). This might be because residence time has little effect on the main decomposition stage, but it significantly influences the char formation stage. Although the kinetic parameters from different researchers differ greatly, the activation energies for cellulose and hemicellulose are usually lower than 200 kJ/mol.

Tab. 3.3: Kinetic parameters for xylan pyrolysis under different heating rates.

Heating rate	Stage	Temperature (°C)	E_i (kJ/mol)	A_i (s^{-1})
5 °C/min	1	215–300	117.76	4.791×10^8
	2	300–530	63.06	40.06
20 °C/min	1	215–310	118.17	6.22×10^8
	2	310–530	65.25	85.79
60 °C/min	1	215–310	119.85	1.15×10^8
	2	310–540	71.77	348.53

3.3.1.3 Semi-global reaction model

Di Blasi et al. [45] developed a semi-global reaction model based on the two-step reaction model, as shown in Fig. 3.7. Xylan pyrolysis firstly forms an intermediate and condensable volatiles, then the intermediate is further transformed into volatiles and char. This model fitted the experimental data well at both high and low temperatures. Compared with a two-step reaction model, the semi-global reaction model reflects the thermally instability of hemicellulose. Both models can simulate the devolatilization behavior at the main reaction stage. Branca et al. [29] employed this model to simulate the pyrolysis of glucomannan and obtained a good result (see Tab. 3.4).

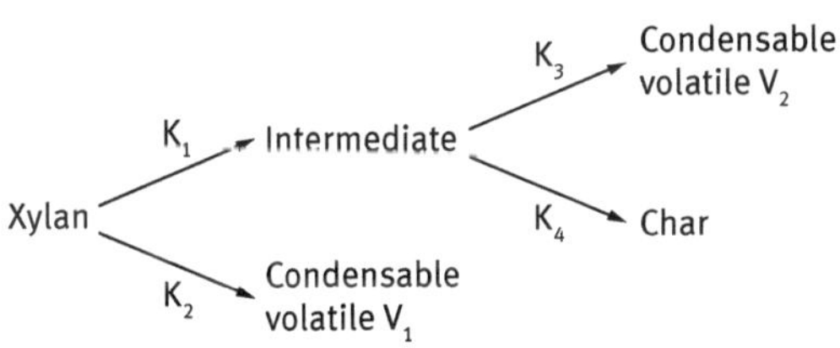

Fig. 3.7: Semi-global reaction model for xylan pyrolysis.

Tab. 3.4: Semi-global reaction model for pyrolysis of xylan and glucomannan.

Sample	Heating rate (°C/min)	Temperature (°C)	Kinetic parameters		Reference
			E (kj/mol)	A (s^{-1})	
Xylan	Isothermal	200–340	$E_1 = 66.1$	$A_1 = 1.74 \times 10^4$	Di Blasi et al. [45]
			$E_2 = 91.4$	$A_2 = 3.31 \times 10^6$	
			$E_3 = 52.5$	$A_3 = 5.87 \times 10^2$	
			$E_4 = 56.3$	$A_4 = 0.43 \times 10^2$	
Glucomannan	300–400	230–320	$E_1 = 54.5$	$A_1 = 2.64 \times 10^3$	Branca et al. [29]
			$E_2 = 98.5$	$A_2 = 1.57 \times 10^7$	
			$E_3 = 48.2$	$A_3 = 5.21 \times 10^1$	
			$E_4 = 70.1$	$A_4 = 2.10 \times 10^4$	

3.3.2 Formation pathway of typical products from hemicellulose pyrolysis

The product species from xylan pyrolysis are in accordance with its chemical structure. The lignin moieties linked with xylan can also decompose into phenols as extra products. As shown in Fig. 3.8, during the pyrolysis of xylan, the linked lignin fragments are firstly cleaved and then decompose into phenols. The O-acetyl groups linked at xylan backbone dissociate as acetic acid. The xylose units in backbone depolymerize and are ring-opened to form small molecular (2–3 carbons) products, such as hydroxyacetone. The ring-opened intermediates can also recyclize to form furfural, furanones and cyclopentenones. Of these, furfural and furanones are recyclized via dehydration of hydroxyls, while cyclopentenones are formed through the combination of C=C bonds.

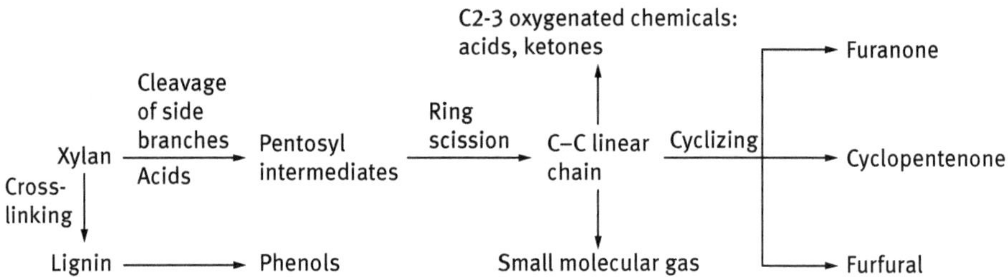

Fig. 3.8: The decomposition pathway of xylan.

3.3.2.1 Formation pathway of acetic acid

Acetic acid is one of the typical products from hemicellulose pyrolysis. The most popular formation pathway of acetic acid is the dissociation of O-acetyl groups linked at C_2 position of xylose residues [22, 46, 47]. In addition, it can also be formed from the decomposition of ring-opened intermediates and uronic acids [48, 49]. The dissociation of O-acetyl groups can occur at a low temperature, thus acetic acid is also produced at low temperature and its yield decreases at elevated temperatures [23]. This is in accordance with the results found by Prins et al. [50]. They observed that acetic acid can be produced at a temperature lower than 300 °C via torrefaction. The other components are only slightly affected at such a low temperature [51]. Thus, two-stage pyrolysis at low temperature can be used to yield acid-rich products and at medium temperatures to produce acid-free bio-oil.

3.3.2.2 Formation pathway of furfural and methanol

Furfural is another typical product from hemicellulose pyrolysis in addition to acetic acid and it is mainly formed from the step-wise dehydration of pentose units, such as D-xylose residue, accompanied with ring contraction [52]. Šimkovic et al. [9] used

TG-MS and found that furfural and methanol were the two typical products during pyrolysis of 4-O-methyl-D-glucurono-D-xylan. Košík et al. [53] found that the main products from 4-O-methyl-D-glucurono-D-xylan pyrolysis were methanol, CO_2, 3-hydroxy-2-penteno-1,5-lactone, and furfural. Methanol was formed from the dissociation of methoxyl linked at C_4 of 4-O-methyl-D-glucuronic acid, while CO_2 was formed through decarboxylation. Jensen et al. [10] also pointed out that methanol was produced via the decomposition of methoxyl. Antal et al. [54] found that xylose could be transformed into anhydride via the dehydration of hydroxyls at the xylosyl pyran ring, which could further dehydrate to form furfural.

3.3.2.3 Formation pathway of gaseous products

The main gaseous products from hemicellulose pyrolysis include CO_2, CO, CH_4, and H_2, of which the yields of CO_2 and CO are usually much higher than those of CH_4 and H_2. The formation of CO_2 is mainly related to the decarboxylation of acetic acid and uronic acid [22, 46]. CO is mainly formed from the decarbonylation of aldehydes, such as glycolaldehyde, formaldehyde and acetaldehyde and its yield increases at elevated temperature. In addition, the further decomposition of volatiles at high temperatures can also contribute to the formation of CO [55, 56]. CH_4 and H_2 are also formed at high temperature (> 500 °C) and their yields are enhanced in the high temperature range. H_2 mainly originates from the cracking and deformation of C=C and C–H bonds, while CH_4 is produced from the decomposition of $O{-}CH_3$ groups [27]. In addition, a small amount of gaseous C_{2+} hydrocarbons can also be yielded due to the secondary degradation of volatiles [6, 9].

3.3.2.4 Formation pathway of ketones

Ketones, including acetone, hydroxyacetone, furanones, pyranones, and alicyclic ketones, can also be detected in the products from hemicellulose pyrolysis and are formed through complex pathways. The formation of linear ketones, such as acetone and hydroxyacetone, might result from the further carbon chain cleavage of ring-opened intermediates [57]. The cyclic ketones, including furanones, pyranones and alicyclic ketones, may originate from the recyclization of ring-opened intermediates, in which furanones and pyranones are recyclized via dehydration of hydroxyls, while alicyclic ketones are formed through the combination of C=C bonds.

3.3.2.5 Formation pathway of char

The char produced from hemicellulose pyrolysis exhibits fusion and soft properties. Fisher et al. [58] pointed out that xylan showed the tendency of fusion under pyrolysis at 200 °C. This might result from the amorphous structure of hemicellulose. Temperature had a great impact on the characteristics of char, whose surface became more smooth and porous as the temperature increased [59]. Ponder et al. [22] studied the pyrolysis of an ash-free synthetic xylan by using a vacuum reactor and obtained a final char yield of 50 %, which confirmed that the complex chemical structure of xylan was the main reason for the high char yield during pyrolysis. Compared with cellulose, whose hydroxyl at C_6 position of glucose units can stabilize the cation produced from the cleavage of glycosidic bonds and form 1,6-anhydrite, which further decomposes to produce levoglucosan, during xylan pyrolysis, which does not have hydroxymethyl at the xylose unit, the cations formed from glycosidic bonds cannot be stabilized via intramolecular dehydration. Instead they might be polymerized with other big molecular radicals, leading to the formation of char. Xin et al. [60] characterized the char produced from xylan pyrolysis at different temperatures by using FTIR and found that compared with xylan, the char obtained at 350 °C showed a lower signal of C–O–C and a higher signal of C=O stretching vibration. This indicated that most glycosidic bonds in xylan had been broken. In the range 350–550 °C, the specific surface area of char was lower than that of xylan. This was due to the deposition of big molecular fragments on the pore canal of char. When the temperature exceeded 650 °C, the specific surface area of char increased rapidly, and the average pore size decreased significantly. This was due to the rapid release of gaseous products, such as CO and CH_4. During this period, the carbon skeleton of char collapsed and merged. As the temperature reached 750–850 °C, the specific surface area of char decreased and the average pore size increased slightly, which might result from the blocking and remerging of the surface pores of char. Yang et al. [61] also found that the amount of pores in char decreased rapidly during pyrolysis at high temperatures.

3.3.3 Mechanism of hemicellulose pyrolysis at molecular scale

Previous studies of the hemicellulose pyrolysis mechanism have usually proposed a reaction pathway by deduction based on product distribution. The specific reaction occurring during pyrolysis is very difficult to obtain by experimental analysis. Density functional theory (DFT) has been introduced to study the elementary reactions in hemicellulose pyrolysis. Xylose, the basic unit comprising xylan, was selected as the model compound for the computational calculation.

Mayer bond order, which can quantitatively assess the strength of each bond in xylose structure [62], was used to judge the initial position of bond breakage during xylose decomposition. The results are shown in Tab. 3.5, the bond order of C_3–O_6 (0.51) is the lowest, thus ring-opening most probably occurs here. The bond O_{10}–H_7 is also very weak (0.72), hence it can be deduced that the ring-opening reaction first starts with the breakage of C_3–O_6, and H_7 moves to O_6 to form a new O_6–H_7 hydroxyl, while the original C_3–O_{10} transforms into aldehyde.

Tab. 3.5: Mayer bond order on each bond in xylose.

Xylose structure	Bond	Value	Bond	Value	Bond	Value
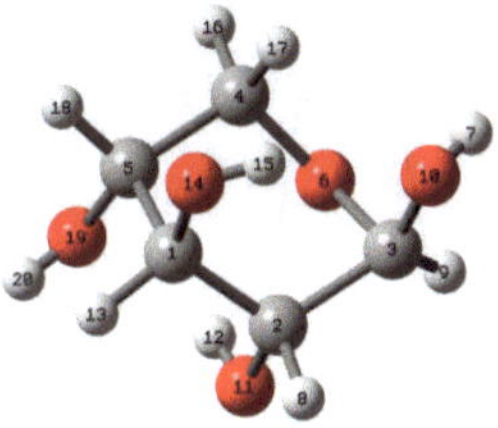	C_3-O_6	0.51	C_2-O_{11}	1.70	C_3-H_9	1.02
	C_4-O_6	0.84	C_1-O_{14}	1.39	C_2-H_8	0.93
	C_3-C_2	0.86	C_5-O_{19}	1.29	C_1-H_{13}	0.96
	C_2-C_1	0.88	$O_{10}-H_7$	0.72	C_5-H_{18}	0.85
	C_1-C_5	0.97	$O_{11}-H_{12}$	0.67	C_4-H_{16}	1.00
	C_5-C_4	0.92	$O_{14}-H_{15}$	0.75	C_4-H_{17}	0.98
	C_3-O_{10}	1.29	$O_{19}-H_{20}$	0.78		

Based on the Mayer bond order assessment and experimental analysis, the decomposition pathway of D-xylose shown in Fig. 3.9 has been proposed. Xylose first decomposes with ring-opening to form a linear intermediate (IM_1), whose further decomposition can be divided into three pathways. Path 1 is the breakage of C_4-C_5 of IM_1 to form formaldehyde and IM_2, IM_2 further dehydrates and recyclizes to form 4-hydroxy-dihydrofuran-2(3H)-one; Path 2 is the decarbonylation of IM_2 to form CO and IM_4, IM_4 further dehydrates and isomerizes to generate hydroxyacetone; Path 3 is the step-wise dehydration of IM_2 accompanied by ring contraction, leading to the formation of furfural.

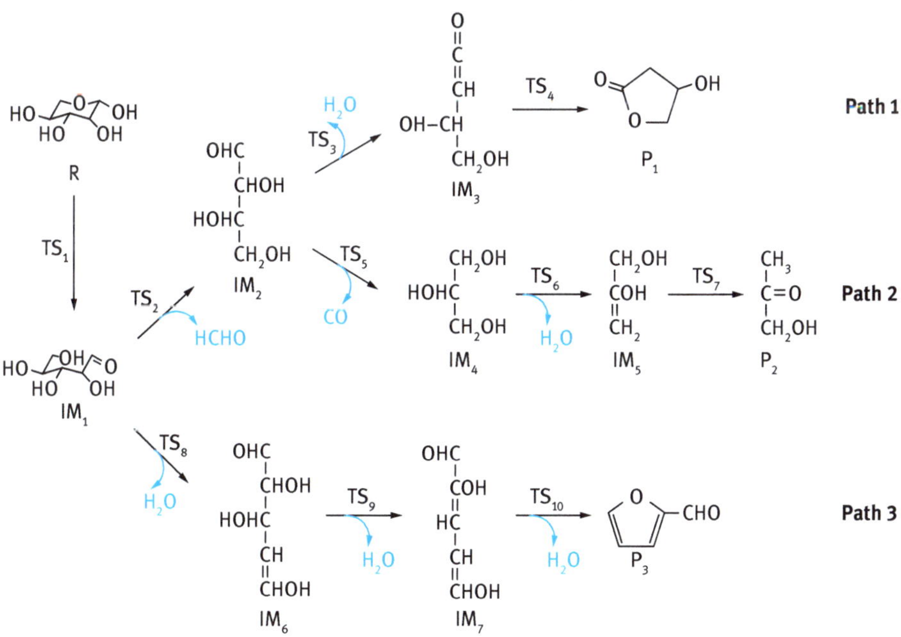

Fig. 3.9: The proposed decomposition pathways of xylose.

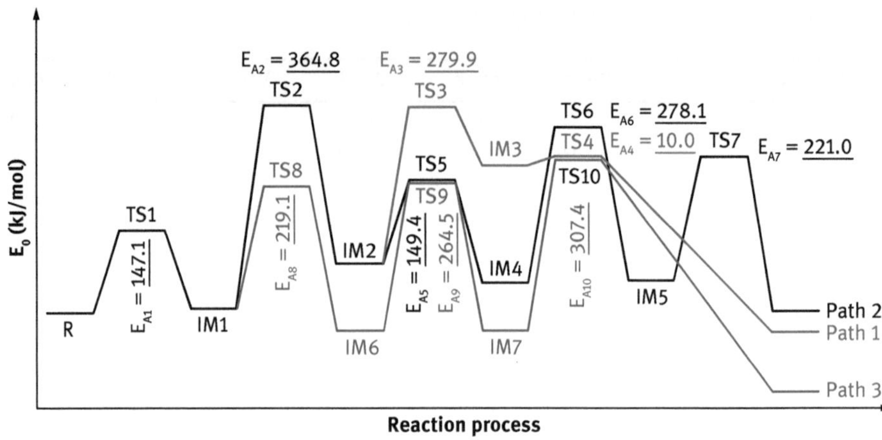

Fig. 3.10: The energy barriers for each elementary reaction in xylose pyrolysis. Reprinted with permission from [63], © 2013 Elsevier.

The corresponding activation energies calculated for each elementary reaction in the three paths are shown in Fig. 3.10. It was indicated that the free energy for Path 3 leading to furfural formation was the lowest, thus the formation of furfural was more favorable than the other two products. Path 1 for 4-hydroxydihydrofuran-2(3H)-one formation exhibited two high energy barriers, namely the dissociation of formaldehyde (364.8 kJ/mol) and the dehydration in the chain end (279.9 kJ/mol), hence it was hard to be formed.

Zhang et al. [64] discussed the formation pathways of furfural, glycolaldehyde, methylglyoxal and formaldehyde from xylose pyrolysis based on DFT study. The Gibbs free energy for each product formation was smaller than zero. During the pyrolytic temperature of 125–525 °C, furfural was the dominant product. When the temperature was higher than 525 °C, the Gibbs free energies for the formation of some smaller molecular products became lower than that for furfural, and the yields of glycolaldehyde and formaldehyde increased. Huang et al. [65] also studied the activation energy distribution during xylose pyrolysis by computational simulation, and found that the ring-opening reaction contained a low energy barrier of 170.4 kJ/mol. The ring-opened intermediates further decomposed to form glycolaldehyde, acetaldehyde, furfural, acetone and CO.

Huang et al. [66] also studied the pyrolysis pathways of O-acetyl-xylose as model compound by DFT method. The decomposition of O-acetyl groups for acetic acid formation contained an energy barrier of 269.4 kJ/mol. The residue anhydroxylose underwent many different decomposition pathways to form small molecular products; the pathways leading to the production of acetaldehyde, glycolaldehyde, acetone, and CO occurred easily, while those for hydroxyacetone and formaldehyde formation showed high energy barriers.

However, the structure of hemicellulose is much more complicated than that of xylose. Thus in order to get better simulated results within restricted computer capability, some researchers also used polysaccharides with several xylose units to perform the theoretical calculation.

Liu et al. [67] built a xylan model with nine xylose units to study the pyrolysis mechanism. The decomposition of xylan occurred during 177–627 °C. The hydroxyls in the end of xylose were first dehydrated at 223 °C. The breakage of glycosidic bonds occurred above 377 °C to form xylosyl oligomer, whose further decomposition led to the production of small molecular compounds, such as H_2O, CO, ethanol, acetic acid, glycolaldehyde, etc. Huang et al. [68] built a more complex model compound as the representative of xylan, which was comprised of ten xylose units as backbone, three groups, O-acetyl, L-arabinosyl, and 4-O-glucuronic acid as branches (L-arabino-4-O-methyl-D-glucurono-O-acetyl-D-xylan). The molecular dynamic simulation showed that hydroxyl underwent cleavage when the temperature reached 177 °C. When the temperature exceeded 277 °C, the glycosidic bonds and branches started to break, and the macromolecule depolymerized to form many kinds of carbohydrate monomers. Meanwhile, the sugar ring was also cracked to generate many fragments. During the main stage within the temperature range of 30–430 °C, most products, including CO_2, acetic acid, glycolaldehyde, hydroxyacetone, furfural, and 1-hydroxy-2-butanone, were formed.

References

[1] Zamora F, Gonzalez MC, Duenas MT, et al. Thermodegradation and thermal transitions of an exopolysaccharide produced by Pediococcus damnosus 2.6. Journal of Macromolecular Science, Part B. 2002;41(3):473–486.

[2] Wang S, Yan Z, Tao L, et al. Catalytic pyrolysis of mannose as a model compound of hemicellulose over zeolites. Biomass & Bioenergy. 2013;57(4):106–112.

[3] Shafizadeh F, Mcginnis GD, Susott RA, et al. Thermal reactions of Alpha-D-xylopyranose and Beta-D-xylopyranosides. The Journal of Organic Chemistry. 1971;36(19):2813–2818.

[4] Räisänen U, Pitkänen I, Halttunen H, et al. Formation of the main degradation compounds from arabinose, xylose, mannose and arabinitol during pyrolysis. Journal of Thermal Analysis and Calorimetry. 2003;72(2):481–488.

[5] Gardiner D. The pyrolysis of some hexoses and derived di-, tri-, and poly-saccharides. Journal of the Chemical Society C: Organic. 1966:1473–1476.

[6] Yang H, Yan R, Chen H, et al. Characteristics of hemicellulose, cellulose and lignin pyrolysis. Fuel. 2007;86(12):1781–1788.

[7] Yang H, Yan R, Chen H, et al. In-depth investigation of biomass pyrolysis based on three major components: hemicellulose, cellulose and lignin. Energy & Fuels. 2006;20(1):388–393.

[8] Biagini E, Barontini F, Tognotti L. Devolatilization of biomass fuels and biomass components studied by TG/FTIR technique. Industrial & Engineering Chemistry Research. 2006;45(13):4486–4493.

[9] Šimkovic I, Varhegyi G, Antal MJ, et al. Thermogravimetric/mass spectrometric characterization of the thermal decomposition of (4-O-methyl-D-glucurono)-D-xylan. Journal of Applied Polymer Science, 1988;36(3):721–728.

[10] Jensen A, Dam-Johansen K, Wójtowicz MA, et al. TG-FTIR study of the influence of potassium chloride on wheat straw pyrolysis. Energy & Fuels. 1998;12(5):929–938.

[11] Beaumont O. Flash pyrolysis products from beech wood. Wood and Fiber Science. 1985;17(2):228–239.

[12] Nowakowski DJ, Woodbridge CR, Jones JM. Phosphorus catalysis in the pyrolysis behaviour of biomass. Journal of Analytical & Applied Pyrolysis. 2008;83(2):197–204.

[13] Nowakowski DJ, Jones JM. Uncatalysed and potassium-catalysed pyrolysis of the cell-wall constituents of biomass and their model compounds. Journal of Analytical & Applied Pyrolysis. 2008;83(1):12–25.

[14] Hosoya T, Kawamoto H, Saka S. Pyrolysis behaviors of wood and its constituent polymers at gasification temperature. Journal of Analytical and Applied Pyrolysis. 2007;78(2):328–336.

[15] Aho A. Pyrolysis of softwood carbohydrates in a fluidized bed reactor. International Journal of Molecular Sciences. 2008;9(9):1665–1675.

[16] Xiao B, Sun XF, Sun R. Chemical, structural, and thermal characterizations of alkali-soluble lignins and hemicelluloses, and cellulose from maize stems, rye straw, and rice straw. Polymer Degradation and Stability. 2001;74(2):307–319.

[17] Yu ZP. Pyrolysis characteristics of hemicelluloses from bamboo stem. China Pulp & Paper. 2012;(11):7–13.

[18] Peng Y, Wu SB. Characteristics and kinetics of sugarcane bagasse hemicellulose pyrolysis by TG-FTIR. Chemical Industry & Engineering Progress. 2009;28(8):1478–1484.

[19] Peng Y, Wu S. Fast pyrolysis characteristics of sugarcane bagasse hemicellulose. Cellulose Chemistry & Technology. 2011;45(9):605–612.

[20] Peng YY, Wu SB. Study on the Characteristics and Kinetics of Sugarcane Bagasse Hemicellulose Pyrolysis . Paper Science & Technology. 2009;3(28):14–18.

[21] Lv G, Wu S, Lou R. Characteristics of corn stalk hemicellulose pyrolysis in a tubular reactor. Bioresources. 2010;5(4):2051–2062.

[22] Ponder GR, Richards GN. Thermal synthesis and pyrolysis of a xylan. Carbohydrate Research. 1991;218:143–155.

[23] Shen DK, Gu S, Bridgwater AV. Study on the pyrolytic behaviour of xylan-based hemicellulose using TG–FTIR and Py–GC–FTIR. Journal of Analytical and Applied Pyrolysis. 2010;87(2):199–206.

[24] Van Putten R, Van der Waal JC, De Jong E, et al. Hydroxymethylfurfural, a versatile platform chemical made from renewable resources. Chemical Reviews. 2013;113(3):1499–1597.

[25] Patwardhan PR, Brown RC, Shanks BH. Product distribution from the fast pyrolysis of hemicellulose. Chemsuschem. 2011;4(5):636–643.

[26] Liu JL, Jiang JC, Huang HT. Study on curie-point pyrolysis of xylan under CP-GC-MS conditions. Chemistry and Industry of Forest Products. 2010;(01):5–10.

[27] Qu TT, Guo WJ, Shen LH, et al. Experimental study of biomass pyrolysis based on three major components: hemicellulose, cellulose, and lignin. Industrial& Engineering Chemistry Research. 2011;50(18):10424–10433.

[28] Zhao K, Xiao J, Shen LH, et al. Experimental study of biomass rapid pyrolysis based on three components. Acta Energiae Solaris Sinica. 2011;(05):710–717.

[29] Branca C, Di Blasi C, Mango C, et al. Products and kinetics of glucomannan pyrolysis. Industrial & Engineering Chemistry Research. 2013;52(14):5030–5039.

[30] Alén R, Kuoppala E, Oesch P. Formation of the main degradation compound groups from wood and its components during pyrolysis. Journal of Analytical and Applied Pyrolysis. 1996;36(2):137–148.
[31] Chen W, Kuo P. Isothermal torrefaction kinetics of hemicellulose, cellulose, lignin and xylan using thermogravimetric analysis. Energy. 2011;36(11):6451–6460.
[32] Wang G, Li W, Li B, et al. TG study on pyrolysis of biomass and its three components under syngas. Fuel. 2008;87(4):552–558.
[33] Sagehashi M, Miyasaka N, Shishido H, et al. Superheated steam pyrolysis of biomass elemental components and Sugi (Japanese cedar) for fuels and chemicals. Bioresource Technology. 2006;97(11):1272–1283.
[34] Giudicianni P, Cardone G, Ragucci R. Cellulose, hemicellulose and lignin slow steam pyrolysis: Thermal decomposition of biomass components mixtures. Journal of Analytical and Applied Pyrolysis. 2013;100:213–222.
[35] Fushimi C, Katayama S, Tasaka K, et al. Elucidation of the interaction among cellulose, xylan, and lignin in steam gasification of woody biomass. Aiche Journal. 2009;55(2):529–537.
[36] Fushimi C, Katayama S, Tsutsumi A. Elucidation of interaction among cellulose, lignin and xylan during tar and gas evolution in steam gasification. Journal of Analytical and Applied Pyrolysis. 2009;86(1):82–89.
[37] Bilbao R, Millera A, Arauzo J. Kinetics of weight loss by thermal decomposition of xylan and lignin. Influence of experimental conditions. Thermochimica Acta. 1989;143:137–148.
[38] Williams PT, Besler S. The pyrolysis of rice husks in a thermogravimetric analyser and static batch reactor. Fuel. 1993;72(2):151–159.
[39] Ramiah MV. Thermogravimetric and differential thermal analysis of cellulose, hemicellulose, and lignin. Journal of Applied Polymer Science. 1970;14(5):1323–1337.
[40] Stamm AJ. Thermal degradation of wood and cellulose. Industrial & Engineering Chemistry. 1956;48(3):413–417.
[41] Min K. Vapor-phase thermal analysis of pyrolysis products from cellulosic materials. Combustion and Flame. 1977;30:285–294.
[42] Varhegyi G, Antal Jr MJ, Szekely T, et al. Kinetics of the thermal decomposition of cellulose, hemicellulose, and sugarcane bagasse. Energy & Fuels. 1989;3(3):329–335.
[43] Koufopanos CA, Lucchesi A, Maschio G. Kinetic modelling of the pyrolysis of biomass and biomass components. The Canadian Journal of Chemical Engineering. 1989;67(1):75–84.
[44] Ward SM, Braslaw J. Experimental weight loss kinetics of wood pyrolysis under vacuum. Combustion and Flame, 1985,61(3):261–269.
[45] Di Blasi C, Lanzetta M. Intrinsic kinetics of isothermal xylan degradation in inert atmosphere. Journal of Analytical and Applied Pyrolysis. 1997;40:287–303.
[46] Shafizadeh F, Mcginnis GD, Philpot CW. Thermal degradation of xylan and related model compounds. Carbohydrate Research. 1972;25(1):23–33.
[47] Demirbaş A. Analysis of liquid products from biomass via flash pyrolysis. Energy Sources. 2002;24(4):337–345.
[48] Beall FC. Thermogravimetric analysis of wood lignin and hemicelluloses. Wood and Fiber Science. 1969;1(3):215–226.
[49] Paine III JB, Pithawalla YB, Naworal JD. Carbohydrate pyrolysis mechanisms from isotopic labeling: Part 2. The pyrolysis of d-glucose: General disconnective analysis and the formation of C_1 and C_2 carbonyl compounds by electrocyclic fragmentation mechanisms. Journal of Analytical and Applied Pyrolysis. 2008;82(1):10–41.
[50] Prins MJ, Ptasinski KJ, Janssen FJ. Torrefaction of wood: Part 2. Analysis of products. Journal of Analytical and Applied Pyrolysis. 2006;77(1):35–40.

[51] Wu Y, Zhao Z, Li H, et al. Low temperature pyrolysis characteristics of major components of biomass. Journal of Fuel Chemistry and Technology. 2009;37(4):427–432.
[52] Shen DK, Gu S, Bridgwater AV. The thermal performance of the polysaccharides extracted from hardwood: Cellulose and hemicellulose. Carbohydrate Polymers. 2010;82(1):39–45.
[53] Košík M, Reiser V, Kováč P. Thermal decomposition of model compounds related to branched 4-(i) O-methylglucuronoxylans. Carbohydrate Research. 1979;70(2):199–207.
[54] Antal MJ, Leesomboon T, Mok WS, et al. Mechanism of formation of 2-furaldehyde from D-xylose. Carbohydrate Research. 1991;217:71–85.
[55] Boroson ML, Howard JB, Longwell JP, et al. Product yields and kinetics from the vapor phase cracking of wood pyrolysis tars. Aiche Journal. 1989;35(1):120–128.
[56] Li S, Lyons-Hart J, Banyasz J, et al. Real-time evolved gas analysis by FTIR method: an experimental study of cellulose pyrolysis. Fuel. 2001;80(12):1809–1817.
[57] Paine III JB, Pithawalla YB, Naworal JD. Carbohydrate pyrolysis mechanisms from isotopic labeling: Part 3. The Pyrolysis of d-glucose: Formation of C_3 and C_4 carbonyl compounds and a cyclopentenedione isomer by electrocyclic fragmentation mechanisms. Journal of Analytical and Applied Pyrolysis. 2008;82(1):42–69.
[58] Fisher T, Hajaligol M, Waymack B, et al. Pyrolysis behavior and kinetics of biomass derived materials. Journal of Analytical and Applied Pyrolysis. 2002;62(2):331–349.
[59] Lv G, Wu S, Lou R. Characteristics of corn stalk hemicellulose pyrolysis in a tubular reactor. Bioresources. 2010;5(4):2051–2062.
[60] Xin S, Yang H, Chen Y, et al. Assessment of pyrolysis polygeneration of biomass based on major components: Product characterization and elucidation of degradation pathways. Fuel. 2013;113(0):266–273.
[61] Yang H, Yan R, Chen H, et al. Mechanism of palm oil waste pyrolysis in a packed bed. Energy & Fuels. 2006;20(3):1321–1328.
[62] Mayer I. Charge, bond order and valence in the ab initio SCF theory. Chemical Physics Letters. 1983;97(3):270–274.
[63] Wang S, Ru B, Lin H, et al. Degradation mechanism of monosaccharides and xylan under pyrolytic conditions with theoretic modeling on the energy profiles. Bioresource Technology. 2013;143:378–383.
[64] Zhang Z, Liu C, Li HJ, et al. Theoretical studies of pyrolysis mechanism of xylan monomer. Acta Chimica Sinica. 2011;69(18):2099–2107.
[65] Huang J, Liu C, Tong H, et al. Theoretical studies on pyrolysis mechanism of xylopyranose. Computational and Theoretical Chemistry, 2012;1001:44–50.
[66] Huang J, Chao L, Hong T, et al. Theoretical studies on pyrolysis mechanism of O-acetyl-xylopyranose. Journal of Fuel Chemistry and Technology. 2013;41(3):285–293.
[67] Liu C, Li HJ, Huang JB. Simulation of molecular dynamics in xylan pyrolysis. Journal of Functional Polymers. 2010;23(3):291–296.
[68] Huang JB, Tong H, Li WM, et al. Molecular dynamic simulation based study on pyrolysis mechanism of hemicellulose. Thermal Power Generation. 2013;(03):25–30.

4 Pyrolysis of lignin

4.1 Lignin pyrolysis process

4.1.1 Fundamental process of lignin pyrolysis

In contrast to the glycan structures of cellulose and hemicellulose, lignin is a highly branched polymer formed by the nonlinear and random combination of three typical phenylpropane units. Although there are only three typical units in lignin, there are various linkages between the units and different functional groups attached to the benzene ring, resulting in the complexity of lignin structure. Compared with the pyrolysis of cellulose and hemicellulose, the pyrolysis of lignin is located in a broader temperature range, about 200–500 °C, and the decomposition is slower, including three weight loss stages: free water release, the formation of main pyrolysis products and the generation of gaseous products from secondary and further decomposition. Further research confirmed that the decomposition of lignin started at 200–275 °C, and a large number of phenolic products were formed at about 400 °C, followed by the decomposition and condensation of aromatic rings to generate some small molecular compounds at temperatures above 500 °C [1]. The products from lignin pyrolysis include pyrolysis oil, light gaseous products and char, and the corresponding yields are 20–30 %, 30–40 % and 30–45 %, respectively.

Lignin pyrolysis oil mainly consists of phenolic compounds as well as some aldehydes and acids. The phenolic compounds include phenols, hydroxylphenols, guaiacols and syringols. Fig. 4.1 shows the typical phenolic products: guaiacols have a methoxyl group at the C_2 position, and a side chain (carbon number ≤ 3) at the C_4 position, including guaiacol, methylguaiacol, vanillin and vinylguaiacol; syringols have two methoxyl groups at the C_2 and C_6 positions, and a side chain (carbon number ≤ 3) at the C_4 position, including syringol, methylsyringol, syringaldehyde, and vinylsyringol. The phenols content is much lower than that of guaiacols and syringols, because the number of H-type units is much smaller than that of G-type and S-type units, and the side chains and methoxyl groups attached to the benzene ring are difficult to remove. Since the phenolic products are mainly generated from the corresponding guaiacyl, p-hydroxylphenyl and syringyl units in the original lignin, the approximate ratio of guaiacyl units to syringyl units in the lignin can be calculated based on the distribution of the phenolic products in the pyrolysis oil. The light gaseous products, mainly including CO, CO_2, CH_4, and H_2, are derived from the cracking of side chains and substituted functional groups attached to the aromatic ring. Moreover, compared with cellulose and hemicellulos, more char is produced during the lignin pyrolysis [2].

DOI 10.1515/9783110369632-006

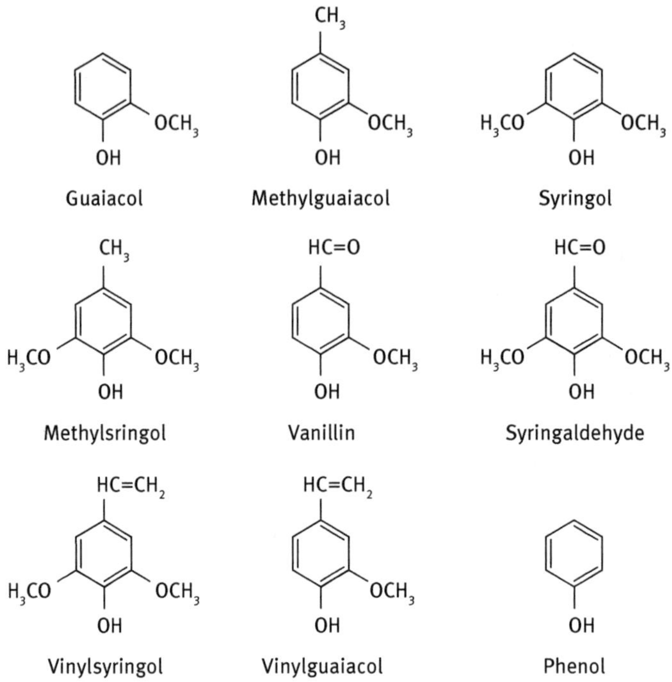

Fig. 4.1: Typical phenolic products from lignin pyrolysis.

4.1.2 Pyrolysis of typical model compounds for lignin

Compared with cellulose and hemicellulose with fixed structural units, the structure of lignin is very complicated because of the variety of the substituted functional groups, the side chains and the linkages between the typical units. Therefore, the study of model compounds containing the typical lignin structure can simplify the pyrolysis behavior of lignin, and this is important for the further study of the detailed mechanism of bond breakage and product generation for lignin pyrolysis. Currently there are two kinds of lignin model compounds: one is the monomeric model compound like phenol derivatives and the other one is the phenolic polymers containing the typical lignin structure.

The selection of monomeric model compounds is based on the structural characteristics of the phenylpropane units, such as guaiacols, syringols, and p-hydroxylphenols. These model compounds are the primary products from lignin pyrolysis, and depending on their different chemical properties, they will subsequently undergo cracking and condensation, or remain stable. Therefore, the study of pyrolysis behavior of these model compounds can estimate their thermal stability during lignin pyrolysis. In the study of guaiacol pyrolysis by Klein [3], guaiacol mainly underwent demethylation to produce methane and catechol, and demethyoxylation also occurred to form phenol and CO. Subsequently, he studied the pyrolysis behavior of a series of monophenolic model compounds, including syringol, isoeugenol, vanillin, anisole, veratrole, acetophenone, cinnamaldehyde and cinnamyl alcohol. The reaction pathways of syringol and isoeugenol were similar to that of guaiacol, which mainly involved demethylation. The pyrolysis of anisole had a complex product distribution and its decomposition temperature was higher than guaiacol, showing that the existence of a phenolic hydroxyl group improved reactivity. The primary pyrolysis product of veratrole was guaiacol, and then catechol and phenol were produced according to the pyrolysis mechanism of guaiacol. In addition, the concerted reaction between the adjacent substituted groups in veratrole could also generate o-cresol. The model compounds that contained carbonyl groups like vanillin, acetophenone, benzaldehyde and cinnamaldehyde, mainly underwent decarbonylation during pyrolysis. Vanillin showed a higher reactivity owing to the electronic effect of the methoxyl and hydroxyl groups. Meanwhile, the pyrolysis of benzaldehyde was easier than that of acetophenone, showing that the terminal C=O group (aldehyde) in the side chain had a higher reactivity than the C=O group in α or β position. The main pyrolysis products of cinnamaldehyde were phenol and allylbenzene, and the Diels–Alder reaction also occurred to form some dimers. The pyrolysis of cinnamyl alcohol generated water, methanol, cinnamaldehyde, allylbenzene, cresol, and some dimers.

Asmadi et al. studied the pyrolysis of guaiacol and syringol as model compounds [4]. The homolysis of $O{-}CH_3$ bonds occurred above 400 °C for these two phenols, followed by some temperature-dependent reactions. At about 400 °C, free-radical-induced rearrangement transformed the aromatic methoxyl group to an aromatic methyl group and then condensation also occurred, which formed some large molecular products. When the temperature was over 450 °C, the intensity of the charring reaction increased obviously. Subsequently, Asmadi et al. [5] studied the thermal reactivities of the primary pyrolysis products of guaiacol and syringol, namely catechols and pyrogallols ($O{-}CH_3$ homolysis products) and cresols and xylenols ($O{-}CH_3$ rearrangement products). The reactivities of catechols and pyrogallols were much higher than those of cresols and xylenols, indicating that the existence of a phenolic hydroxyl group significantly improves reactivity. Meanwhile, pyrogallols were easier to convert than catechols, suggesting that more substituted groups on the aromatic ring would lead to a higher reactivity. Moreover, at the primary pyrolysis stage, catechols and pyrogallols were effectively converted to CO (additional CO_2 in the case of pyrogallols), while a long-term pyrolysis was needed for the conversion of cresols and xylenols to H_2 and CH_4.

There are two kinds of linkages between the phenylpropane units in lignin, namely ether bond linkages (α-O-4, β-O-4, 4-O-5, etc.) and C–C bond linkages (β-1, α-1, β-5, 5-5, etc.). β-ether bonds are the most common linkage in lignin. In a pyrolysis study of Phenethyl Phenyl Ether (PPE), the simplest dimeric model compound, Klein et al. found that it mainly underwent cracking of β-ether bonds to form phenol and styrene [6]. Although the cleavage of β-ether bonds predominated during the pyrolysis of the dimers with the β-ether linkage, the functional groups on the side chain (aliphatic hydroxyl group and carbonyl group) and attached to the aromatic ring (phenolic hydroxyl group and methoxyl group) influenced the reactivity of β-ether bonds. Domburg et al. [7] studied the pyrolysis behaviors of different dimmers with β-ether linkages which contained an additional methoxyl group, hydroxyl group or carbonyl group. The result showed that when the $C_\alpha-OH$ bond was oxygenated to a $C_\alpha=O$ bond, the cleavage of β-ether linkages was strongly suppressed. Additionally, the change of substituted functional groups on the aromatic ring from C_4-OH to C_4-O-CH_3 slightly affected the cleavage of β-ether linkages. Savinykh et al. [8] further studied the influence of phenyl and methoxyl groups on the reactivity of β-ether linkages, and found that the removal of phenyl significantly suppressed the cleavage of β-ether linkages, while the methoxyl group had less influence.

In addition to the β-ether linkage, there are some other linkages in lignin, such as α-ether, 5-O-4 and β-1. Although these linkages are less common than β-ether linkages in lignin structure, they affect the integral pyrolysis behavior of lignin. In general, the ether linkages, such as α-ether, 5-O-4, and β-ether, are easily cleaved to form phenolic products, and the linkages between the alkyl group (side chain) and the carbon atom in the aromatic ring have relatively lower reactivities, while the depolymerization of linkages combining the carbon atoms in different aromatic rings is difficult. Kawamoto et al. [9] investigated the reactivities of dimers with different linkages, including α-O-4, β-O-4, β-1, and 5-5. The different linkages significantly affected the reactivities of dimers, and the substituted functional groups (phenolic hydroxyl group and methoxyl group) also had some influence. The reactivities of dimers followed this sequence: α-O-4 (phenolic and non-phenolic (with methoxyl group)), β-O-4 (phenolic) > β-O-4 (non-phenolic), β-1 (phenolic and non-phenolic) > 5-5 (phenolic and non-phenolic). For compounds with the same linkage, the phenolic type was more reactive than the non-phenolic type. Moreover, based on the distribution of pyrolysis products, β-O-4 type dimers were found to mainly undergo the cleavage of $C_\beta-O$ bonds to form cinnamyl alcohols and guaiacol, and the cleavage of $C_\beta-C_\gamma$ bonds to form vinyl ethers. There were two competing reaction pathways for the pyrolysis of β-1 type dimers, namely the cleavage of $C_\alpha-C_\beta$ to form benzaldehydes and styrenes, and the cleavage of $C_\beta-C_\gamma$ to generate formaldehyde and stilbenes. In addition, the phenolic hydroxyl group promoted the cleavage of $C_\beta-O$ bonds in β-O-4 type dimers and the cleavage of $C_\beta-C_\gamma$ bonds in β-1 type dimers.

The trimers and other model compounds with a higher degree of polymerization are closer to natural lignin, and their study better reflects the pyrolysis behavior of lignin. Faix et al. [10] studied the pyrolysis of two trimers which contained both α-ether linkages and β-ether linkages. The detailed pyrolysis mechanism of trimers was difficult to conclude, but similar to the pyrolysis of dimers: the syringol, guaiacol and acetoguaiacone in the final products were derived from the cleavage of α- or β-ether linkages, while the benzaldehyde was produced from the cleavage of $C_\alpha-C_\beta$ linkages. Liu et al. [11] investigated the pyrolysis of β-O-4 model polymers (guaiacyl type, as

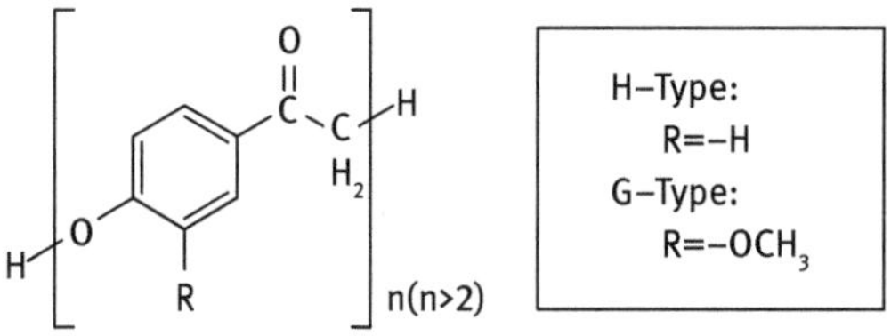

Fig. 4.2: Lignin β-O-4 model polymers [11].

shown in Fig. 4.2) with an average molecular weight of 4,832. Because of the complex molecular structure, the cleavage of different linkages occurred, including C_α–C_β, C_β–O, Ar–C_α, and O-4. However, on the basis of the product distribution, the cleavage of C_β–O bonds still dominated during depolymerization, suggesting similarity to the depolymerization mechanism of the β-O-4 model compounds.

Chu et al. [12] used t-butoxycarbonlymethyl vanillin to synthesize β-O-4 type model polymers, as shown in Fig. 4.3. During the pyrolysis of these polymers, the β-O-4 linkage mainly decomposed in the temperature range of 250–350 °C. A free radical reaction mechanism was deduced from the product distribution. The main monophenolic products, namely vanillin and 4-methylguaiacol, were produced from the cleavage of β-O-4 linkages followed by proton abstraction from weak C–H or O–H bonds. Meanwhile, the collision of free radicals from the cleavage of C–O bonds also generated 1,4-butanediol diacetate. Some other phenolic products, such as 4-propylguaiacol and eugenol, were derived from the secondary reactions of the radicals which were formed from the cleavage of β-O-4 linkages. The secondary reactions included H-abstraction, double bond formation, rearrangement, isomerization and concerted reaction. In addition, char was also produced, and the elemental analysis showed that pyrolytic char accounted for 50 % of the carbon content in the original sample. This pyrolytic char was likely formed by the polymerization of small radicals, such as aromatics, alkanes, and alkenes.

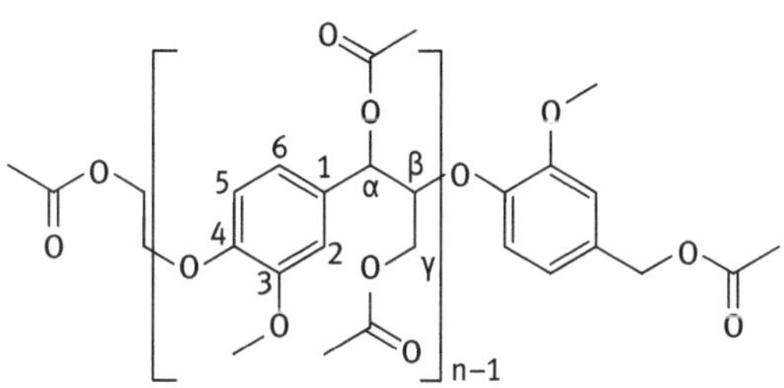

Fig. 4.3: β-O-4 type model polymers. Reprinted with permission from [12], © 2012 Royal Society of Chemistry.

4.1.3 Pyrolysis of different lignin model compounds

Lignin can be isolated from biomass by certain methods. However, because of the effects of isolation methods on the structure of the derived lignin, the pyrolysis behaviors of these lignin samples differ.

4.1.3.1 Pyrolysis of pickling lignin

During the Van Soest analysis of biomass, 72 % sulfuric acid solution was used to dissolve the cellulose and hemicellulose and solid residue was obtained, namely strong acid detergent fiber (SADF). SADF is mainly composed of lignin and a little ash. Fig. 4.4 presents the TG/DTG curves of SADF from fir and birch. The weight losses of these two lignin samples both covered a wide temperature range. The residue yields are both about 40 %. The DTG curves show that there were three weight loss stages, and the maximum weight loss rate for each stage increased in sequence. The slight weight loss in the first stage was attributed to the release of water and the main weight loss occurred in the second and third stages. In the second stage, primary depolymerization and cracking of side chains occurred, which generated a lot of volatiles, such as phenols, water, CO and CO_2. The third weight loss stage corresponded to the further depolymerization of lignin and some secondary reactions, which released CO, CH_4, methanol, and some other small molecular gases. The pyrolysis study of SADF from pine cone and corncob by Xie et al. also showed that the weight loss continued in a wide temperature range, and the final char yield was above 40 % [14]. The DTG curve had several weight loss peaks. Besides the weight loss peak for the release of water at a temperature lower than 120 °C, there were two and three weight loss peaks for the pyrolysis of SADF from pine cone and corncob, respectively, which could be attributed to the difference in the feedstock and the catalysis by ash in SADF.

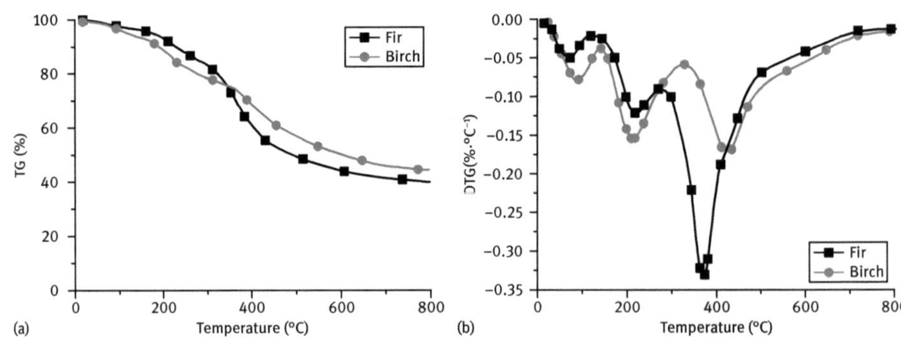

Fig. 4.4: TG/DTG curves of SADF from fir and birch at the heating rate of 20 K/min. Reprinted with permission from [13], © 2008 Elsevier.

The evolution of pyrolysis products can be analyzed by an on-line infrared spectrometer combined with a thermobalance. In the primary stage, physically absorbed water was released first, and the elimination of aliphatic hydroxyl groups could also generate some water. When pyrolysis progressed into the major pyrolysis stages, more products were released. Fig. 4.5 shows the FTIR spectra at maximum volatile release in the subsequent two major pyrolysis stages. In the first major pyrolysis stage, the bands at 3,964–3,500 cm^{-1} and 1,800–1,300 cm^{-1} correspond to the release of water.

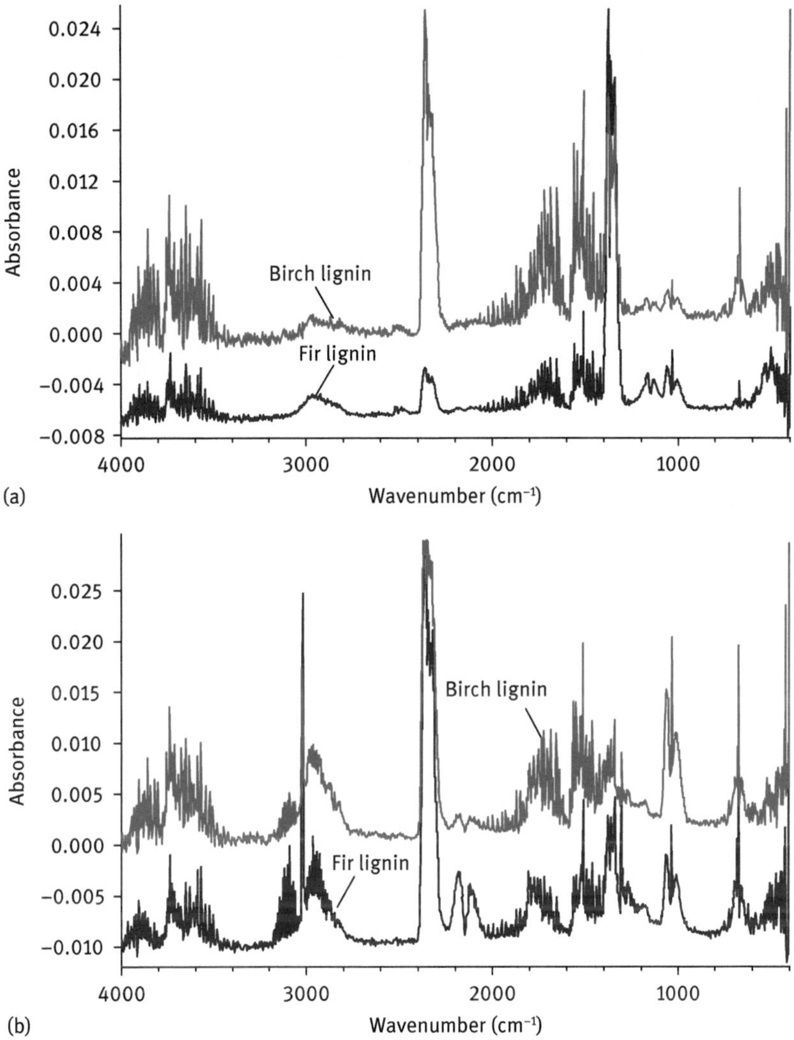

Fig. 4.5: FTIR spectra at maximum volatile release from the pyrolysis of SADF; (a) the first weight loss peak, (b) the second weight loss peak. Reprinted with permission from [13], © 2008 Elsevier.

The bands at 2,112 cm^{-1} and 2,180 cm^{-1} indicate the formation of CO and the bands at 2,391–2,217 cm^{-1} and 726–586 cm^{-1} show the generation of CO_2. Water, CO, and CO_2 can be produced by elimination of aliphatic hydroxyl groups in the side chains and cleavage of C–C bonds. The bands at 2,801 cm^{-1} and 1,747 cm^{-1} showed the presence of formaldehyde, which was mainly generated from the fragmentation of hydroxymethyl groups in the side chains. The most significant band is at 1,400–1,300 cm^{-1}, which represents the vibration of O–H bonds, mainly from the phenols formed by the cleavage of ether bonds. In the second major pyrolysis stage, the release of CO and CO_2 was enhanced. Meanwhile, the band at 3,200–2,850 cm^{-1} indicates the existence of hydrocarbons, especially methane, and the strong absorption band at 3,200–2,850 cm^{-1} suggests the continuous production of methanol. Both methane and methanol can be generated by the fragmentation of methoxyl groups attached to the benzene ring.

The Klason method is a common way to determine the content of lignin in biomass, and the derived lignin is called Klason lignin. This lignin has some similarity to SADF because they are both treated by 72 % sulfuric acid solution. However, there are still some differences, since the preparation of SADF has extra steps compared with neutral and weak acid detergent treatment. The TG/DTG curves of Klason lignin samples from Mongolian Scots pine and Manchurian ash are shown in Fig. 4.6. The final char yields for the pyrolysis of these two lignin samples were both above 40 %, close to that of SADF from fir and birch, because the strong acid treatment causes the condensation of lignin structure and therefore its reactivity decreased. There are two weight loss stages for the pyrolysis of Klason lignin samples from Mongolian Scots pine and Manchurian ash, including the primary stage for the release of physically absorbed water as well as some water generated by chemical reaction at a temperature below 150 °C, and the major pyrolysis stage involving the formation of main pyrolysis products with the main weight loss peak at about 400 °C. The study by Caballero et al. also showed that when the treatment time of sulfuric acid solution increased, the pyrolysis of the treated lignin had a higher residual char yield [15]. Similar results were observed by Haykiri-Acma et al. in the pyrolysis study of Klason lignin from Turkish hazelnuts [16]. When the pyrolysis temperature was lower than 150 °C, the release of water occurred. Within the major pyrolysis stage at 150–900 °C, the maximum weight loss rate was at 409 °C, while the final char yield was 36.9 %. Jiang et al. [17] studied the pyrolysis behaviors of Klason lignin samples from beech, willow, mixed softwood, cassava stalk and cassava rhizome, and found that the corresponding char yields were 40–45 %. The DTG curve showed only one strong weight loss peak during pyrolysis, and this was owing to the drying pretreatment at 105 °C which led to the disappearance of the weight loss peak in the primary pyrolysis stage. In a pyrolysis study of

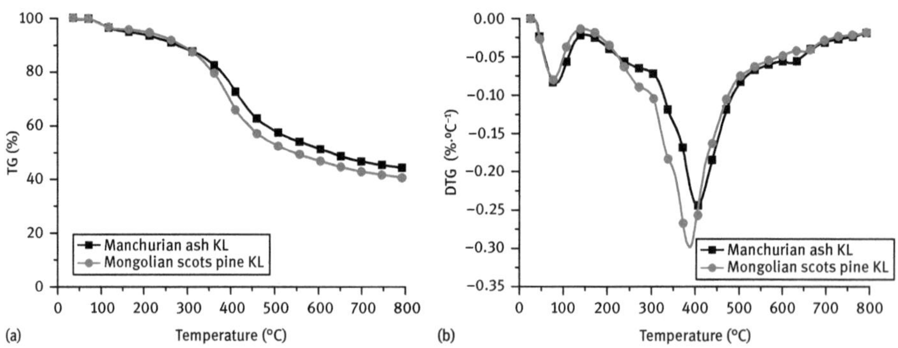

Fig. 4.6: TG/DTG curves of Klason lignin samples from Mongolian Scots pine and Manchurian ash.

Klason lignin from Chinese fir, Zhao et al. [18] observed two weight loss stages, including the initial stage for the release of water and partial volatiles, and the major pyrolysis stage at 180–650 °C. However, their study on the pyrolysis of Klason lignin from maple showed an extra weight loss stage at 180–300 °C before the major pyrolysis stage, indicating that the biomass feedstock affects the pyrolysis characteristics of the corresponding Klason lignin. Furthermore, the char yields of these two Klason lignin samples were both high, reaching 46.4 % and 47.5 %, respectively, proving that the strong acid treatment lowers the reactivity of lignin.

4.1.3.2 Pyrolysis of Milled Wood Lignin (MWL)

MWL can be isolated from biomass by the Bjokman method. Compared with Klason lignin, MWL maintains the structure of original lignin and shows a higher reactivity [19]. The TG/DTG curves of MWL from Mongolian Scots pine (softwood) and Manchurian ash (hardwood) are shown in Fig. 4.7. The final char yields for the pyrolysis of Mongolian Scots pine MWL and Manchurian ash MWL were 37 % and 26 %, respectively, which were lower than that of SADF and Klason lignin. The DTG curves indicate two weight loss stages, including the primary stage for the release of physically absorbed water below 120 °C, and the major pyrolysis stage at 150–650 °C with only one strong weight loss peak, especially for Manchurian ash. This reflects that MWL has a higher reactivity and can generate volatiles more easily. Meanwhile, the DTG curves show that for the pyrolysis of Mongolian Scots pine MWL, the maximum weight loss rate was at a higher temperature, and a weak shoulder peak existed at about 300 °C. It is considered that the methoxyl group content in lignin significantly affects the final pyrolytic char yield, and a higher methoxyl group content leads to a lower char yield. Compared with softwood lignin, hardwood lignin has a higher methoxyl group content, so the char yield during its pyrolysis is lower. Moreover, the guaiacyl units in softwood lignin have a higher tendency to undergo condensa-

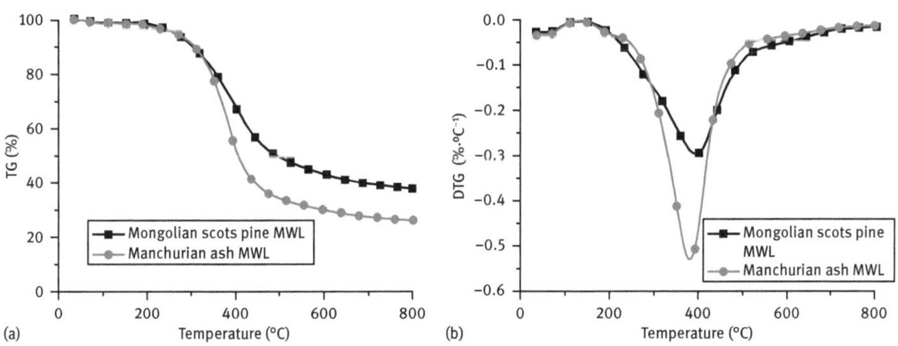

Fig. 4.7: TG/DTG curves of MWL from Mongolian Scots pine and Manchurian ash. Reprinted with permission from [22], © 2009 Elsevier.

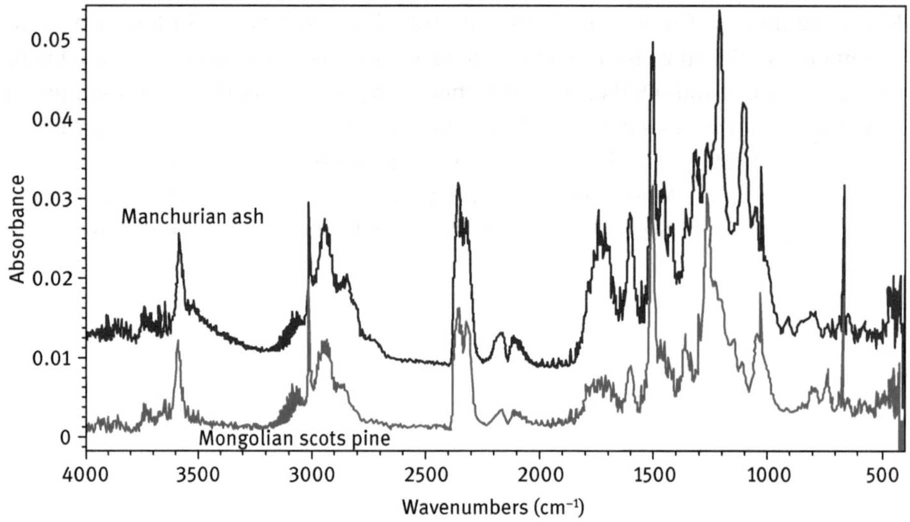

Fig. 4.8: FTIR spectra at maximum volatile release for the pyrolysis of MWL. Reprinted with permission from [22], © 2009 Elsevier.

tion than the syringyl units in hardwood lignin. Similar results were obtained by Faix et al. [20], who found that in the pyrolysis of MWL from spruce, beech and bamboo, the weight losses started at 180–200 °C, and the residual char yields were 38 %, 27 %, and 32 %, respectively. The TG study of 16 kinds of MWL by Jakab et al. showed that the char yield was 26–39 %, and the char yield of hardwood was lower than that of softwood [21].

Fig. 4.8 presents the FTIR spectra at maximum volatile release for the pyrolysis of MWL from Mongolian Scots pine and Manchurian ash. The band at 4,000–3,500 cm^{-1} shows the formation of water. The band at 3,200–2,800 cm^{-1} corresponds to the production of hydrocarbons, and it can be deduced that CH_4 is the main hydrocarbon product on the basis of the band at 3,016 cm^{-1}. The band at 2,400–2,230 cm^{-1} shows the presence of CO_2, and bands of CO at 2,110 cm^{-1} and 2,170 cm^{-1} also exist. The band at 1,900–900 cm^{-1} indicates the release of several organic compounds, including alcohols, aldehydes, phenols, and acids. Phenolic compounds with methoxyl groups are the typical pyrolysis products of lignin, such as guaiacol, syringol and their derivatives. Alcohols are another typical product, and the strong absorbance at 1,057 cm^{-1} shows the existence of methanol.

4.1.3.3 Pyrolysis of Kraft lignin

Kraft lignin can be extracted from the black liquor produced by pulping factories, and the TG/DTG curves of a typical Kraft lignin sample are presented in Fig. 4.9. The pyrolysis of lignin covered a wide temperature range, and the final char yield was about 40 %. The DTG curve indicates several weight loss peaks, and the main weight loss happened at about 375 °C. Meanwhile, a slight shoulder peak was observed at 270–300 °C, which could be attributed to the pyrolysis of some residual hemicellulose structure combined with the main lignin structure [23]. In the TG analysis of four kinds of black liquors, Alén et al. [24] found that in addition to the first weight loss stage for water release at 100–150 °C, there were also several weight loss peaks in the major pyrolysis stage at 150–700 °C, with a high final char yield above 60 %. The difference in the pyrolysis behaviors of different Kraft lignin samples is due to variations in biomass feedstock and the pulping process, which result in the different structures of Kraft lignin samples. A TG-FTIR study by Fenner et al. showed that there were two weight loss stages for the pyrolysis of Kraft lignin: one was at 120–300 °C and the other one was at 300–480 °C; the latter accounted for more than 50 % of the total weight loss [25]. Additionally, SO_2 was detected in the first stage, and given that the pyrolysis of natural lignin does not generate SO_2, the presence of sulfur was related to the pulping process. In a pyrolysis study of Kraft lignin samples from different biomass by Brodin et al. [26], two weight loss peaks were observed, including a weak weight loss peak at 100–200 °C and the main weight loss peak at 300–500 °C. However, Brebu et al. [27] did not observe an obvious weight loss peak for water release during the pyrolysis of Kraft lignin. The major pyrolysis stage included a weak weight loss peak at 150–300 °C and a main weight loss peak at 200–500 °C. The maximum weight loss rate was at about 385 °C, and the final char yield was as high as 54 %. The pyrolysis research on Kraft lignin by Zhang et al. suggested that in addition to the weight loss peak for water release at 35–200 °C and the main weight loss peak at 200–500 °C, two subsequent weak weight loss peaks also existed, and the residual char yield was about 45 % [23]. Based on the analysis by TG-FTIR, the weight loss peak at 500–700 °C corresponded to the generation of CO_2 and some aromatic compounds. The weight loss peak at 700–900 °C was related to the further release of CO formed by the cleavage of ether bonds between aromatic rings and the generation of CO_2. Hu et al. [28] used benzene/ethanol solution and acetone solution to extract lignin from black liquor. They found that there were

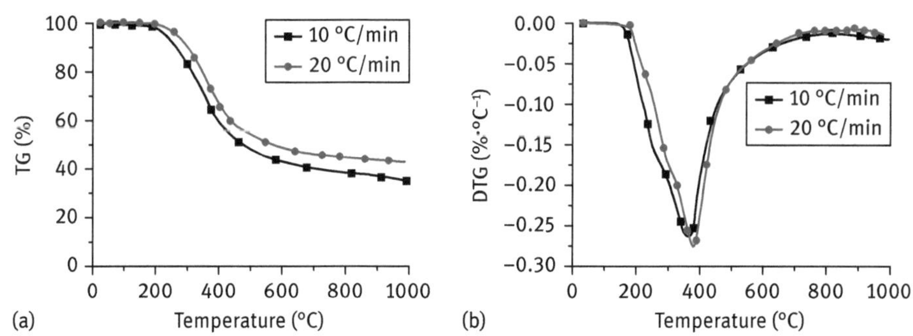

Fig. 4.9: TG/DTG curves of a typical Kraft lignin sample.

two pyrolysis stages. The first pyrolysis stage at 110–180 °C was for the volatilization of solution, and the second stage at 200–500 °C was the major pyrolysis stage, with a maximum weight loss rate at 390–400 °C and a final char yield of 35 %.

4.1.3.4 Pyrolysis of organosolv lignin

Lignin can be dissolved in some solvents under certain temperature and pressure and the derived lignin is called organosolv lignin. The common solvents include ethanol, formic acid, acetic acid and acetone. The TG/DTG curves of organosolv lignin from beech wood are shown in Fig. 4.10. Weight loss occurred in the temperature range from 50 °C to 800 °C and the maximum weight loss rate was at about 360 °C, with a residual char yield of 40 %. The DTG curve of organosolv lignin is similar to that of MWL, including a weak weight loss peak for water release in the primary stage below 150 °C, and a significant weight loss peak in the major pyrolysis stage. Similar pyrolysis behavior of organosolv lignin from hardwood was found by Jiang et al., in which only one strong weight loss peak appeared in the major pyrolysis stage, and the char yield was about 35 % [17]. However, owing to variations in feedstock and extraction processes, the corresponding pyrolysis behaviors appear to be very different. Several weight loss peaks were observed by Brebu et al. in their pyrolysis study of organosolv lignin from hardwood using an acetic acid/phosphinic acid solvent, and the first weight loss peak at 110 °C corresponded to the removal of free water, while the major pyrolysis stage was located at 150–600 °C. In addition to the strong weight loss peak at 380 °C, there was another slight weight loss peak at about 200 °C, and the char yield at 575 °C was as high as 54 % [27]. De Wild et al. [29] extracted organosolv lignin from wheat straw by ethanol, and the pyrolysis result showed a high final char yield of 50–60 %, and in the major pyrolysis stage at 150–500 °C, a maximum weight loss rate at 365 °C was achieved. Domínguez et al. [30] carried out a pyrolysis study of organosolv lignin extracted from *Eucalyptus globulus* by methanol-water solvent, and they found that there were three weight loss stages: in the first weight loss stage below 200 °C, in addition to the release of free water, the generation of small molecular products like CO and CO_2 also occurred; the second weight loss stage at 200–450 °C was the major pyrolysis stage, corresponding to the formation of main phenolic products; the third weight loss stage above 450 °C involved the decomposition and condensation of the aromatic rings. Zhao et al. [31] studied the pyrolysis behavior of organosolv lignin from Siam weed using acetic acid-water solution and formic acid-water solution, and four pyrolysis stages were clarified: the first weight loss stage represented the release of water, and the temperature was located in the range of 30–130 °C; in the second weight loss stage at 130–200 °C, the glass transition of lignin occurred; the third weight loss stage at 200–500 °C was the major pyrolysis stage, and the maximum weight loss rate was at 370 °C, involving the formation of monophenols, and some cracking of aromatic rings; the last weight loss stage was at 500–900 °C, in which aromatic rings underwent further decomposition and condensation.

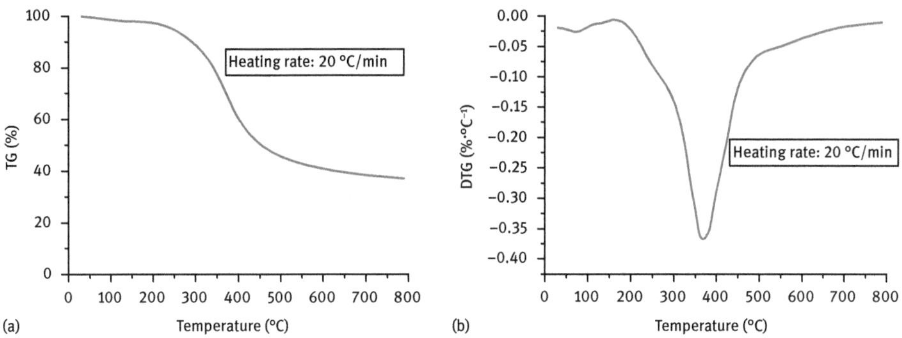

Fig. 4.10: TG/DTG curves of organosolv lignin. Reprinted with permission from [32], © 2012 Elsevier.

Limited by the extraction solvents and techniques, the terminal of the molecular structure of organosolv lignin is often combined with the glucosidic bond. A Py-GC/MS study of organosolv lignin from beech at 600 °C showed the weight loss ratio of lignin was 61.9 %, and the main pyrolysis products included phenols, furans, some straight-chain acids and esters ($C > 16$), and a little ethanol and acetic acid. The typical pyrolysis products of hemicellulose like acetic acid were also detected by Domínguez et al. in their pyrolysis study of organosolv lignin from *Eucalyptus globulus* [30]. Meanwhile, a similar product distribution was obtained by Zhang et al. in the pyrolysis of organosolv lignin, in which phenols were the main pyrolysis products and a few furans were also formed, such as furfural, 5-methylfurfural, and 5-hydroxymethylfurfural [23].

4.1.3.5 Pyrolysis of pyrolytic lignin

Pyrolytic lignin is not similar to natural lignin, and it is a mixture of incomplete depolymerization products from lignin. Pyrolytic lignin can be extracted from pyrolysis oil by using solvents such as water, ethanol, acid, alkali etc. In recent years, pyrolytic lignin has drawn wide attention because it lowers the stability of pyrolysis oil storage and also causes severe catalyst deactivation during catalytic upgrading processes. Therefore, researchers improved the technical route for upgrading pyrolysis oil by adopting a pretreatment to remove pyrolytic lignin before the catalytic process. Furthermore, because pyrolytic lignin is an incomplete depolymerization product from lignin and contains some of the original structure of lignin, employing secondary pyrolysis or depolymerization of pyrolytic lignin to obtain some light fraction may be a promising way to achieve the high-grade utilization of pyrolytic lignin.

The TG and DTG curves of pyrolytic lignin extracted from lauan pyrolysis oil are presented in Fig. 4.11. There were three weight loss stages. The first weight loss stage below 180 °C was for the release of free water and some small molecular gases. The second weight loss stage at 180–500 °C was the major pyrolysis stage, in which a lot of volatiles were generated. The third weight loss stage was at 500–800 °C, corresponding to secondary decomposition and further carbonization. The char yield at 800 °C was 38.83 %. In the major pyrolysis stage, there was a shoulder peak at about 230 °C and the maximum weight loss rate was at 320 °C. Compared with the pyrolysis of other lignin samples, the major pyrolysis stage of pyrolytic lignin pyrolysis moved towards the low temperature region, because pyrolytic lignin is derived from partial depolymerization of natural lignin which makes it more thermo-instable than natural lignin samples. The pyrolysis of pyrolytic lignin samples with different molecular weights by Jiang et al. exhibited three weight loss stages [33]. For the pyrolysis of low molecular weight pyrolytic lignin, the first weight loss stage was at 25–160 °C, in which water and some volatile compounds were released. The second weight loss stage at 160–600 °C was the major pyrolysis stage and the pyrolytic lignin underwent depolymerization. The third weight loss stage at 600–900 °C corresponded to the further cracking of some relatively stable structures, and the final char yield was 27.1 %. The pyrolysis behavior of high molecular weight pyrolytic lignin was similar to that of low molecular weight pyrolytic lignin, while the temperature range of the major pyrolysis stage (180–620 °C) was a little different and the char yield was higher at 36.4 %. Chang et al. found that pyrolytic lignin had its major pyrolysis stage in the temperature range of 150–550 °C, and was more thermo-instable compared with other lignin samples such as alkali lignin [34].

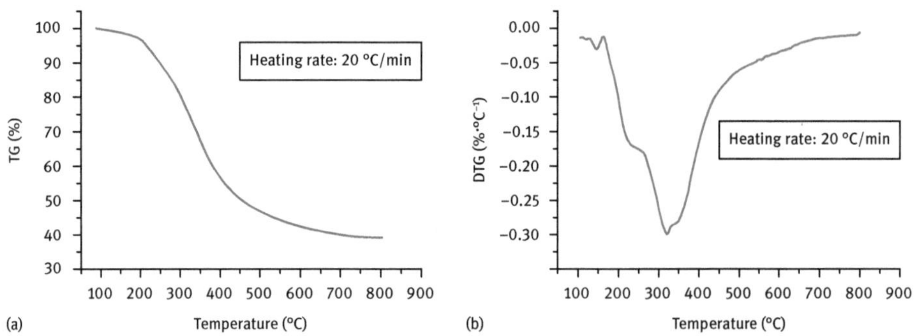

Fig. 4.11: TG/DTG curves of pyrolytic lignin. Reprinted with permission from [35], © 2012 Elsevier.

Fig. 4.12 presents the FTIR spectrum of the maximum volatile release at 320 °C. The bands at 4,000–3,500 cm^{-1}, 2,400–2,230 cm^{-1}, and 2,230–2,030 cm^{-1} show the formation of water, CO_2, and CO. A band of CH_4 at 3,016 cm^{-1} also exists, and the band at 2,946 cm^{-1} indicates the presence of compounds with unsaturated C=C bonds. The band at 2,000–800 cm^{-1} is for the production of organic compounds such as phenols, aldehydes, acids, ketones, and alcohols. Meanwhile, there is a weak absorbance at 1,612 cm^{-1}, corresponding to the stretching of C=O bonds conjugating to the aromatic ring. The bands at 1,274 cm^{-1} and 1,114 cm^{-1} show the existence of guaiacols and syringols.

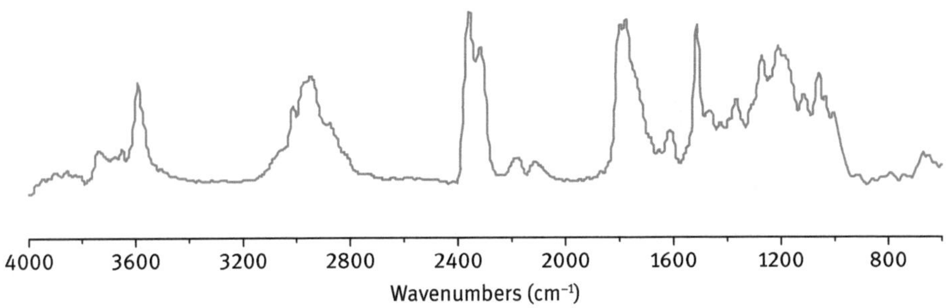

Fig. 4.12: FTIR spectrum at the maximum volatile release for the pyrolysis of pyrolytic lignin. Reprinted with permission from [35], © 2012 Elsevier.

4.2 Effect of reaction parameters on the pyrolysis behavior of lignin

Using the infrared radiation-heating reactor mentioned above, the pyrolysis product distributions of MWL from Mongolian Scots pine and Manchurian ash were obtained and are shown in Fig. 4.13. The pyrolysis oil yields of MWL from Manchurian ash and Mongolian Scots pine were 25.6 % and 19.1 %, respectively, while the corresponding char yields were 34.3 % and 42.8 %. The gas yields of these two lignin samples were similar, and the concentrations of CO, CO_2, and CH_4 in the gaseous products were above 60 %, nearly 20 %, and 9 %, respectively. The pyrolysis of MWL from Mongolian Scots pine generated more char but fewer volatiles, which could be attributed to the higher amount of guaiacyl units in this lignin and thus crosslinking reactions easily occurred to form char.

Pyrolysis oil mainly consists of phenolic compounds, such as phenol, 2,4-dimethyl-phenol, 2-methylphenol, catechol, and 3-methylcatechol. Many phenolic compounds contain a methoxyl group and an alkyl group, showing the structural characteristics of original lignin. The phenolic compounds containing a methoxyl group include guaiacol, 4-methylguaiacol, 4-ethylguaiacol, syringol, and 4-propenylguaiacol. Small quantities of acetic acid, benzyl alcohol, and 3,5-dimethyl-benzyl alcohol have also been detected. In addition, some aldehydes are found in pyrolysis oil. In addition to the phenolic aldehydes like vanillin and its derivatives, some typical pyrolysis products of hemicellulose like furfural are generated, which is related to the preparation of MWL. In contrast to the Klason method which uses strong acid to dissolve the cellulose and hemicellulose, MWL is isolated by ball milling and solvent extraction, and thus the side chains of the phenylpropane units may be combined with some residual glucosidic bonds. This is in agreement with the results obtained by Ucar et al., showing that a few polysaccharides were detected in MWL [36].

During the pyrolysis of lignin, devolatilization competes with char formation and some subsequent secondary reaction also occurs. Therefore, the reaction parameters,

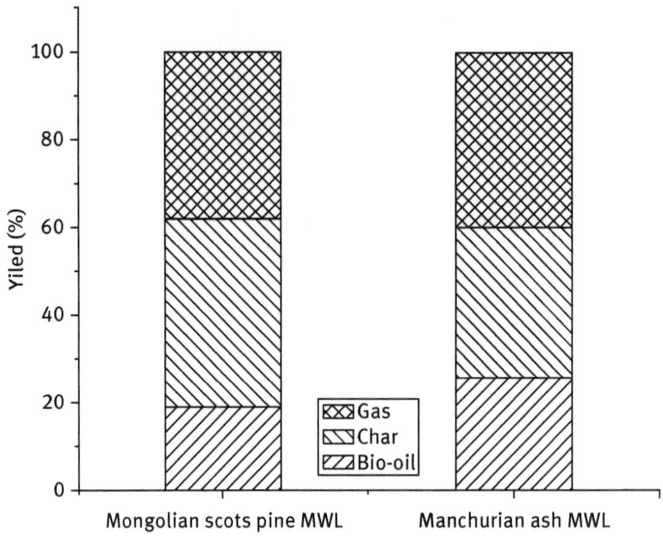

Fig. 4.13: Pyrolysis product distribution of MWL.

such as reaction temperature, reaction atmosphere, residence time and heating rate, will influence the intensities of devolatilization, char formation and secondary reaction, and then affect the distribution of pyrolysis products.

4.2.1 Effect of reaction temperature

The chemical bonds in the functional groups of lignin have different dissociation energies, so the distribution of pyrolysis products is significantly influenced by reaction temperature. The TG analysis of lignin shows that with increasing pyrolysis temperature, the weight of solid residue (char) decreases, while volatiles are generated continuously. Some volatiles can be condensed to liquid (pyrolysis oil), and the others are non-condensable gases, such as CO, CO_2, and CH_4.

4.2.1.1 Effect of reaction temperature on the yields of pyrolysis products

The yields of pyrolysis products under different reaction temperatures show the following tendency: as reaction temperature increases, the yield of gas rises continuously. Meanwhile, the yield of pyrolysis oil first increases and then decreases and the char yield goes down. It has been found that during lignin pyrolysis the production of volatiles requires the cleavage of the linkages between basic structural units, 40 % of which remain stable at temperatures lower than 300 °C. Therefore, when the reaction temperature is low, crosslinking reactions predominate, which favors the formation of char. As reaction temperature increases, more side chains as well as C–C and

C–O bonds in the phenylpropane units are cracked, which increases the devolatilization reaction intensity and thus promotes the production of pyrolysis oil and gaseous products. When the reaction temperature is even higher, due to the occurrence of a secondary reaction, the yield of pyrolysis oil decreases while the yield of gas further increases. In their study of the secondary pyrolysis of primary pyrolysis oil, Hosoya et al. [37] found that the aldehydes in pyrolysis oil undergo decarbonylation to produce CO, and some phenolic compounds containing side chains and methoxyl groups are decomposed to generate catechol and phenol, as well as some light gases such as CO, CH_4, and methanol.

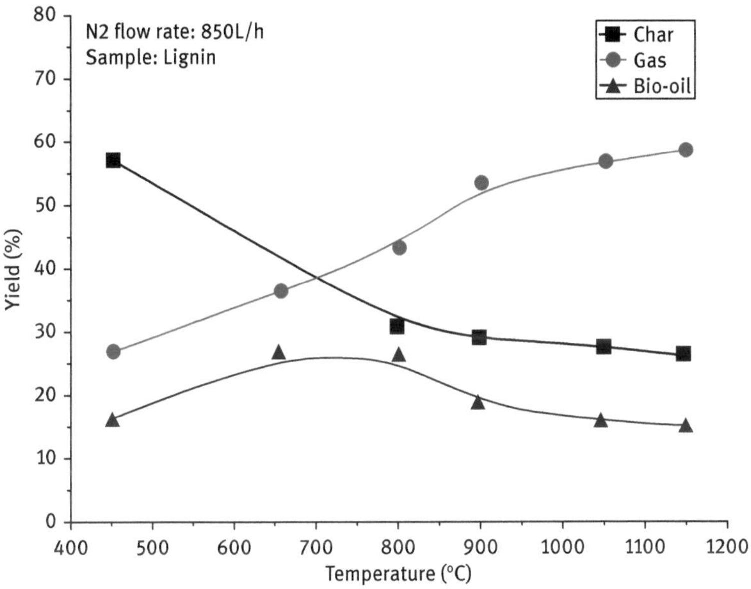

Fig. 4.14: Distribution curves of products from lignin pyrolysis with temperature.

Fig. 4.14 shows the distribution of products from Kraft lignin pyrolysis carried out in an infrared radiation heating reactor where the temperature of the radiation source is 450–1,150 °C. The results show that in the reaction temperature range studied, the yield of char decreased along with increasing temperature and this tendency was more obvious at temperatures lower than 900 °C. When the temperature increased further, the yield of char reached a stable value of about 26 %. The variation in gas yield was the opposite of char yield; gas yield increased from 26 % at 450 °C to 57 % at 1,150 °C. The yield of pyrolysis oil first increased and then decreased. At low reaction temperature, char formation predominates and the release of volatiles was not obvious. As reaction temperature increased, far more condensable volatiles were released, which increased the yield of pyrolysis oil. The highest yield of pyrolysis oil was 27 % at 655 °C.

The further increase in reaction temperature caused the secondary decomposition of some condensable compounds, and thus the yield of pyrolysis oil decreased while the gas yield increased further. At 1,150 °C, the yield of pyrolysis oil was only about 15 %.

Similar influences of reaction temperature on the yields of pyrolysis products were observed in other studies of lignin pyrolysis, and the main difference is the temperature for the maximum pyrolysis oil yield. Nunn et al. [38] studied the pyrolysis of MWL at 325–1,125 °C under N_2 atmosphere. They found that the weight loss of MWL increased as the temperature rose, and it reached a stable value of 86 % at 775 °C. The maximum pyrolysis oil yield was 53 % at 627 °C. The increase in gas yield continued until the temperature reached 877 °C, remaining at a stable value of about 36 %. Ferdous et al. [39] studied the pyrolysis of Alcell lignin (organosolv lignin extracted from hardwood by ethanol/water solvent) at 350–800 °C in a fixed bed reactor. The char yield decreased continuously to 34 % at 800 °C. In the temperature range of 350–550 °C, the yields of gas and pyrolysis oil both increased with rising temperature and the gas yield increased slightly, because in this stage gaseous products were mainly derived from some primary decomposition of lignin, which was not intensive. In the temperature range of 550–650 °C, both the primary decomposition of lignin and the secondary decomposition of volatiles were enhanced and caused the significant increase in gas yield, which rose from 20 % to 40 %. When the temperature was above 650 °C, the increase in gas yield was not obvious because the secondary decomposition of volatiles was weakened. Lou et al. [40] studied the pyrolysis behavior of enzymatic/mild acidolysis lignin (EMAL) isolated from bamboo, and found that the yield of char decreased from 43 % at 400 °C to 28 % at 900 °C, while the yield of gas increased continuously from 6 % to 26 %. The yield of pyrolysis oil first rose and then fell. The maximum pyrolysis oil yield of 55 % was achieved at 600 °C and afterwards the yield of pyrolysis oil decreased because of the secondary decomposition of some condensable volatiles at higher temperature. The research on Kraft lignin pyrolysis by Ben et al. also showed that as temperature rose, pyrolysis oil yield first increased and then decreased, while the gas yield kept on increasing [41].

4.2.1.2 Effect of reaction temperature on gaseous product composition

As mentioned above, the yield of gas increases with rising reaction temperature during lignin pyrolysis. Furthermore, the composition of gaseous products also changes, because the reaction temperature significantly affects the intensities of the different devolatilization and secondary decomposition processes. In general, CO, CO_2, CH_4, and H_2 are typical gaseous products during lignin pyrolysis. When the temperature is low, CO and CO_2 account for a high proportion of the gases. As temperature increases, the increasing tendencies of CO and H_2 production are significant.

The influence of temperature on gas composition in a study of Kraft lignin pyrolysis is shown in Tab. 4.1. CO, CO_2 and CH_4 were the main gaseous products, and their total concentration reached above 70 %. The concentration of CO_2 was the highest among the gaseous products at low temperature, and it decreased continuously to 27.27 % when the temperature increased to 1,150 °C. The concentration of H_2 showed the opposite varying tendency and increased from 1.99 % at 450 °C to 16.90 % at 1,150 °C. The concentration of CO first increased and then decreased and the maximum value was 35.82 % at 800 °C. The concentration of CH_4 increased with rising temperature and finally remained stable at about 21 %. C_{2+} gases were only formed at high temperature, and their concentrations increased slightly with increasing temperature.

Tab. 4.1: Distribution of gaseous products from Kraft lignin pyrolysis.

Temperature (°C)	H_2 (vol%)	CO (vol%)	CH_4 (vol%)	CO_2 (vol%)	C_2H_4 (vol%)	C_2H_6 (vol%)	C_3H_8 (vol%)
450	1.99	18.94	15.54	63.53	0.00	0.00	0.00
656	4.79	32.59	16.34	45.97	0.16	0.15	0.00
800	5.16	35.82	21.56	34.36	1.99	1.06	0.05
900	6.90	34.99	20.19	35.12	1.99	0.77	0.04
1050	9.80	33.26	22.56	30.68	2.73	0.96	0.01
1150	16.90	29.98	20.96	27.27	3.34	1.51	0.05

A similar influence of reaction temperature on gaseous product distribution was observed by other researchers. In a study of Kraft lignin pyrolysis, Iatridis et al. found that when the reaction temperature increased from 400 °C to 650 °C, the increase in CO and CH_4 yields was remarkable, while the yield of CO_2 increased slowly [42]. The research on EMAL pyrolysis by Lou et al. [40] showed that in the temperature range of 400–900 °C, the yields of H_2 and CO both increased with rising temperature and the maximum yields of CO_2 and CH_4 were achieved at 600 °C and 700 °C, respectively. C_2H_4 and C_2H_6 started to appear at 700 °C, and their yields remained low. Ferdous et al. [39] studied the distribution of gaseous products from the pyrolysis of Alcell lignin and Kraft lignin. For the pyrolysis of Alcell lignin, as the temperature increased from 350 to 800 °C, the concentration of H_2 in the gaseous products increased significantly to 31.5 %. The concentration of CO first decreased and then increased, with a minimum concentration at 550 °C. The increase of CO concentration at high temperature was attributed to the enhanced secondary decomposition of volatiles. The maximum concentration of CO_2 was observed at 500 °C, and then it decreased. The concentration of CH_4 also decreased at high temperature, whereas the concentration of C_{2+} hydrocarbons remained low. For the pyrolysis of Kraft lignin, the variation of gaseous product distribution was similar to Alcell lignin, while the decrease in CO_2 concentration and the increase in H_2 concentration with increasing temperature were more significant.

4.2.1.3 Effect of reaction temperature on the composition of pyrolysis oil

As the pyrolysis temperature increases, the devolatilization reaction becomes more intensive and the secondary decomposition of volatiles also occurs. These reactions significantly affect the composition of pyrolysis oil. When the reaction temperature increases, the yields of typical products corresponding to the original structural units of lignin, mainly guaiacols and syringols, show a tendency to increase and then decrease, while the yields of phenols and catechols increase continuously. The increase in guaiacols and syringols yields is attributed to the promotion of the cleavage of ethers bonds between structural units and some C–C bonds with rising temperature. When pyrolysis temperature increases further, demethoxylation and demethylation reactions are enhanced, resulting in the secondary decomposition of guaiacols and syringols to form phenols and catechols. In general, because guaiacols and syringols account for a large proportion of the phenolic products, the total yield of phenolic products also first increases and then decreases with increasing temperature, which is in accordance with the variation in pyrolysis oil yield. In addition, some products from the pyrolysis of glucosidic bonds combined with the lignin structure, such as aldehydes and ketones, will be released completely at low reaction temperature. When the temperature rises further, decarbonylation and condensation will decrease their yields.

The composition of pyrolysis oil from Kraft lignin pyrolysis obtained under different temperatures was compared by GC-MS analysis. A few furans like furfural were found in the pyrolysis oil, and they were derived from the pyrolysis of residual saccharide structure. The yield of furans decreased with increasing temperature. The comparison of the phenolic compounds contents in pyrolysis oil under different pyrolysis temperatures showed the contents of phenolic compounds containing a methoxyl group, such as guaiacol, 4-propenylguaiacol and syringol, decreased with increasing temperature, indicating that demethoxy and demethylation reactions were enhanced, while the contents of the phenolic compound without a methoxyl group, like phenol and catechol, increased correspondingly. The influence of pyrolysis temperature on the distribution of organosolv lignin pyrolysis products was also studied on a Py-GC/MS system in the temperature range of 450–700 °C, as shown in Fig. 4.15. The amounts of furans like furfural decreased with rising temperature. The quantities of other products, such as guaiacol, 4-propylguaiacol and syringol, all showed the tendency of increasing first and then decreasing, and the maximum contents are achieved at 600 °C.

A similar influence of pyrolysis temperature on the composition of pyrolysis oil from lignin pyrolysis was also obtained by other researchers. In their study of Kraft lignin pyrolysis at 400–650 °C, Iatridis et al. [42] found that the total yield of main phenolic products, including phenol, guaiacol, 4-methylguaiacol, 4-ethylguaiacol and methylphenol, increased from 2.18 % at 400 °C to 2.95 % at 650 °C. The pyrolysis of lignin from acid hydrolysis of corncob by Huang et al. showed that the quantities of 4-vinylguaiacol and guaiacol reached the maximum value of 5.97 % and 3.57 % at

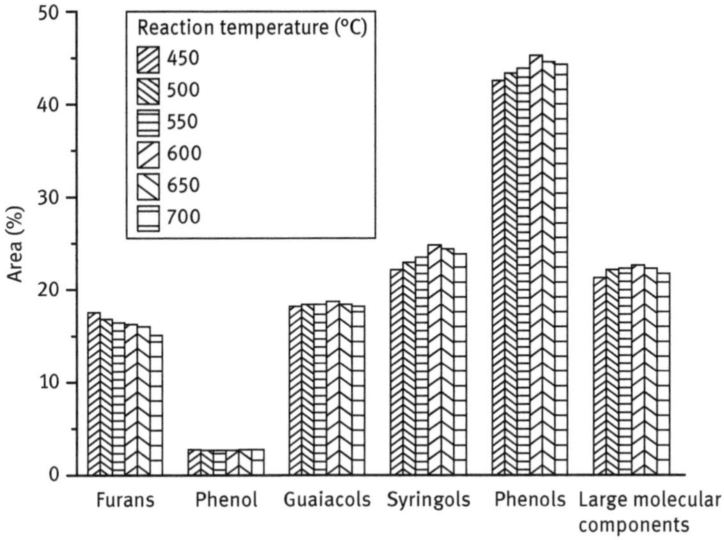

Fig. 4.15: Influence of temperature on the distribution of lignin pyrolysis products.

600 °C, respectively, while the maximum syringol content of 2.58 % was obtained at 520 °C. Additionally, the phenol content increased continuously [43]. Jiang et al. [44] analyzed pyrolysis oil from the pyrolysis of Alcell lignin and Asian lignin, and compared the yields of phenolic products at 400–800 °C. For the pyrolysis of Alcell lignin, the total yields of phenolic compounds increased from 13.4 % at 400 °C to 17.2 % at 600 °C, and then decreased to 12.4 % at 800 °C. The yields of phenolic compounds containing methoxyl groups displayed a similar tendency to the variation of the total phenol yield. However, the phenolic compounds which did not contain a methoxyl group, such as phenol, catechol and 3-methylcatechol, could not be detected at 400 °C but could be detected at 800 °C. The variation in pyrolysis oil composition from the pyrolysis of Asian lignin was similar to that from the pyrolysis of Alcell lignin. The maximum yields of 4-vinylguaiacol, guaiacol and syringol were achieved at 600 °C, while the yields of phenol and catechol increased with rising temperature. In their study of EMAL pyrolysis Lou et al. found that phenol, methylphenol and ethylphenol started to be generated at 600 °C, and their relative contents reached maximum values of 2.99 %, 5.26 %, and 1.59 % at 800 °C, respectively. Guaiacol, ethylguaiacol, syringol and isoeugenol were found to be formed at above 320 °C. Their maximum yields were obtained at 400–600 °C, and afterwards the yields of these phenolic compounds decreased [45].

4.2.1.4 Effect of reaction temperature on pyrolytic char

The solid residue (pyrolytic char) from lignin pyrolysis is a complex mixture of organics, and its composition changes under different temperatures. At low reaction temperature, because of the low intensity of devolatilization, the char contains a lot of original lignin structure, including some aliphatic structures like the side chains in the phenylpropane units and some hydroxyl and methoxyl groups. With increasing pyrolysis temperature, devolatilization is enhanced, and pyrolytic char exhibits higher carbonization, stability and aromaticity.

Pasquali et al. [46] characterized the pyrolysis residue of Kraft lignin using IR and found that the original lignin sample was rich in guaiacyl and syringyl structures at the ambient temperature. After pyrolysis at 226 °C, the absorption peaks for these two structures were significantly weakened, indicating that because of the cleavage of ether bonds, a lot of guaiacols and syringols were released from lignin. Sharma et al. [47] conducted a characterization of char samples derived from pyrolysis of alkali lignin at different temperatures by IR and ^{13}C-CP/MAS-NMR. With increasing pyrolysis temperature, the typical structure in lignin disappeared gradually, and hydrogen and oxygen were removed resulting in a higher aromaticity of the pyrolytic char. From 250 °C to 350 °C, the aromatic, phenolic, methoxyl, hydroxymethyl, and aliphatic carbon atoms gradually diminished to generate char with higher aromaticity, which contained many hydroxyl groups (mainly phenolic hydroxyl groups) and few methoxyl groups. When the temperature was above 350 °C, the O/C ratio in the char was further lowered. The aromatic-bonded oxygen atoms decreased due to the dehydration and decarboxylation reactions. In the char from pyrolysis at a high temperature, aromatic carbon accounted for more than 90 % of the total carbon.

Li et al. [48] investigated the characteristics of char from the pyrolysis of larch lignin and Manchurian ash lignin. By comparing the residue from lignin pyrolysis at 25–460 °C using FTIR, they found that with increasing temperature the absorption peaks for the C–O bond in ether bonds (C–O–C), aliphatic –OH and C–H bonds became weaker, showing that ether bonds and aliphatic structures were easily decomposed during pyrolysis. In contrast, the peaks for aromatic C–O bonds (guaiacyl and syringyl) did not significantly change. Therefore, the aromatic character of char increased as the temperature rose. The XPS (X-ray photoelectron spectroscopy) characterization showed that the relative intensity of C–C bonds on the char surface increased with rising temperature, while the relative intensity of C–O and C=O bonds decreased, indicating gradual carbonization of the lignin sample. Lou et al. [40] investigated the charring of EMAL from bamboo and found that the C/O ratio increased in the temperature range of 400–700 °C, indicating that the charring process mainly involved condensation and COx release. However, when the temperature increased to above 700 °C, the C/O ratio decreased, because the residual oxygen existed in the form of stable inorganic oxides. The FTIR analysis of char showed that the aliphatic hydroxyl group in the side chains of phenylpropane units, as well as the phenolic hydroxyl group, methyl group and methoxyl group attached to the aromatic ring, progressively disappeared when the temperature increased owing to the further condensation of the aromatic rings.

Diehl et al. [49] characterized the chemical structure of residue from pyrolysis of pine softwood and eucalyptus hardwood at different temperatures. ^{13}C-CP/MAS-NMR showed these two lignin samples were stable when the temperature was below 300 °C. When the temperature was higher than 300 °C, the amounts of Ar–O bonds, methyl groups and alkyl side chains began to decrease, while Ar–C bonds and Ar–H bonds increased remarkably, leading to a high aromaticity. The results from FTIR analysis showed that when the pyrolysis temperature reached 250 °C, the amount of hydroxyl groups slightly decreased, while the amount of side chains attached to the aromatic ring remained almost the same, suggesting the elimination of terminal hydroxyl groups in the side chains. When the temperature was higher than 500 °C, the amount of hydroxyl groups decreased obviously. Meanwhile, C–H bonds started to decrease at 400 °C, and a significant decrease was observed in the temperature range of 500–600 °C. Then they almost disappear at 700 °C. Substituted groups attached to the benzene ring also began to decrease at 400 °C and nearly disappear at 800 °C and 900 °C. In addition, the residue of lignin pyrolysis at high reaction temperature was also characterized by Raman spectrum. The result showed that the pyrolytic char was rich in amorphous carbon structure, as well as some graphite structure, which was similar to the Raman spectrum of coal.

4.2.2 Effect of residence time

During lignin pyrolysis, residence time also affects the distribution of pyrolysis products. Too short a residence time will cause the incomplete depolymerization of lignin, which not only lowers the yields of gas and pyrolysis oil, but also leads to a heterogeneous liquid product because of the random cleavage of chemical bonds and interaction. A proper prolongation of the residence time will favor devolatilization, which increases the yields of pyrolysis oil and gas. However, too long a residence time will cause secondary decomposition of primary pyrolysis products, resulting in a decrease in pyrolysis oil yield and a further increase in gas yield.

For lignin pyrolysis on a fixed bed reactor, due to the influence of heat and mass transfer efficiency, prolonging residence time remarkably affects the distribution of pyrolysis products. Iatridis et al. [42] studied the influence of residence time on Kraft lignin pyrolysis at 400–700 °C and found that with increasing residence time, the yields of gaseous and liquid products significantly increased. At a temperature of 650 °C, when residence time increased from 10 s to 120 s, the yields of CH_4, CO, and CO_2 increased from 1.80 %, 4.50 %, and 5.48 % to 4.83 %, 9.20 %, and 7.20 %, respectively. Furthermore, the total yield of phenolic compounds, including phenol, guaiacol, 4-methylguaiacol, 4-ethylguaiacol, and cresol, increased from 1.13 % to 3.22 %. In addition, prolonging the residence time can also affect the intensity of secondary reactions. Jegers et al. [50] investigated the effect of residence time on pyrolysis of Kraft lignin and MWL at 400 °C. For the pyrolysis of Kraft lignin, the yield of CO was stable after 20 min, while the yield of CH_4 and CO remained stable after 60 min, indicating that CH_4 and CO were important products of secondary reactions. Furthermore, the yields of phenolic compounds without a methoxyl group, such as phenol, cresol, and 4-ethylphenol, increased with extending residence time and became stable after 100 min. However, the yields of guaiacols, such as guaiacol, 4-methylguaiacol, 4-ethylguaiacol, and 4-propylguaiacol, reached their maximum value at 7.5 min. Afterwards the yields of these guaiacols decreased continuously and some guaiacols even disappeared, suggesting the occurrence of secondary reactions (demethylation and demethoxylation). The yields of catechols, such as catechol, 4-methylcatechol and 4-ethylcatachol, also showed the tendency of increasing followed by decreasing, indicating the thermo-instability of these catechols. The influence of residence time on MWL pyrolysis was similar to that on Kraft lignin pyrolysis, and the main difference was that MWL pyrolysis had a higher reaction rate of primary decomposition.

Ferdous et al. [39] adjusted the reaction residence time by varying the carrier gas flow rate, and studied the influence of residence time on Alcell lignin pyrolysis at 800 °C. They found that the conversion of lignin was not obviously affected by residence time, while the yields of pyrolysis oil and gas were significantly affected. A short residence time could suppress the secondary decomposition of pyrolysis oil, and thus the pyrolysis oil yield increased. Furthermore, a decreased residence time lowered the yields of CO and CO_2 but raised the yield of H_2, while the yields of C_{2+} hydrocarbons nearly remained unchanged. Rutherford et al. [51] found that during the pyrolysis of organosolv lignin, because of the relatively high dissociation energy of chemical bonds in lignin structure, a long-time pyrolysis under a low temperature would lead to insufficient devolatilization, while higher pyrolysis temperature made the conversion of lignin much more intensive. In addition, when the pyrolysis time was longer, the degree of aromatization for char was even higher, and some original phenolic structure was transformed into heterocyclic ring structure.

4.2.3 Effect of other reaction parameters

In addition to temperature and residence time, some other conditions, such as the pyrolysis atmosphere and heating rate, also influence the behavior of lignin pyrolysis. The pyrolysis atmosphere significantly affects the reaction pathway. For example, during pyrolysis under air atmosphere, the oxidation reaction predominates, but under inert gas atmosphere the release of volatiles predominates. Increasing the heating rate can lower the char yield, because a high heating rate can decrease the residence time in the low temperature region, which prohibits the formation of char at low temperature with low reactivity and crosslinking structure.

Li et al. [48] compared the pyrolysis of larch lignin and Manchurian ash lignin under N_2 and air atmospheres. During larch lignin pyrolysis under N_2 atmosphere, the maximum weight loss rate was obtained at 343 °C, and the char yield at 700 °C was 50 %. Under air atmosphere, a violent oxidation reaction occurred, and the larch lignin was burnt out at 500 °C. Manchurian ash lignin displayed similar pyrolysis behavior. Furthermore, the FTIR measurement of these two lignin samples showed that under air atmosphere, the carbonyl group content increased with increasing temperature, indicating slow oxidation at low temperature. Ferdous et al. [39] studied the influence of heating rate on the pyrolysis of Alcell lignin and Kraft lignin, and found that a high heating rate facilitated the conversion of lignin. They also reported that pyrolysis oil yield of Alcell lignin increased much faster. In addition, with increasing heating rate the H_2 concentration in the gaseous products increased significantly, while the CO concentration slightly decreased.

4.3 Mechanism of lignin pyrolysis

4.3.1 Pyrolysis kinetic model for lignin pyrolysis

In early studies of lignin pyrolysis, a one-step global kinetic model, which supposes that lignin is converted into gaseous, liquid and solid products in one step, was usually used to explain the weight loss in lignin pyrolysis. The corresponding kinetic calculation mainly aims at the single activation energy and pre-exponential factor. For example, the kinetic calculation of MWL pyrolysis by Nunn et al. showed that the activation energy and pre-exponential factor were 82.3 kJ/mol and $3.39 \times 10^5\ s^{-1}$, respectively [38]. However, this kinetic model cannot reflect the complex and continuous reaction in lignin pyrolysis, and it also does not match the common phenomenon that two weight loss peaks exist during lignin pyrolysis, so it is not widely used in current studies.

In view of the two main weight loss peaks during lignin pyrolysis, researchers tried to improve the one-step global kinetic model, and thus proposed a modified model of double-reaction competition (Fig. 4.16). In this model, two weight loss stages are studied individually, corresponding to two competing reactions, and the kinetic parameters are also calculated individually. Based on TG curves of fir and black birch SADF pyrolysis, the activation energies for two weight loss stages of fir SADF pyrolysis are 79.2 kJ/mol and 136.9 kJ/mol, respectively, while those of black birch SADF

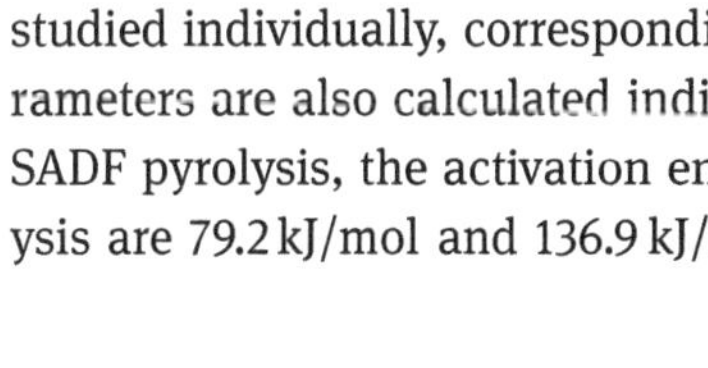

Fig. 4.16: Model of double-reaction competition.

pyrolysis are 87.2 kJ/mol and 141.7 kJ/mol, respectively. In addition, the calculation of the TG curve of Kraft lignin pyrolysis shows the activation energies of these two stages are 58.4 kJ/mol and 120.0 kJ/mol, respectively. This difference in the activation energy indicates that two kinds of reaction exist for these two weight loss stages. When the reaction temperature is low, the aromatic ring is stable, while the side chains attached to the aromatic ring are easily decomposed, including the cleavage of C–C bonds and the elimination of terminal hydroxyl groups in the side chains. Some light gaseous products and char are generated in this stage, and the corresponding activation energy is relatively low. In their TG-FTIR study, Fenner et al. found that CO and CO_2 were released at 50 °C [25]. The FTIR analysis of lignin pyrolysis residue by Li et al. showed that at low temperature, the intensity of peaks for some ether bonds and aliphatic hydroxyl groups decreased, while no obvious change was observed for the aromatic C–O bond. When the reaction temperature was high, the main structure of lignin started to be depolymerized and condensable products like phenolic compounds were formed. Meanwhile, some secondary reactions also occurred, such as demethylation, demethoxylation, and the decomposition and condensation of aromatic rings. At this stage, the activation energy was relatively high [48].

Based on the double-reaction competition model, some researchers proposed modified kinetic models to better match experimental results. Considering that some primary solid products undergo secondary decomposition, Cho et al. [52] proposed a kinetic model based on the pyrolysis product distribution. The products included CO, CO_2, pyrolysis oil, and solid char. In addition to some components that could be detected by GC-MS, such as guaiacols, syringols, and vanillin, the pyrolysis oil also contained some GC-MS-undetectable heavy components, which accounted for 60 % of the carbon content in the pyrolysis oil. Adam et al. [53] studied the pyrolysis of Kraft lignin and proposed the corresponding kinetic model. Three parallel reaction pathways were proposed for lignin pyrolysis with a short residence time, corresponding to the formation of gas, pyrolysis oil and char, respectively; while for a relatively long residence time, taking the secondary decomposition of pyrolysis oil into consideration, an extra reaction pathway of pyrolysis oil decomposition into gaseous products was added. The modified kinetic model involving the secondary decomposition of pyrolysis oil was used to predict the product distribution of lignin pyrolysis and the prediction was in good agreement with the experimental results. Farag et al. [54] made a detailed classification of the condensable volatiles into five groups, namely water, GC-MS-detectable phenolic compounds, heavy molecular weight compounds, non-phenolic aromatic compounds and aliphatic compounds, and proposed an improved model of double-reaction competition, which successfully predicted the yields of main pyrolysis products.

In the pyrolysis process of lignin, although there may be several main weight loss peaks, the weight loss rate is still much slower than that of cellulose and hemicellulose and the weight loss curve shows a gentle variation tendency. This is due to the wide distribution of bond energy in lignin, which leads to the wide distribution of the corresponding activation energies. On account of this, some models involving the variation of activation energy have been proposed. In their study of lignin pyrolysis at 150–900 °C, Caballero et al. [55] proposed a component function kinetic model, which supposed that lignin was composed of a large amount of components and these components started to decompose only at the characteristic temperature. The distributed activation energy model is another kinetic model that considers the variation in activation energy. In this model, reactions are assumed to consist of a series of independent first-order reactions with different activation energies, and the activation energies for all parallel reactions satisfy a probability distribution function. Ferdous et al. [56] used this model to calculate the activation energy of Alcell lignin and Kraft lignin pyrolysis, and obtained the distribution of activation energy.

Klein et al. [3] proposed a free radical mechanism in their research on the pyrolysis of PPE, a model compound of lignin. In the subsequent study, Afifi et al. [57] indicated that the cleavage of ether bonds in lignin generated free radicals with high reactivity and instability. These radicals then underwent rearrangement, electron exchange and some interaction to form some more stable products. Sarkanen et al. [58] considered that the cleavage of C–C bonds would form many free radicals, and the recombination of these radicals could produce guaiacols and syringols. On the basis of previous research work, Faravelli et al. [59] proposed a kinetic model of lignin pyrolysis based on a free radical reaction mechanism. A syringyl dimer was used to represent the typical structure of lignin. After the estimation of dissociation energy of different chemical bonds, it was found that the C_β–O bond was first cleaved to form a phenoxy radical and an alkylaromatic radical, and this process was the initiation of free radicals. During the propagation of free radicals, hydroxyl radical was the most reactive and it could quickly activate the dimer to form the active intermediate, which could undergo β-decomposition to produce coumaryl and synapyl alcohols. Furthermore, the decomposition of some active intermediates could generate hydroxyl radicals and some other radicals. In the addition reaction of free radicals, the phenoxy radical could be added to the lignin structure via the release of a methoxyl group to form a C–O–C bond. Moreover, some reactions like condensation, char formation, CO release and radical recombination also occurred.

4.3.2 Lignin pyrolysis mechanism based on product distribution

4.3.2.1 Mechanism of small molecular product formation

Many small molecular products are generated during lignin pyrolysis. The evolution behavior of these products with temperature can be obtained by TG-FTIR analysis. Combined with structure analysis and dissociation probability of chemical bonds in the original lignin, the corresponding formation mechanism can be deduced. Fig. 4.17 presents the specific FTIR profiles of small molecular product evolution from the pyrolysis of Mongolian Scots pine MWL and Manchurian ash MWL. In general, the

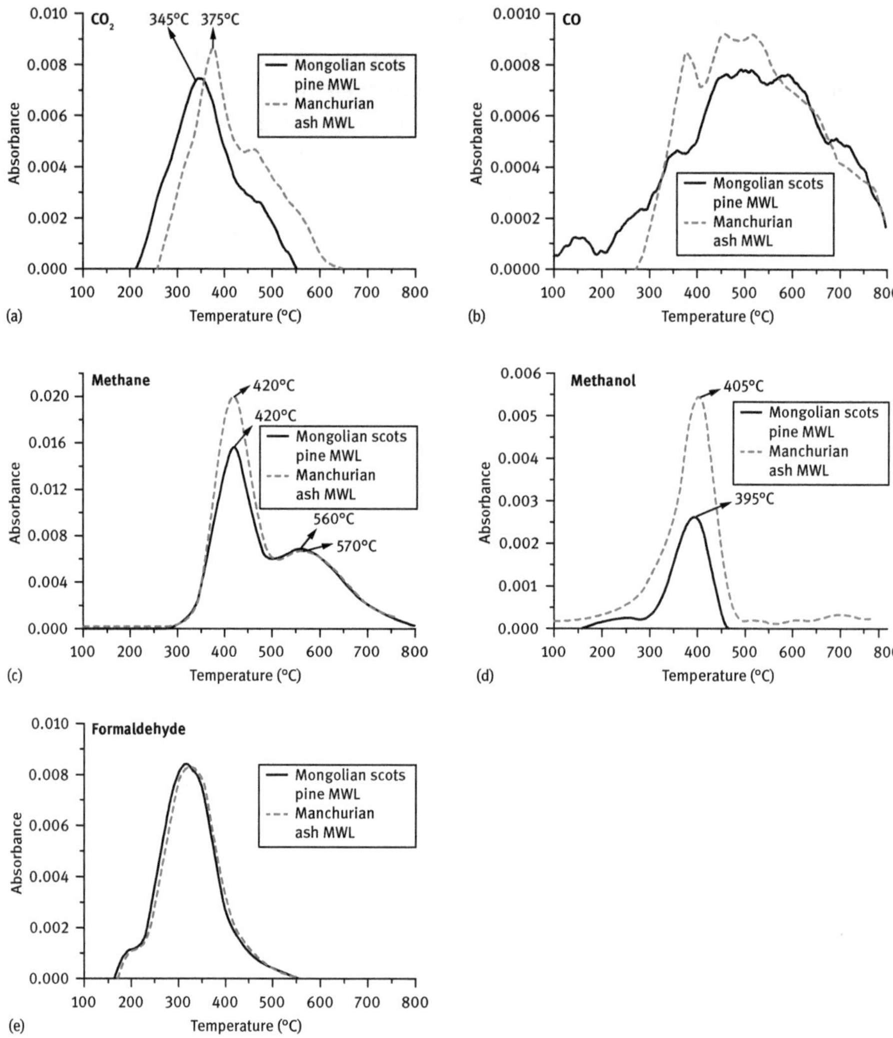

Fig. 4.17: FTIR profiles of some small molecular products' evolution from MWL pyrolysis. Reprinted with permission from [22], © 2009 Elsevier.

small molecular products are mainly generated from the cracking of side chains in the phenylpropane units and the removal of the functional groups attached to the aromatic ring.

As shown in Fig. 4.17, the production of CO covers a wide temperature range. At low temperature, CO is mainly produced from the cleavage of ether bonds in the side chains, because these ether bonds have relatively low dissociation energies. At high temperature, the generation of CO is also obvious owing to the cleavage of ether bonds between aromatic rings and secondary decomposition of some volatiles. The production of CO_2 is located in a relatively narrow temperature range, from 220–250 °C to 600 °C. CO_2 is mainly derived from the cracking and reforming of the active functional groups in the side chains like the carbonyl group and the carboxyl group. The evolution of CH_4 shows two obvious absorption peaks. The first is strong at 420 °C while the second is relatively weak at 560 °C. When the temperature is lower than 500 °C, CH_4 is mainly formed by the cracking of the side chains and the demethylation of the methoxyl group attached to the benzene ring. Due to its higher methoxyl group content, Manchurian ash MWL gave a higher CH_4 yield than Mongolian Scots pine MWL by pyrolysis. CH_4 generation at a higher temperature corresponding to the second absorption peak was attributed to the further rupture of aromatic rings.

Methanol is released in a narrow temperature range of 300–450 °C and it is mainly generated from the hydroxymethyl group in the γ position of the side chain and the methoxyl group attached to the aromatic ring. Because of the higher methoxyl group content, more methanol was produced in the pyrolysis of Manchurian ash MWL than Mongolian Scots pine MWL. Formaldehyde is mainly produced in the temperature range of 250–400 °C by the cleavage of C_β–C_γ bonds in the side chains containing a hydroxymethyl group or γ bond decomposition of the carboxyl group. Moreover, formaldehyde can also be formed by C_γ elimination in the pyrolysis of β-1 type dimer as the model compound of lignin.

In addition, some other small molecular products are also generated, such as water, H_2, and C_2–C_3 gaseous hydrocarbons. Nunn et al. [38] found that water formation covered the whole of lignin pyrolysis, 80 % of which was produced when the temperature was below 427 °C. Most water was derived from the elimination of the hydroxyl group in the aliphatic side chains, and only little water was generated from the secondary reaction of the volatiles. The production of H_2 started at 500 °C and occurred in the subsequent heating process [21, 60]. Fisher et al. [60] proposed that primary H_2 generation was attributed to the rearrangement and condensation of the aromatic rings. At a higher temperature, Avni et al. [61] pointed out that H_2 was released from the cracking and rearrangement of the strong bonds in the aromatic ring. Moreover, Lou et al. [40] observed the formation of C_2–C_3 gaseous hydrocarbons at a high temperature, mainly from the secondary decomposition of some intermediates, char, and some condensable volatiles.

4.3.2.2 Mechanism of phenolic compound formation

Phenolic compounds are the main products of lignin pyrolysis, because the original lignin contains many guaiacyl and syringyl structures. Although the content of hydroxylphenyl structure is not high in original lignin, the amount of phenols without a methoxyl group, like phenol and catechol, cannot be neglected in the final products, especially for pyrolysis at high temperatures and long residence times.

Guaiacols and syringols are mainly produced from the corresponding guaiacyl and syringyl structures in lignin. The thermal stability of the linkages between the typical structural units, such as the α-O-4 and β-O-4 structure, is much lower than that of functional groups attached to the aromatic rings, such as the phenolic hydroxyl group and the methoxyl group. Therefore, during lignin pyrolysis, the C–O bond in the side chain can be cleaved at a low temperature, while demethylation and demethoxylation occur at higher temperatures. As a result, at the primary pyrolysis stage, the functional groups attached to the aromatic rings cannot be removed easily and the phenolic compounds generated mainly contain the original methoxyl group. Generally speaking, the side chains in the phenylpropane units start to crack at 230–260 °C to form guaiacols and syringols with substituted functional groups such as methyl group, ethyl group and vinyl group. Domburg et al. [7] found that the formation of phenolic compounds began with the elimination of the hydroxyl group in the side chains, followed by the cleavage of the ether bond linkages.

Although the C–O bond in the α-O-4 linkage has proved to be easily cleaved, the corresponding products are complex, so the detailed pyrolysis mechanism of α-O-4 structure has not yet been proposed. Current study is mainly focusing on the transformation of β-O-4 structure to phenolic products. Here, based on the distribution of guaiacols and syringols from organosolv lignin pyrolysis and the cleavage mechanism of the C_β–O bond, the formation mechanism of typical phenolic products is proposed.

As shown in Fig. 4.18, when R_1 is a methoxyl group and R_2 is hydrogen, the model compound is ascribed to the guaiacyl type. The products include guaiacol, vanillin, 4-propenylguaiacol and 4-vinylguaiacol. The amount of phenolic compounds which have a saturated alkyl group in the C_4 position like 4-methylguaiacol and 4-ethylguaiacol is small, while the amount of phenolic compounds which have an unsaturated alkyl group in the C_4 position is large. Therefore, it can be deduced that after the cleavage of a C_β–O bond in the side chain, some subsequent reactions occur. When the C_α=C_β bond is cleaved, the C_α easily undergoes oxidation and dehydrogenation to form vanillin instead of 4-methylguaiacol. When the C_β–C_γ bond is cleaved, the C_α=C_β bond is maintained, which will not be hydrogenated to form 4-ethylguaiacol. The hydroxyl group in the γ position is also easily eliminated to produce 4-propenylguaiacol. The removal of the whole side chain from the aromatic ring is relatively more difficult, so that less guaiacol is generated.

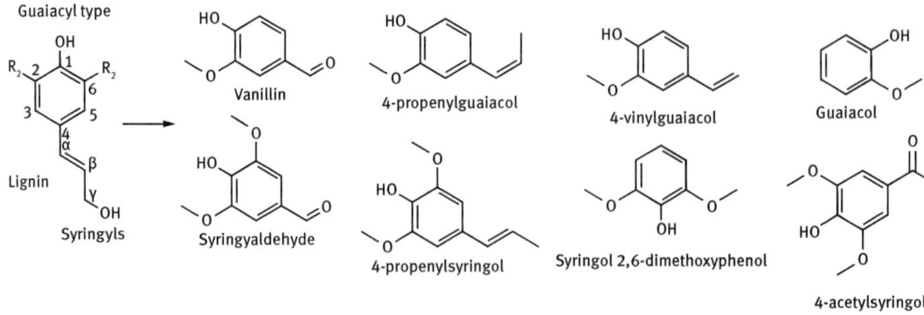

Fig. 4.18: Mechanism of phenolic compound formation from lignin pyrolysis.

When R_1 and R_2 are both a methoxyl group, the model compound is ascribed to the syringyl type. The pyrolysis products are mainly syringols, such as syringaldehyde, 4-propenylsyringol, syringol, and 4-acetylsyringol. The relative content of these products shows that the cracking of side chains still predominates during pyrolysis. The cleavage of the $C_\alpha=C_\beta$ bond followed by the oxidation and dehydrogenation of C_α can easily form syringaldehyde. The elimination of a hydroxyl group in the γ position also can generate 4-propenylsyringol. This reaction mechanism is similar to that of the guaiacyl type compound, in which the cleavage of the $C_\alpha=C_\beta$ bond to generate an aldehyde group is the most active, followed by the cleavage of the C_γ–OH bond, while the complete removal of the side chain is relatively difficult.

Similar to the reaction pathway mentioned above, Hu et al. [62] proposed a radical mechanism for the formation of phenolic compounds during lignin pyrolysis. For a typical β-O-4 structure compound, the homolysis of the C_β–O bond occurred first, then the removal of the side chain, the elimination of hydroxyl groups in α and γ positions, the cleavage of C_α–C_β, C_β–C_γ bonds, and hydrogen abstraction reaction occurred, to generate guaiacol, 4-propenylguaiacol, vanillin, and vinylguaiacol. For a compound containing both β-5 and α-O-4 linkages, the occurrence of homolysis of the C–O bond and the C–C bond can produce benzofuran. Meanwhile, this compound can also first undergo homolysis of the C_α–O bond, then the cleavage of C_β–Ar, C_α–C_β, and C_β–C_γ bonds, the elimination of hydroxyl groups in γ positions and hydrogen abstraction reaction, to produce methylguaiacol, ethylguaiacol, and some other guaiacol derivatives.

Phenols and catechols are also found in the pyrolysis products of lignin. Although some phenols are generated from the hydroxylphenol units in the original lignin, most phenols and catechols are produced from the secondary decomposition of primary pyrolysis products, such as guaiacols and syringols, via demethylation and demethoxylation in which the methoxyl group can be converted into a hydroxyl group or be eliminated.

Branca et al. [63] found that during biomass pyrolysis, guaiacols and syringols were important intermediates, and their yields decreased at higher temperature. Petrocelli et al. [64] observed that the generation of guaiacols was favored at low temperature, and much more phenol and catechol were produced when the temperature increased. Because the original lignin did not contain the catechol type structure, they concluded that catechol was derived from the secondary decomposition of guaiacol during a long residence time. The study of 4-propylguaiacol pyrolysis by Masuku et al. also showed high production of 4-propylcatechol, indicating that the alkyl side chain in the C_4 position of the aromatic ring was more stable than the C–O bond in the methoxyl substituted group [65]. Vuori et al. [66] used guaiacol as model compound to study lignin pyrolysis and found that when the temperature reached 400 °C, catechol was generated quickly, while phenol was also detected. They proposed that the pyrolysis of guaiacol was a free-radical and synergistic reaction, while phenol was produced from the further decomposition of catechol.

In addition to the monophenolic products, there are also some phenolic oligomers such as dimers, trimers and tetramers. Two reaction pathways are considered for the generation of these oligomers. The first is the incomplete depolymerization of lignin and the second is the reoligomerization of the monophenolic compounds produced from the complete depolymerization of lignin. Bai et al. [67] found that some monophenolic compounds, such as 4-vinylguaiacol, guaiacol and syringol, had high reactivity and can undergo reoligomerization to produce oligomers. They also found that some structures in original lignin, such as phenylcoumaran, stilbene and biphenol, had high stability during pyrolysis, so these compounds were probably derived from direct depolymerization of lignin. However, Kotake et al. [68] indicated that most oligomers were produced from the incomplete depolymerization of lignin, because the C–C bonds in these oligomers were thermo-stable.

4.3.2.3 Mechanism of char formation

It is generally believed that during lignin pyrolysis, pyrolytic char, which has a complex three-dimensional network and heterocyclic structure with high thermostability, is produced from the crosslinking reaction of the inner fractions in lignin. As mentioned above, the aliphatic structure in lignin is mainly in the form of propyl side chains, which are relatively thermo-instable. The cracking of the side chain can generate some small molecular products, such as H_2O, CO, and CO_2. Meanwhile, some residual side chains that have double bonds after dehydration can react with other molecules to undergo cyclization and form heterocyclic compounds. With increasing carbonization degree, the aliphatic character of char disappears gradually and the aromatic character is strengthened. Moreover, different aromatic rings can undergo condensation to form polycyclic aromatics. Based on the characterization of pyrolytic char, Cao et al. [69] proposed a simplified mechanism of char formation. The side chains of the phenolic intermediate first form double bonds which undergo further cyclization with the radicals generated from other phenolic intermediates to produce heterocyclic aromatic compounds. Hosoya et al. [70] found that the presence of a methoxyl group was necessary for char formation, and O-quinone was an important intermediate. In their study of char formation from Kraft lignin pyrolysis, Rodríguez-Mirasol et al. [71] concluded that the introduction of some inorganic salts during lignin preparation also affects char formation. Therefore, the mechanism of char formation is very complicated and needs further investigation.

4.3.2.4 Mechanism of other product formation

Some aromatic hydrocarbons and benzyl alcohol are also found in lignin pyrolysis products. Benzyl alcohol may be generated from the rearrangement of aromatics with methoxyl groups. Alén et al. [72] observed the production of aromatic hydrocarbon at 800–1,000 °C during lignin pyrolysis, reporting that a high pyrolysis temperature could completely remove the oxygen-containing functional groups on the aromatic ring. In the study of EMAL pyrolysis by Lou et al., some hydrocarbons like xylene started to be generated at 400 °C [40]. In addition, the generation of acetic acid, furfural, 5-methylfurfural, and 5-hydroxyfurfural is sometimes observed during lignin pyrolysis. Considering that these compounds are typical products of cellulose and hemicellulose pyrolysis, the production of acetic acid and furans is attributed to the pyrolysis of residual glucosidic bonds combined with the lignin structure.

4.3.3 Mechanism of lignin pyrolysis at the molecular scale

In research on the lignin pyrolysis mechanism, theoretical calculation is a newly developed and efficient method to describe the pyrolysis mechanism at molecular level. Through the simulation of lignin structure, the cleavage of chemical bonds and the formation of products, the easily cleaved chemical bonds and the possible reaction pathways can be obtained. Recent research mainly focuses on the simple dimer model compound, and the simulation of β-O-4 structure decomposition is the hot topic.

Huang et al. [73] studied the mechanism of guaiacol pyrolysis by theoretical calculation. The calculation of the dissociation energies of different chemical bonds in guaiacol showed the homolysis of the CH_3–O bond occurred first to produce catechol and guaiacol can also undergo homolysis of the O–H bond to generate o-cresol and 2-hydroxybenzaldehyde. Parthasarathi et al. [74] calculated the dissociation energies of the different linkages in lignin, and estimated the thermal stability of the linkages. They classified the linkages into two groups: one was the ether bond linkages, including α-O-4, β-O-4, and 5-O-4 structures and the other was the C–C bond linkages, including β-1, α-1, β-5, and 5-5 structures. After the optimization of geometry based on quantum chemistry calculation, it was found that the ether bond linkages of α-O-4 and β-O-4 structures had much lower dissociation energies than all C–C bond linkages, which proved the low thermal instability of these two linkages. Beste et al. [75] investigated the free radical chain reaction mechanism during the pyrolysis of PPE, a simple β-O-4 structure model compound, and predicted the behaviors of phenoxyl free radical and benzyl free radical participating in the hydrogen abstraction from PPE. The relative rate constant of reaction was obtained by Density Functional Theory calculation and it was then used in the kinetic analysis of free radical intermediates to predict the selectivity of products, which was in good agreement with the experimental results. Moreover, in their subsequent study of substituted PPE derivatives pyrolysis, the selectivity of pyrolysis products was also predicted based on the theoretical calculation, which also matched well with the experimental results. Huang et al. [76] also simulated the pyrolysis of PPE and designed 10 reaction pathways, including the free radical reactions involving the cleavage of β-O-4 linkages and C_α–C_β bonds as well as two concerted reactions. The thermodynamic and kinetic calculation showed that all the reactions were endothermic and concerted reactions with relatively low activation energies were the dominating reactions.

The theoretical calculation of PPE and its derivatives pyrolysis has been widely studied. However, the detailed simulation of more complex model compound pyrolysis is very difficult. Compared with the calculation of activation energy, the calculation of thermodynamic enthalpy change in different reaction pathways is easier and currently more appropriate for the deduction of the mechanism for complex model compound pyrolysis. Wang et al. [77] used 1-(4-hydroxy-phenyl)-2-phenoxy-propane-1,3-diol as the model compound of lignin, and inferred the reaction pathway and main products by comparing the thermodynamic enthalpy change of different reaction steps. The thermodynamic calculation showed that the first step was the cleavage of β-O-4 and $C_\alpha-C_\beta$ bonds, and the subsequent simulation of free radical reactions suggested that p-hydroxybenzylalcohol, phenol and ethanol were the possible products. Elder et al. [78] also adopted the method of thermodynamic enthalpy change calculation to study the radical propagation reaction after the homolysis of the substituted dimer model compound of lignin. The results showed that for the initiation reactions, the bonds of $C_\alpha-C_\beta$ and β-O-4 had similar dissociation energies, indicating that the probabilities of the homolysis of these two bonds were nearly equal. In the propagation reaction stage, the free radicals generated from homolysis reaction had different delocalization of the unpaired electron and therefore their stabilities differed. The cleavage of the β-O-4 bond can generate phenoxyl free radical and 1-phenyl-2-propyl free radical. Because of the lower spin density of unpaired electrons and thus the lower delocalization, phenoxyl radical was more stable. Therefore, when 1-phenyl-2-propyl radical reacted with the dimer through hydrogen abstraction, more energy was released. Similarly, for the hydroxybenzyl radical and β-phenoxyethanol radical generated from the cleavage of $C_\alpha-C_\beta$ bonds, the former had a lower spin density of unpaired electrons in the α position than the latter had in the β position. Hence, when reacting with the dimer through hydrogen abstraction reaction, the former was endothermic while the latter was exothermic.

Although research on the lignin pyrolysis mechanism on the basis of theoretical calculation is still at an early stage and mainly concentrates on simple dimer pyrolysis, as understanding of the lignin pyrolysis mechanism is furthered and computer technology develops, in the future it will be possible to simulate the pyrolysis of more complex lignin model compounds and even real lignin.

References

[1] Brebu M, Vasile C. Thermal degradation of lignin – A review. Cellulose Chemistry & Technology. 2010;44(9):353–363.

[2] Yang H, Yan R, Chen H, et al. Characteristics of hemicellulose, cellulose and lignin pyrolysis. Fuel. 2007;86(12-13):1781–1788.

[3] Klein MT. Model pathways in lignin thermolysis. Massachusetts Institute of Technology; 1981.

[4] Asmadi M, Kawamoto H, Saka S. Thermal reactions of guaiacol and syringol as lignin model aromatic nuclei. Journal of Analytical & Applied Pyrolysis. 2011;92(1):88–98.

[5] Asmadi M, Kawamoto H, Saka S. Thermal reactivities of catechols/pyrogallols and cresols/xylenols as lignin pyrolysis intermediates. Journal of Analytical & Applied Pyrolysis. 2011;92(1):76–87.

[6] Klein MT, Virk PS. Model pathways in lignin thermolysis. 1. Phenethyl phenyl ether. Industrial & Engineering Chemistry Fundamentals. 1983;22(1):35–45.

[7] Domburg G, Rossinskaya G, Sergseva V. Study of thermal stability of b-ether bonds in lignin and its models: Therm Anal Proc Int Conf, 4th, Budapest; 1974;2:221.

[8] Savinykh VI, Kislitsyn AN, Rodionova ZM, et al. On thermal stability of monoarylglycol ethers. Khim Drev. 1975;(5):100–102.

[9] Kawamoto H, Horigoshi S, Saka S. Pyrolysis reactions of various lignin model dimers. Journal of Wood Science. 2007;53(2):168–174.

[10] Faix O, Meier D, Fortmann I. Pyrolysis-gas chromatography-mass spectrometry of two trimeric lignin model compounds with alkyl-aryl ether structure. Journal of Analytical and Applied Pyrolysis. 1988;14(2–3):135–148.

[11] Liu J, Wu S, Lou R. Chemical structure and pyrolysis response of beta-O-4 lignin model polymer. Bioresources. 2011;6(2):1079–1093.

[12] Chu S, Subrahmanyam AV, Huber GW. The pyrolysis chemistry of a β-O-4 type oligomeric lignin model compound. Green Chemistry. 2013;15(1):125–136.

[13] Liu Q, Wang S, Zheng Y, et al. Mechanism study of wood lignin pyrolysis by using TG–FTIR analysis. Journal of Analytical and Applied Pyrolysis. 2008;82(1):170–177.

[14] Xie H, Yu Q, Qin Q, et al. Study on pyrolysis characteristics and kinetics of biomass and its components. Journal of Renewable and Sustainable Energy. 2013;5(1):13122.

[15] Caballero JA, Marcilla A, Conesa JA. Thermogravimetric analysis of olive stones with sulphuric acid treatment. Journal of Analytical and Applied Pyrolysis. 1997;44(1):75–88.

[16] Haykiri-Acma H, Yaman S, Kucukbayrak S. Comparison of the thermal reactivities of isolated lignin and holocellulose during pyrolysis. Fuel Processing Technology. 2010;91(7):759–764.

[17] Jiang G, Nowakowski DJ, Bridgwater AV. Effect of the temperature on the composition of lignin pyrolysis products. Energy & Fuels. 2010;24(8):4470–4475.

[18] Zhao J, Xiuwen W, Hu J, et al. Thermal degradation of softwood lignin and hardwood lignin by TG-FTIR and Py-GC/MS. Polymer Degradation and Stability. 2014;108:133–138.

[19] Rencoret J, Marques G, Gutiérrez A, et al. Isolation and structural characterization of the milled-wood lignin from Paulownia fortunei wood. Industrial Crops and Products. 2009;30(1):137–143.

[20] Faix O, Jakab E, Till F, et al. Study on low mass thermal degradation products of milled wood lignins by thermogravimetry-mass-spectrometry. Wood Science and Technology. 1988;22(4):323–334.

[21] Jakab E, Faix O, Till F. Thermal decomposition of milled wood lignins studied by thermogravimetry/mass spectrometry. Journal of Analytical and Applied Pyrolysis. 1997;40–41:171–186.

[22] Wang S, Wang K, Liu Q, et al. Comparison of the pyrolysis behavior of lignins from different tree species. Biotechnology Advances. 2009;27(5):562–567.

[23] Zhang M, Resende FLP, Moutsoglou A, et al. Pyrolysis of lignin extracted from prairie cordgrass, aspen, and Kraft lignin by Py-GC/MS and TGA/FTIR. Journal of Analytical and Applied Pyrolysis. 2012;98:65–71.

[24] Alén R, Rytkönen S, Mckeough P. Thermogravimetric behavior of black liquors and their organic constituents. Journal of Analytical and Applied Pyrolysis. 1995;31:1–13.

[25] Fenner RA, Lephardt JO. Examination of the thermal decomposition of kraft pine lignin by Fourier transform infrared evolved gas analysis. Journal of Agricultural and Food Chemistry; 1981;29(4):846–849.

[26] Brodin I, Sjöholm E, Gellerstedt G. The behavior of Kraft lignin during thermal treatment. Journal of Analytical and Applied Pyrolysis. 2010;87(1):70–77.
[27] Brebu M, Tamminen T, Spiridon I. Thermal degradation of various lignins by TG-MS/FTIR and Py-GC-MS. Journal of Analytical and Applied Pyrolysis. 2013;104:531–539.
[28] Hu J, Xiao R, Shen D, et al. Structural analysis of lignin residue from black liquor and its thermal performance in thermogravimetric-Fourier transform infrared spectroscopy. Bioresource Technology. 2013;128:633–639.
[29] De Wild P J, Huijgen W, Heeres HJ. Pyrolysis of wheat straw-derived organosolv lignin. Journal of Analytical and Applied Pyrolysis. 2012;93:95–103.
[30] Domínguez JC, Oliet M, Alonso MV, et al. Thermal stability and pyrolysis kinetics of organosolv lignins obtained from Eucalyptus globulus. Industrial Crops and Products. 2008;27(2):150–156.
[31] Zhao X, Liu D. Chemical and thermal characteristics of lignins isolated from Siam weed stem by acetic acid and formic acid delignification. Industrial Crops and Products. 2010;32(3):284–291.
[32] Luo Z, Wang S, Guo X. Selective pyrolysis of Organosolv lignin over zeolites with product analysis by TG-FTIR[J]. Journal of Analytical and Applied Pyrolysis. 2012;95:112–117.
[33] Jiang X, Ellis N, Shen DK, et al. Thermogravimetry – FTIR analysis of pyrolysis of pyrolytic lignin extracted from bio-oil. Chemical Engineering & Technology. 2012;35(5):827–833.
[34] Chang S, Zhao Z, Zheng A, et al. Properties of pyrolytic lignin from bio-oil. Nongye Jixie Xuebao [Transactions of the Chinese Society for Agricultural Machinery]. 2011;42(11):99–105.
[35] Wang S, Lin H, Ru B, et al. Comparison of the pyrolysis behavior of pyrolytic lignin and milled wood lignin by using TG–FTIR analysis. Journal of Analytical and Applied Pyrolysis. 2014;108:78–85.
[36] Ucar G, Meier D, Faix O, et al. Analytical pyrolysis and FTIR spectroscopy of fossil Sequoiadendron giganteum (Lindl.) Wood and mwls isolated hereof. Holz als Roh- und Werkstoff. 2005;63(1):57–63.
[37] Hosoya T, Kawamoto H, Saka S. Secondary reactions of lignin-derived primary tar components. Journal of Analytical and Applied Pyrolysis. 2008;83(1):78–87.
[38] Nunn TR, Howard JB, Longwell JP, et al. Product compositions and kinetics in the rapid pyrolysis of sweet gum hardwood. Industrial & Engineering Chemistry Process Design and Development. 1985;24(3):836–844.
[39] Ferdous D, Dalai AK, Bej SK, et al. Production of H 2 and medium Btu gas via pyrolysis of lignins in a fixed-bed reactor. Fuel Processing Technology. 2001;70(1):9–26.
[40] Lou R, Wu S. Products properties from fast pyrolysis of enzymatic/mild acidolysis lignin. Applied Energy. 2011;88(1):316–322.
[41] Ben H, Ragauskas AJ. Comparison for the compositions of fast and slow pyrolysis oils by NMR characterization. Bioresource Technology. 2013;147:577–584.
[42] Iatridis B, Gavalas GR. Pyrolysis of a precipitated kraft lignin. Industrial & Engineering Chemistry Product Research and Development. 1979;18(2):127–130.
[43] Huang Y, Wei Z, Qiu Z, et al. Study on structure and pyrolysis behavior of lignin derived from corncob acid hydrolysis residue. Journal of Analytical and Applied Pyrolysis. 2012;93:153–159.
[44] Jiang G, Nowakowski DJ, Bridgwater AV. Effect of the temperature on the composition of lignin pyrolysis products. Energy & Fuels. 2010;24(8):4470–4475.
[45] Lou R, Wu S, Lv G. Fast pyrolysis of enzymatic/mild acidolysis lignin from moso bamboo. Bioresources. 2010;5(2):827–837.
[46] Pasquali CL, Herrera H. Pyrolysis of lignin and IR analysis of residues. Thermochimica Acta, 1997;293(1):39–46.
[47] Sharma RK, Wooten JB, Baliga VL, et al. Characterization of chars from pyrolysis of lignin. Fuel. 2004;83(11):1469–1482.

[48] Li J, Li B, Zhang X. Comparative studies of thermal degradation between larch lignin and Manchurian ash lignin. Polymer Degradation and Stability. 2002;78(2):279–285.
[49] Diehl BG, Brown NR, Frantz CW, et al. Effects of pyrolysis temperature on the chemical composition of refined softwood and hardwood lignins. Carbon. 2013;60:531–537.
[50] Jegers HE, Klein MT. Primary and secondary lignin pyrolysis reaction pathways. Industrial & Engineering Chemistry Process Design and Development. 1985;24(1):173–183.
[51] Rutherford DW, Wershaw RL, Rostad CE, et al. Effect of formation conditions on biochars: compositional and structural properties of cellulose, lignin, and pine biochars. Biomass and Bioenergy. 2012;46:693–701.
[52] Cho J, Chu S, Dauenhauer PJ, et al. Kinetics and reaction chemistry for slow pyrolysis of enzymatic hydrolysis lignin and organosolv extracted lignin derived from maplewood. Green Chemistry. 2012;14(2):428–439.
[53] Adam M, Ocone R, Mohammad J, et al. Kinetic investigations of Kraft lignin pyrolysis. Industrial & Engineering Chemistry Research. 2013;52(26):8645–8654.
[54] Farag S, Kouisni L, Chaouki J. Lumped approach in kinetic modeling of microwave pyrolysis of kraft lignin. Energy & Fuels. 2014;28(2):1406–1417.
[55] Caballero JA, Font R, Marcilla A. Study of the primary pyrolysis of Kraft lignin at high heating rates: yields and kinetics. Journal of Analytical and Applied Pyrolysis. 1996;36(2):159–178.
[56] Ferdous D, Dalai AK, Bej SK, et al. Pyrolysis of lignins: experimental and kinetics studies. Energy & Fuels. 2002;16(6):1405–1412.
[57] Afifi AI, Hindermann JP, Chornet E, et al. The cleavage of the aryl–O–CH_3 bond using anisole as a model compound. Fuel. 1989;68(4):498–504.
[58] Sarkanen KV, Ludwig CH. Lignins: occurrence, formation, structure and reactions. Lignins: occurrence, formation, structure and reactions. Wiley, New York; 1971.
[59] Faravelli T, Frassoldati A, Migliavacca G, et al. Detailed kinetic modeling of the thermal degradation of lignins. Biomass and Bioenergy. 2010;34(3):290–301.
[60] Fisher T, Hajaligol M, Waymack B, et al. Pyrolysis behavior and kinetics of biomass derived materials. Journal of Analytical and Applied Pyrolysis. 2002;62(2):331–349.
[61] Avni E, Coughlin R W, Solomon P R, et al. Mathematical modelling of lignin pyrolysis. Fuel. 1985;64(11):1495–1501.
[62] Hu J, Shen D, Xiao R, et al. Free-radical analysis on thermochemical transformation of lignin to phenolic compounds. Energy & Fuels. 2012;27(1):285–293.
[63] Branca C, Giudicianni P, Di Blasi C. GC/MS characterization of liquids generated from low-temperature pyrolysis of wood. Industrial & Engineering Chemistry Research. 2003;42(14):3190–3202.
[64] Petrocelli FP, Klein MT. Simulation of kraft lignin pyrolysis. In: Fundamentals of thermochemical biomass conversion. Springer, Netherlands; 1985. pp. 257–273.
[65] Masuku CP. Thermal reactions of the bonds in lignin. IV. Thermolysis of dimethoxyphenols. Thermolysis of dimethoxyphenols. 1991;45(3):181–190.
[66] Vuori AI, Bredenberg JB. Thermal chemistry pathways of substituted anisoles. Industrial & Engineering Chemistry Research. 1987;26(2):359–365.
[67] Bai X, Kim KH, Brown RC, et al. Formation of phenolic oligomers during fast pyrolysis of lignin. Fuel. 2014;128:170–179.
[68] Kotake T, Kawamoto H, Saka S. Mechanisms for the formation of monomers and oligomers during the pyrolysis of a softwood lignin. Journal of Analytical and Applied Pyrolysis. 2014;105:309–316.
[69] Cao J, Xiao G, Xu X, et al. Study on carbonization of lignin by TG-FTIR and high-temperature carbonization reactor. Fuel Processing Technology. 2013;106:41–47.

[70] Hosoya T, Kawamoto H, Saka S. Role of methoxyl group in char formation from lignin-related compounds. Journal of Analytical and Applied Pyrolysis. 2009;84(1):79–83.
[71] Rodríguez-Mirasol J, Cordero T, Rodríguez JJ. High-temperature carbons from kraft lignin. Carbon. 1996;34(1):43–52.
[72] Alén R, Kuoppala E, Oesch P. Formation of the main degradation compound groups from wood and its components during pyrolysis. Journal of Analytical and Applied Pyrolysis. 1996;36(2):137–148.
[73] Huang J, Li X, Wu D, et al. Theoretical studies on pyrolysis mechanism of guaiacol as lignin model compound. Journal of Renewable and Sustainable Energy. 2013;5(4):43112.
[74] Parthasarathi R, Romero RA, Redondo A, et al. Theoretical study of the remarkably diverse linkages in lignin. The Journal of Physical Chemistry Letters. 2011;2(20):2660–2666.
[75] Beste A, Buchanan AC, Britt PF, et al. Kinetic analysis of the pyrolysis of phenethyl phenyl ether: Computational prediction of α/β-selectivities. The Journal of Physical Chemistry A. 2007;111(48):12118–12126.
[76] Huang X, Liu C, Huang J, et al. Theory studies on pyrolysis mechanism of phenethyl phenyl ether. Computational and Theoretical Chemistry. 2011;976(1):51–59.
[77] Wang H, Zhao Y, Wang C, et al. Theoretical study on the pyrolysis process of lignin dimer model compounds. Acta Chim Sin. 2009;67(9):893–900.
[78] Elder T. A computational study of pyrolysis reactions of lignin model compounds. Holzforschung. 2010;64(4):435–440.

5 Cross coupling pyrolysis of biomass components

Biomass is a high-molecular organic substance, whose main framework is cellulose, filled with hemicellulose and lignin. It also contains a small amount of extractives and inorganic salts. The pyrolysis behavior of biomass is significantly affected by the quantity and the form of three major components, namely cellulose, hemicellulose and lignin. Some researchers have concluded that biomass pyrolysis behavior as a whole could be regarded as the integration of the individual pyrolysis behaviors of cellulose, hemicellulose, and lignin, indicating that the interactions between the three major components can be ignored. Alén et al. [1] studied the pyrolysis products of pine and its three major components and found that the pyrolysis behavior of pine was almost the integration of the three components. Svenson et al. [2] carried out a study of birch and the pyrolysis behaviors of its major components, and found that the yield of char from birch could be calculated from the proportion of the three major components according to linear superposition. The calculated result was in good agreement with the experimental result. Yang et al. [3] studied the TG curves of a mixed biomass sample and found that the weight loss during biomass pyrolysis could be obtained by the linear superposition of the individual weight losses of the three major components and that the interactions between the three major components were small enough to be ignored. This was also confirmed by the calculation of IR spectra [4].

However, there are also many research results showing that the pyrolysis processes of the individual components are not independent. The three major components have overlapping temperature ranges and component interactions exist during pyrolysis. Caballero et al. [5] carried out kinetic studies on olive and almond shell pyrolysis and found that the pyrolysis processes were well fitted by three independent parallel reactions. Due to the interactions between components, the real kinetic parameters cannot simply be obtained from those of extracted components' pyrolysis. Worasuwannarak et al. [6] analyzed biomass pyrolytic products and found an interaction between lignin and cellulose, which resulted in a decreased tar yield and an increased char yield. In addition, hemicellulose also influences the pyrolysis of cellulose and lignin. Hosoya et al. [7] found that the existence of lignin dramatically suppressed the tendency of thermal polymerization and carbonization from the macromolecular organics, such as levoglucosan, which was produced from cellulose pyrolysis. It promoted the decomposition of cellulose to small molecular products. Furthermore, the presence of cellulose suppressed char formation from lignin, but promoted the conversion of lignin to phenols, like guaiacol and 4-methyl guaiacol.

DOI 10.1515/9783110369632-007

5.1 Influence of component interaction on pyrolysis

The influence of component interaction on biomass pyrolysis is complicated. Therefore, it is more reasonable to investigate the influence of the interaction between each pair of components on each other's pyrolysis behaviors. A good understanding of the interactions between component pairs is needed to analyze the coupling effect of components on biomass pyrolysis. The amounts of cellulose, hemicelluloses and lignin are about 40–80 %, 15–30 %, and 10–25 % respectively [8]. As a result, the amounts of these three major components could be mixed in these ranges accordingly. Therefore, the interactions between components can be observed by comparing the experimental TG/DTG curves with the calculated TG/DTG curves, which are summed according to the quantities of their three major components in the mixed sample.

5.1.1 Effect of the ratio of hemicellulose to cellulose

Microcrystalline cellulose and xylan (Sigma, extracted from beech) are selected as model compounds for cellulose and hemicellulose, respectively. The pyrolysis samples are composed of microcrystalline cellulose and xylan in different proportions. The corresponding DTG curves are presented in Fig. 5.1. It shows that the TG curves of mixtures still keep the pyrolysis characteristics of each component rather than displaying integration. Two weight loss peaks are assigned to the decomposition of hemicellulose (200–327 °C) and cellulose (327–450 °C). With increasing hemicellulose content in the mixture, the maximum decomposition rate of hemicellulose increases, while that for cellulose decreases. Due to the low temperature for hemicellulose decomposition, the hemicellulose can decompose into the melted organics which cover the cellulose surface when the reaction temperature is reached. This suppresses the release of cellulose pyrolysis volatiles and makes the weight loss strength of cellulose in the mixture weaker than that of pure cellulose.

According to TG curves for the mixture samples (Fig. 5.2), weight loss occurring at temperatures below 327 °C can be ascribed to hemicellulose pyrolysis. In this case, the experimental and calculated TG/DTG curves are in good agreement, indicating that the influence of cellulose on hemicellulose pyrolysis is slight. When the temperature is higher than 327 °C, the weight loss rate of experimental DTG is lower than that predicted by the calculated DTG curve, resulting in a higher final residue yield compared with the calculated value. Therefore, it can be concluded that the pyrolysis of cellulose is affected by hemicellulose, suppressing the formation of volatiles but promoting the generation of char.

The pyrolytic products of the mixture of cellulose and hemicellulose are diverse, including almost all of the products of the individual pyrolysis of cellulose and hemicellulose respectively. The interaction between hemicellulose and cellulose during pyrolysis was examined by studying the formation of two typical cellulose pyrolytic prod-

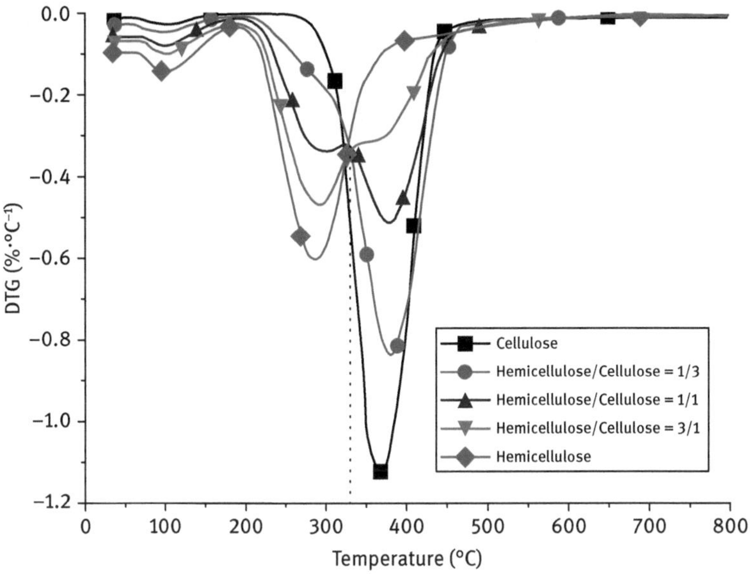

Fig. 5.1: DTG curves for the pyrolysis of cellulose and hemicellulose mixtures. Reprinted with permission from [9], © 2011 Elsevier.

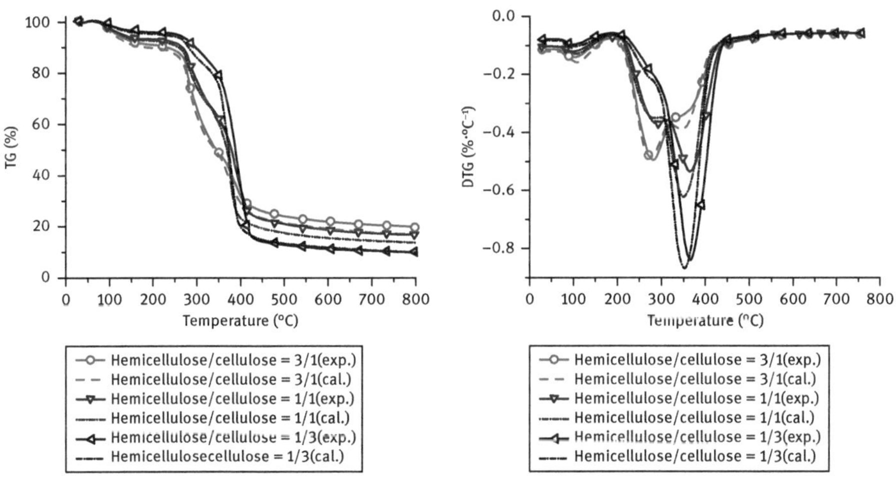

(a) (b)

Fig. 5.2: Experimental and calculated TG/DTG curves for the pyrolysis of cellulose and hemicellulose mixtures. Reprinted with permission from [9], © 2011 Elsevier.

ucts, LG and hydroxyacetaldehyde [10, 11]. The formation curves of LG and hydroxyacetaldehyde are shown in Fig. 5.3. They show that the production of LG decreases as the proportion of hemicellulose increases and that LG production in this case is even lower than during pure cellulose pyrolysis, indicating that LG formation is suppressed by the presence of hemicellulose. Furthermore, hydroxyacetaldehyde production is also influenced by hemicellulose content. When hemicellulose content is between 25 % and 50 %, hydroxyacetaldehyde production is higher than that from pure cellulose, suggesting that the formation of hydroxyacetaldehyde can be promoted by the appropriate addition of hemicellulose. This phenomenon might be caused by the secondary decomposition of LG to hydroxyacetaldehyde [12].

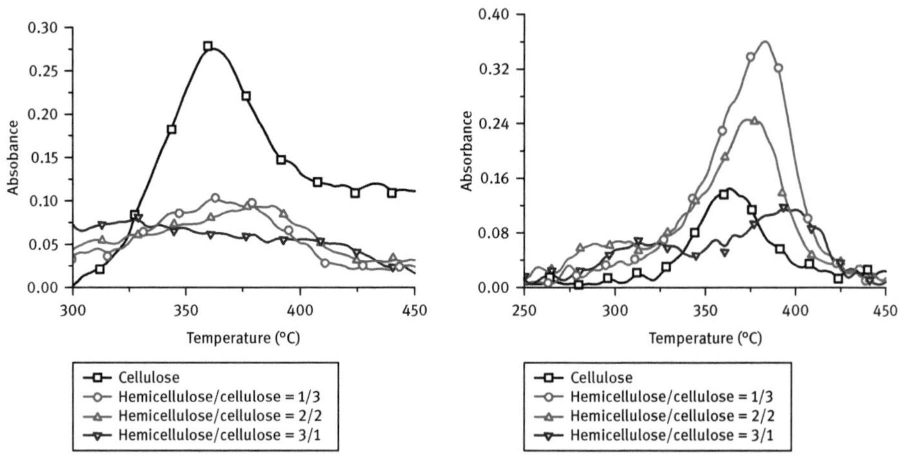

(a) (b)

Fig. 5.3: Evolution profiles for typical compounds during cellulose pyrolysis with the ratio of cellulose/hemicellulose; (a) levoglucosan, (b) hydroxyacetaldehyde. Reprinted with permission from [9], © 2011 Elsevier.

5.1.2 Effect of the ratio of cellulose to lignin

The composition, structure and pyrolysis behavior of lignin are different from those of cellulose. The pyrolysis behavior of a mixture of cellulose and lignin stays in good accordance with that of the single components in TG analysis. The main difference is that weight loss in the mixture sample is slightly lower than that in the single components, as shown in Fig. 5.4. However, at a high temperature, the production and composition of small molecular gases, tar and char are significantly affected by the interaction between cellulose and lignin, leading to an increased tar yield and a decreased char yield [7]. Hosoya et al. [13] considered that solid-liquid and gas-gas phase interactions for the pyrolytic products of cellulose and lignin result in the increase of tar and the suppression of char. They pointed out that the volatiles from cellulose are hydrogen donors, while the volatiles from lignin are hydrogen acceptors. The presence of lignin suppresses the formation of LG from cellulose pyrolysis, but it promotes the production of small molecular compounds. In addition, the presence of cellulose accelerates the conversion of lignin to phenols [6]. Haensel et al. [14] studied the pyrolysis characteristics of a mixture of cellulose and lignin at high temperature and analyzed the structure of pyrolytic char by XPS, UPS (ultraviolet photoelectron spectrometer), and SEM. They found that a graphite type structure was formed in the char, because the lignin underwent strong deoxygenation during the mixture's pyrolysis. On the other hand, the structure of the char from the pure lignin pyrolysis was found to be disorganized.

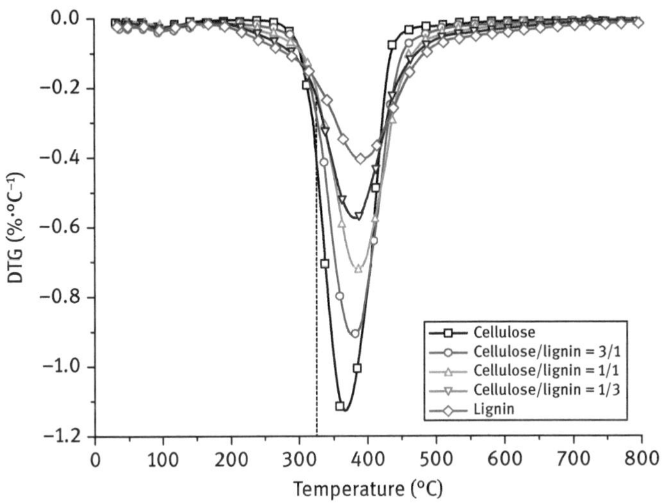

Fig. 5.4: DTG curves for the pyrolysis of cellulose and lignin mixtures. Reprinted with permission from [9], © 2011 Elsevier.

The TG curves for the mixture samples of cellulose and lignin are presented in Fig. 5.5. The calculated and the experimental TG/DTG curves are in good agreement when the ratio of cellulose to lignin is 1/3. When the proportion of cellulose increases to 1/2 or 3/4, the experimental TG/DTG curves shift slightly towards a high temperature region and their weight loss rates slow down in comparison with the calculated TG/DTG curves.

The interaction between cellulose and lignin is also confirmed by the pyrolysis of acid detergent fiber (ADF), which is obtained by the Van Soest method. In the case of the pyrolysis of poplar, in which the cellulose and lignin content are 60.7 % and 14.8 %, respectively, a cellulose/lignin ratio of about 4 is adopted. The corresponding TG/DTG curves are shown in Fig. 5.6. The maximum weight loss rate is at about 350 °C

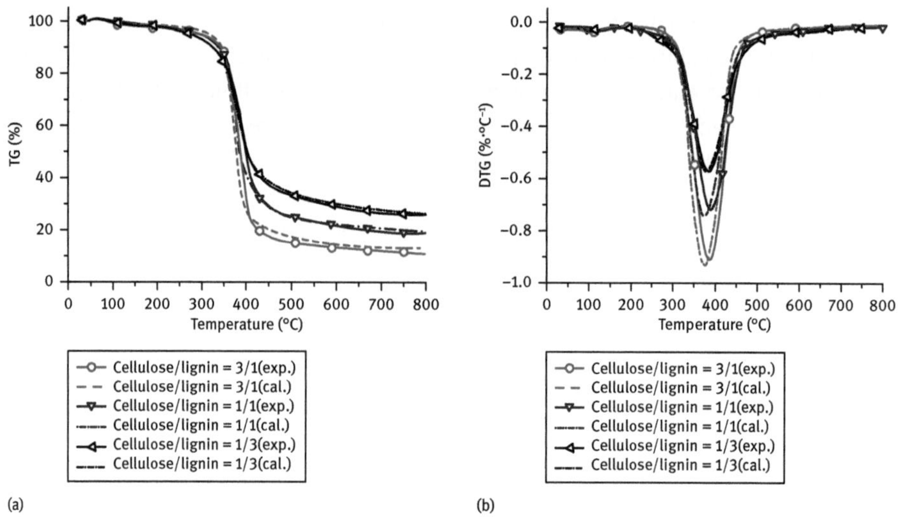

Fig. 5.5: Experimental and calculated TG/DTG curves for the pyrolysis of cellulose and lignin mixtures. Reprinted with permission from [9], © 2011 Elsevier.

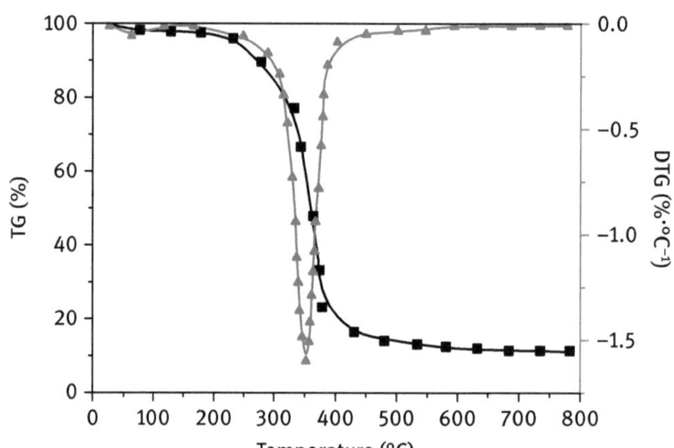

Fig. 5.6: TG/DTG curves of poplar ADF. Reprinted with permission from [17], © 2009 Springer.

due to the abundant cellulose. Furthermore, the characteristics and tendencies of the TG/DTG curves correspond with those found in pure cellulose pyrolysis. Its pyrolytic products contain abundant aldehydes, acetones, and acids, which mainly come from cellulose pyrolysis. Meanwhile, due to the existence of a small crosslinking structure between cellulose and lignin, ADF starts to decompose slowly at a low temperature, which corresponds to the pyrolysis characteristic of lignin. Furthermore, the curves for the mixture of cellulose and lignin are close to those of pure cellulose, suggesting that the presence and the form of a crosslinking structure will influence the pyrolysis process of the mixture. Xie et al. [15] also investigated the pyrolytic behaviors of cellulose, hemicellulose, and lignin by TG analysis on NDF, ADF, and SADF, and confirmed that the different components interacted in the pyrolysis process. Zhang et al. [16] found that a cellulose and lignin mixture sample containing the natural linking structure showed obvious interactions in the pyrolysis process. The yield of LG decreased due to the interaction between cellulose and lignin in the herbaceous biomass, while the yields of small molecular chemicals and furans increased. However, these effects could not be found in woody biomass. The differences between pyrolysis in herbaceous and woody biomass might be caused either by the differences in the connection way between cellulose and lignin, or by the abundance of non-acid-solved ash and minerals in the herbaceous biomass. Giudicianni et al. [18] also found that the interactions between different biomass components in their experimental studies could not be ignored. The interactions had a significant influence on the yields and characteristics of the products and they also led to a decrease of specific BET surface of char. The interaction between cellulose and lignin had the most obvious impact, with the presence of lignin suppressing the decomposition of cellulose to LG, while promoting the production of small organics from cellulose.

5.1.3 Effect of the ratio of hemicellulose to lignin

The DTG curves for the mixture of hemicellulose and lignin are shown in Fig. 5.7. Two weight loss peaks in the main reaction region are assigned to hemicellulose and lignin pyrolysis respectively. With a rising proportion of hemicellulose, weight loss increases at the first peak while decreasing at the second peak. The interaction between hemicellulose and lignin is strong. The presence of lignin increases the rate of hemicellulose pyrolysis, while the presence of hemicellulose decreases the initial decomposition temperature and the weight loss rate for lignin pyrolysis. The strong interaction can also be confirmed by comparing the experimental TG curves for the hemicellulose and lignin mixture samples (Fig. 5.8) to the calculated TG curves through the linear superposition of components.

Fig. 5.9 shows the evolution profiles of furfural, a key product of hemicellulose pyrolysis. The formation of furfural was found to be evidently inhibited by lignin. This tendency is caused by the decomposition of lignin at low temperatures, affecting the pyrolysis behavior of hemicellulose. Some typical chemicals from lignin pyrolysis are also identified in the corresponding products.

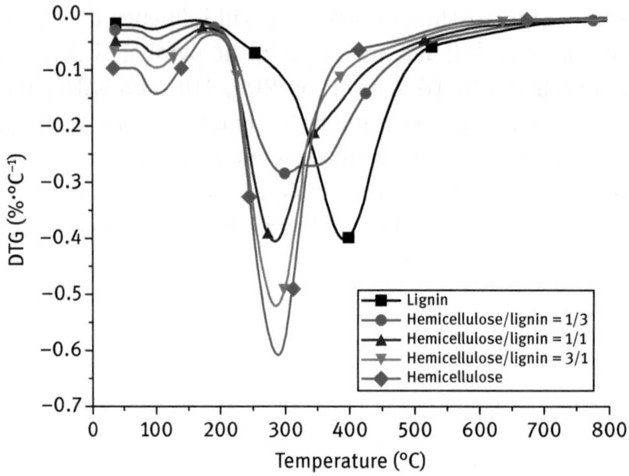

Fig. 5.7: DTG curves for the pyrolysis of hemicellulose and lignin mixtures. Reprinted with permission from [9], © 2011 Elsevier.

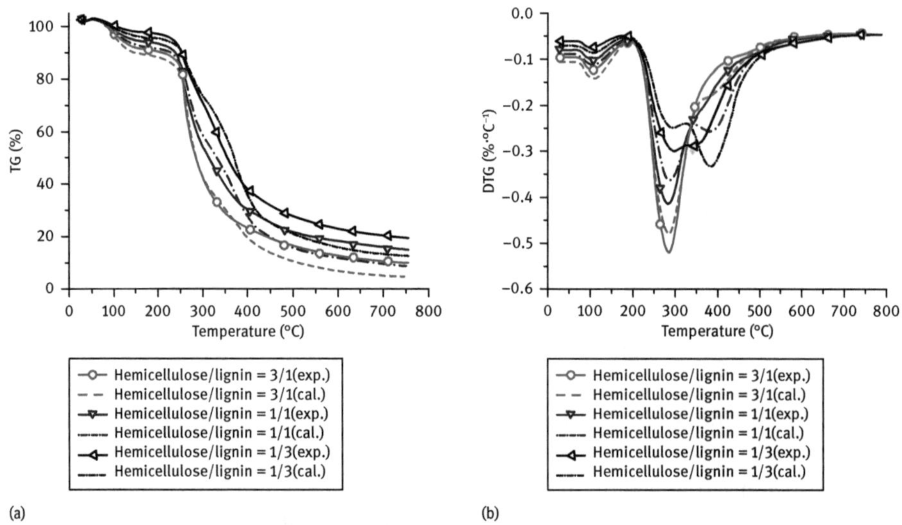

Fig. 5.8: Experimental and calculated TG/DTG curves for the pyrolysis of hemicellulose and lignin mixtures. Reprinted with permission from [9], © 2011 Elsevier.

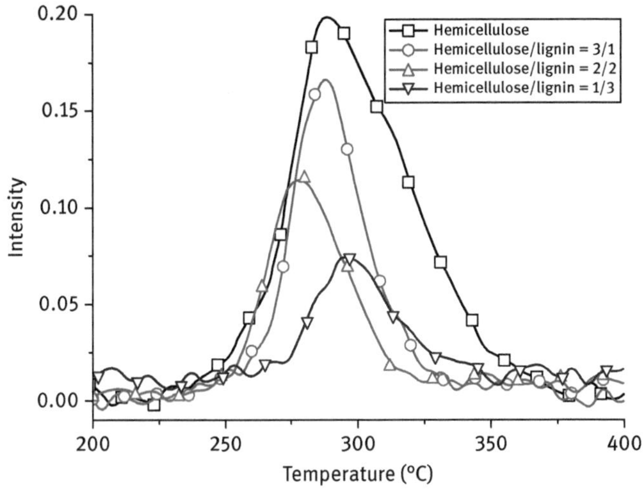

Fig. 5.9: Evolution profiles for furfural with varying hemicellulose/lignin ratio. Reprinted with permission from [9], © 2011 Elsevier.

5.2 Coupled pyrolysis of components

5.2.1 Pyrolysis behavior of a mixture of biomass components

In order to reveal the effect of component interaction, typical proportions for three components can be selected based on their abundance in the biomass. Lignin has a great influence on hemicellulose pyrolysis at low temperatures, while lignin and hemicellulose have a joint influence on cellulose pyrolysis. As a result, the experimental TG curves for mixed samples are different from calculated TG curves. Therefore, complex interactions exist among the three major components. Microcrystalline cellulose, xylan and organosolv lignin were selected as the model compounds for cellulose, hemicellulose, and lignin, respectively. They were mixed in the ratios of 6 : 1.5 : 1.5, 5 : 3 : 1, 5 : 2 : 2, and 5 : 1 : 3, respectively. The corresponding experimental and calculated DTG curves for these four samples are presented in Fig. 5.10. The differences between the experimental and calculated DTG curves are slight and the only differences revealed between the two temperature ranges of 200–327 °C and 327–410 °C are caused by the differences in component content. The weight loss peak at 200–327 °C corresponds to hemicellulose decomposition and the peak at 327–410 °C is caused by cellulose pyrolysis. As a result, the decrease of cellulose in the mixed samples can lead to a slower maximum weight loss rate. In addition, according to an analysis of the TG curves of various mixed biomass samples, the difference in weight loss at temperatures below 327 °C can be ascribed to the influence of lignin on hemicellulose pyrolysis, while the difference at temperatures above 327 °C can be attributed to the influence of hemicellulose on cellulose pyrolysis. Due to the small contribution of

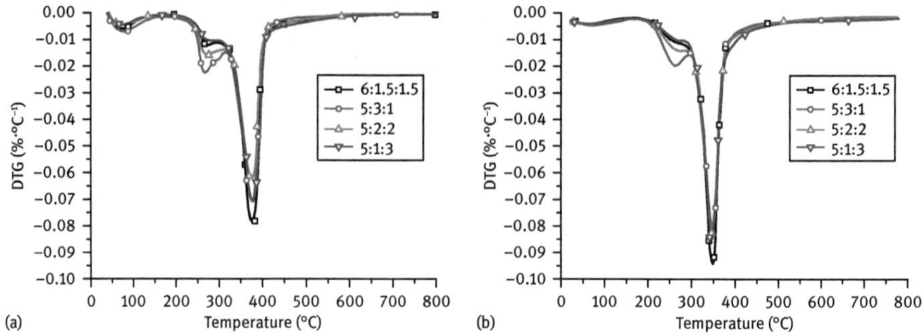

Fig. 5.10: DTG curves for the mixed biomass samples; (a) experimental results, (b) calculated results. Reprinted with permission from [19], © 2011 Elsevier.

lignin to total weight loss, the influence of hemicellulose on lignin pyrolysis makes only a small contribution to the difference.

An analysis of IR spectra for the different mixed biomass samples has found that their pyrolytic products contain LG, furfural, acetic acid, 2,6-dimethoxylphenol, and others, which are typical products of cellulose, hemicellulose and lignin pyrolysis. The primary factor influencing the distribution of pyrolytic products is the quantity of each component in the sample. The evolution profiles for typical products from mixed biomass with varied component proportions are shown in Fig. 5.11. The formation of LG, which is an intermediate product of cellulose pyrolysis, is promoted by the presence of hemicellulose and lignin. The formation of furfural and acetic acid are different from their formation in pure hemicellulose pyrolysis. The formation of furfural and acetic acid in the temperature range of 200–350 °C is related to the decomposition of hemicellulose, while the subsequent formation of these two products at higher temperature is closely related to the interactions among components. When the ratio of cellulose/hemicellulose/lignin is 5 : 3 : 1, the decomposition of hemicellulose to acetic acid is promoted, while the formation of furfural is inhibited. When the temperature elevates, acetic acid is also formed from cellulose pyrolysis due to the influence of hemicellulose and lignin. The formation of 2,6-dimethoxylphenol is based exclusively on lignin and is also promoted to some extent by the presence of cellulose and hemicellulose.

5.2.2 Influence of component proportions on the distribution of pyrolytic products

In order to obtain a better understanding of the interactions among the three main components, an infrared radiation reactor is used to study the effect of component proportions on the pyrolytic products of biomass. The yields of bio-oil, small molecular gases and char from cellulose pyrolysis at 550 °C are 81.41 %, 12.51 %, and 6.08 %, respectively. Those from hemicellulose are 44.22 %, 36.73 %, and 19.05 % respectively,

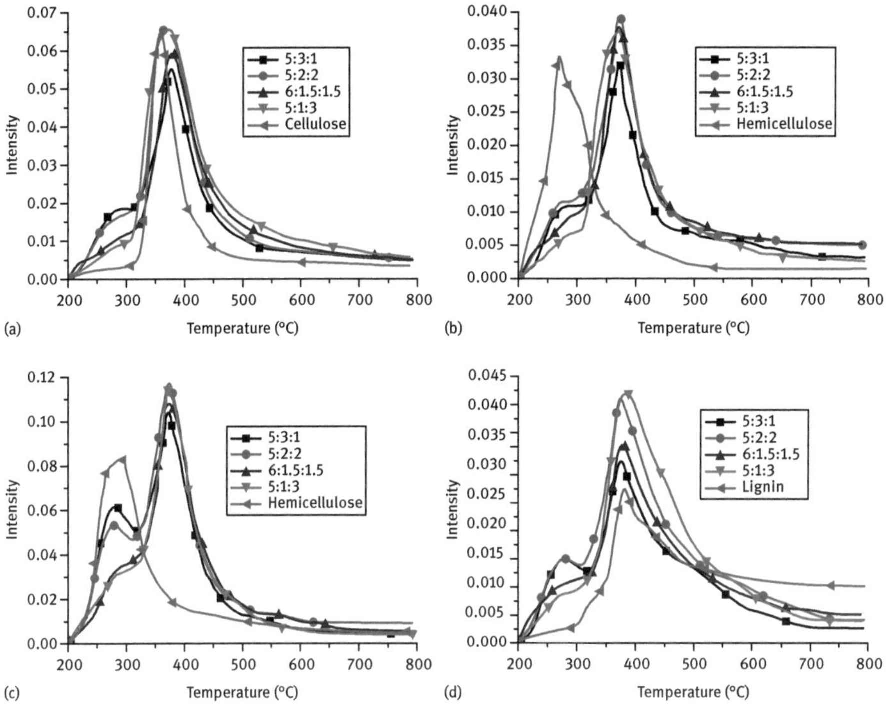

Fig. 5.11: Evolution profiles for typical products of the pyrolysis of mixed biomass samples and components; (a) levoglucosan, (b) furfural, (c) acetic acid, and (d) 2,6-dimethyl phenol. Reprinted with permission from [19], © 2011 Elsevier.

while those from lignin are 21.77 %, 37.90 %, and 40.33 % respectively. The highest yield of bio-oil is from cellulose pyrolysis, while small molecular gases are most abundant in hemicellulose pyrolysis and char is most common in lignin pyrolysis. After mixing the components in different proportions, the product yields for four mixed biomass samples under the same pyrolysis temperature are presented in Fig. 5.12. There is an obvious difference between the experimental and calculated values for the four mixed biomass samples. The interactions between components reduce the yield of bio-oil, while increasing the yields of char and small molecular gases. Zhao et al. [20] prepared samples according to the proportion of the three major components in rice husks and corn stalks, respectively. Based on the analysis of pyrolytic products, they found that the calculated and experimental results were in good accordance, but that the yields of specific products were different. This indicates that the three main components undergo coupled physical as well as chemical interactions. Cellulose is covered by melted hemicellulose and lignin at a low reaction temperature, suppressing the release of volatiles and providing sufficient residence time for the secondary reaction. As a result, the yield of bio-oil under experimental conditions is lower than that given by the calculation.

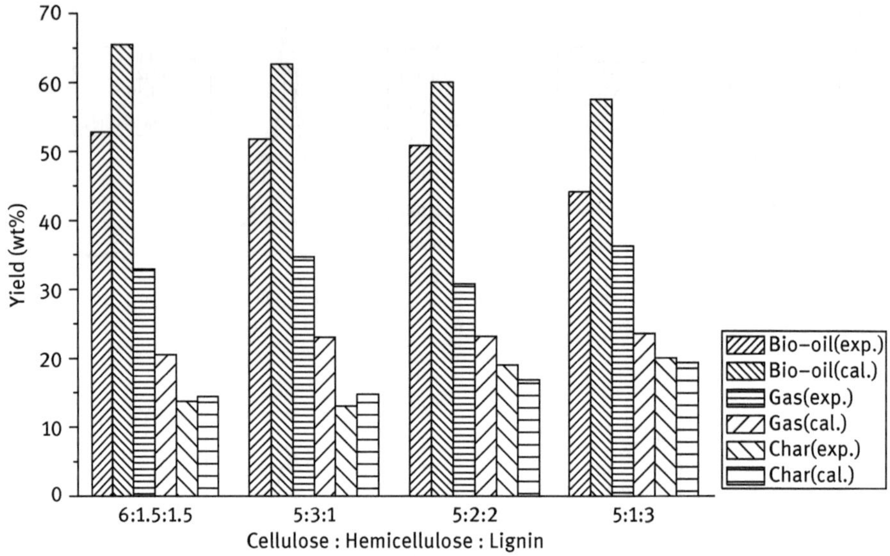

Fig. 5.12: Experimental and calculated product distributions from the pyrolysis of mixed biomass samples. Reprinted with permission from [19], © 2011 Elsevier.

The composition and content of small molecular gases are shown in Tab. 5.1. The major components are CO and CO_2 and there are also a few small molecular alkanes such as CH_4, C_2H_4, and C_2H_6, as well as some H_2. The gas yield is high for hemicellulose, but relatively low for cellulose. Furthermore, the formation of CO and CO_2 is slightly affected by the interactions between different components. The interactions have a significant effect on the production of hydrocarbon gases, especially in the case of the influence of cellulose and hemicellulose on the pyrolysis of lignin, which leads to a decreased yield of CH_4 and C_2H_6. Caballero et al. [21] used the Pyroporbe 1000 instrument to study the distribution of the pyrolytic products of almond shells, their holocellulose and lignin. They found that due to the cross interactions of components, the amounts of CH_4, C_2H_6, C_3H_6, methanol, and acetaldehyde were not equal to the integrated amounts from holocellulose and lignin pyrolysis.

The bio-oil obtained from mixed biomass pyrolysis contains the all same products as yielded by pure component pyrolysis, such as alkanes, acids, alcohols, ketones, furans, aldehydes, phenols, and sugars, representing the characteristics of the pyrolysis of the three major components. 2,5-diethoxy tetrahydrofuran, 3,4-dehydration altrose and LG are typical products of cellulose pyrolysis; acetic acid and furfural are mainly formed from hemicellulose; and phenol, 2,6-dimethoxylphenol as well as 2,6-dimethoxy-4-(2-propenyl)phenol are typical products of lignin pyrolysis. In addition, ketones, such as hydroxyacetone, hydroxy-2-cyclopenten-1-one, and 3-methyl-2-hydroxy-2-cyclopenten-1-one, are generated by the pyrolysis of cellulose and hemicellulose.

Tab. 5.1: Gas yields from the individual components and mixed biomass (wt%).

Sample			H_2	CH_4	CO	CO_2	C_2H_4	C_2H_6	C_3H_6	C_3H_8
Component	Cellulose		0.09	0.07	8.46	2.84	0.40	0.06	0.23	0.01
	Xylan		0.42	0.13	10.85	24.05	0.56	0.36	0.28	0.08
	Lignin		0.27	3.95	20.53	8.47	2.47	1.53	0.50	0.17
Mixed biomass	6 : 1.5 : 1.5	Cal.	0.26	0.71	19.77	9.90	1.15	0.38	0.53	0.05
		Exp.	0.28	0.40	21.65	8.20	1.40	0.25	0.54	0.04
	5 : 3 : 1	Cal.	0.29	0.52	17.99	12.31	1.01	0.34	0.48	0.05
		Exp.	0.24	0.36	16.71	13.85	1.02	0.32	0.44	0.06
	5 : 2 : 2	Cal.	0.25	0.83	17.55	9.96	1.10	0.42	0.47	0.06
		Exp.	0.16	0.46	16.54	11.97	0.84	0.28	0.35	0.05
	5 : 1 : 3	Cal.	0.26	1.29	20.16	9.31	1.40	0.58	0.53	0.07
		Exp.	0.27	0.75	20.44	9.73	1.45	0.37	0.54	0.05

The interactions between components can be identified by comparing typical product distributions. As shown in Tab. 5.2, the experimental quantities of 3,4-dehydration altrose and LG obtained from cellulose pyrolysis are lower than those obtained by calculation. Conversely, the experimental value for 2,5-diethoxytetrahydrofuran is higher than the calculated one and this tendency is strengthened by an increase in hemicellulose content while cellulose content remains constant. Consequently, hemicellulose significantly improves the formation of 2,5-diethoxytetrahydrofuran, while suppressing the formation of 3,4-dehydrationaltrose and LG. The impact of lignin on these products is weak. As mentioned above, acetic acid and furfural are typical products of hemicellulose pyrolysis. Comparing their yields from the mixed biomass samples with the ratios of 6 : 1.5 : 1.5 and 5 : 2 : 2 shows that formation of acetic acid and furfural is promoted by an increase in cellulose content. Conversely, when the lignin content increases in the mixed samples, the relative yields for these two chemicals from the experiment and the calculation gradually converge and the experimental yield is even lower than the calculated one. Therefore, it can be concluded that cellulose promotes the formation of acetic acid and furfural, while lignin suppresses their formation. Phenol, 2,6-dimethoxylphenol and 2,6-dimethoxy-4-(2-propenyl)phenol are typical products of lignin pyrolysis and their experimental values are all higher than their calculated values, especially for 2,6-dimethoxylphenol. Furthermore, the yield of phenols decreases with the rising hemicellulose content when cellulose content remains constant. It can be asserted that cellulose strongly promotes phenol production, while hemicellulose hinders it. When the cellulose content is higher than the hemicellulose content, the yield of phenols increases.

Tab. 5.2: Relative yields of main products from the individual components and mixed biomass.

Compound	Component (%)		Mixed biomass (%)				
	Yield	Source		6:1.5:1.5	5:3:1	5:2:2	5:1:3
2,5-Diethoxytetra-hydrofuran	7.93	Cellulose	Cal.	16.14	13.87	8.30	5.63
			Exp.	5.29	4.41	4.41	4.41
Altrose	4.71		Cal.	1.20	0.20	0.75	1.00
			Exp.	3.14	2.61	2.61	2.61
Levoglucosan	13.23		Cal.	2.58	0.30	2.09	3.92
			Exp.	8.82	7.35	7.35	7.35
Acetic acid	5.26	Xylan	Cal.	4.57	5.15	1.33	0.66
			Exp.	0.88	1.75	1.17	0.58
Furfural	4.88		Cal.	2.71	2.77	0.76	0.00
			Exp.	0.81	1.63	1.08	0.54
Phenol	1.36	Lignin	Cal.	0.44	0.36	0.83	1.39
			Exp.	0.23	0.15	0.30	0.45
2,6-Dimethoxylphenol	3.16		Cal.	2.16	0.61	4.73	7.80
			Exp.	0.53	0.35	0.70	1.05
2,6-Dimethoxy-4-(2-propenyl)phenol	0.67		Cal.	0.42	0.26	0.64	0.81
			Exp.	0.11	0.07	0.15	0.22

5.3 Pyrolysis behaviors of detergent fibers

The inherent crosslinked structure of components is also an important factor. The corresponding investigation of the pyrolysis behaviors of detergent fibers can be beneficial in revealing the component interactions from another perspective. The detergent fibers, mainly NDF, ADF, and SADF, are obtained through the Van Soest method. The difference between these fibers and simple mixed biomass is that part of the natural crosslinked structure between components is still retained in these fibers.

NDF, whose structure is rarely damaged, was selected for comparison with the pyrolysis behaviors of mixed biomass in order to determine the interactions among the three major components of the crosslinked structure. In fir the cellulose content is 48.95 % and those of hemicellulose and lignin are 11.94 % and 31.21 %, respectively. The corresponding ratio is about 5 : 1 : 3. The NDF obtained from fir retains the original proportions and structures of the three major components.

The comparison between pyrolysis curves for the fir NDF and the mixed biomass sample with a cellulose/hemicellulose/lignin ratio of 5 : 1 : 3 are shown in Fig. 5.13. It shows that the existence of crosslinked structure has little effect on the initial pyrolysis and the char formation stage, while it has a great influence on the main pyrolysis stage ranging from 200 °C to 450 °C. The weight loss rate in the main pyrolysis stage for fir NDF is much greater than that of the mixed biomass sample. The existence of

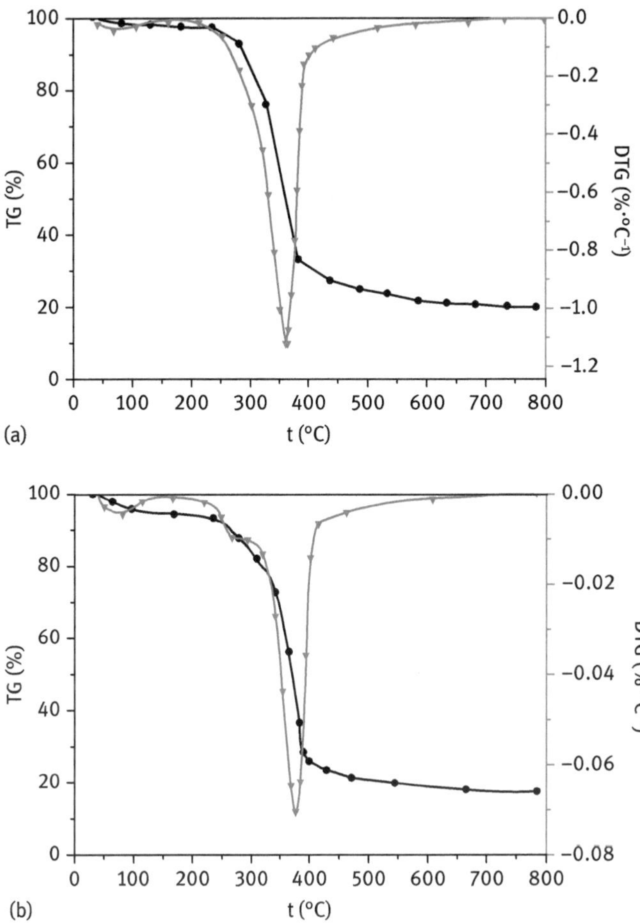

Fig. 5.13: TG/DTG curves for (a) fir NDF and (b) mixed biomass sample. (a) Reprinted with permission from [17], © 2009 Springer. (b) Reprinted with permission from [19], © 2011 Elsevier.

a crosslinked structure makes the weight loss of the three major components in the main stage merge together without obvious subpeaks. However, during the pyrolysis of mixed biomass samples, weak weight loss peaks of hemicellulose and the strong peaks of cellulose are identified. As a result, the crosslinked structure makes the component interactions stronger.

5.3.1 Pyrolysis behaviors of different detergent fibers

Fir and poplar are two widely used types of biomass. The lignin content and extractives content in fir are significantly higher than in poplar, while the cellulose content, hemicellulose content and acid-insoluble ash content in fir are lower than in poplar.

The TG/DTG curves for fir NDF and poplar NDF are shown in Fig. 5.14. The large difference in component contents leads to the obvious distinctions between TG curves for fir and poplar raw biomass, while if soluble minerals and extractives are excluded, the corresponding NDF/TG curves are similar. The slight difference between the TG curves for fir NDF and poplar NDF is caused by the different amounts of the three major components in the two samples. The presence of soluble minerals and extractives in the samples has a significant influence on the pyrolysis behaviors of the three major components in comparison with raw biomass and its NDF pyrolysis. The pyrolysis of raw fir starts at a relatively low temperature compared to the fir NDF sample and it has a high yield of residue. Therefore, it can be inferred that extractives can make pyrolysis initiate early and also dramatically catalyze cross reactions. Compared to ADF, the thermal weight loss for NDF starts earlier and its residue yield is higher, since it contains hemicellulose. Furthermore, due to the high hemicellulose content in poplar, a weak shoulder peak, a characteristic of hemicellulose pyrolysis, appears in the DTG curve for poplar NDF.

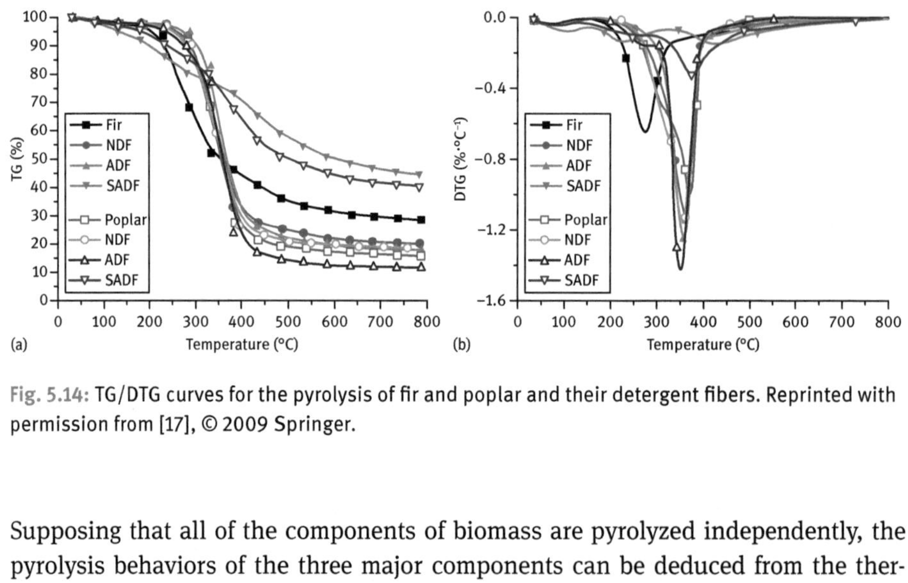

Fig. 5.14: TG/DTG curves for the pyrolysis of fir and poplar and their detergent fibers. Reprinted with permission from [17], © 2009 Springer.

Supposing that all of the components of biomass are pyrolyzed independently, the pyrolysis behaviors of the three major components can be deduced from the thermal weight loss curves of raw biomass and its detergent fibers. The TG/DTG curves of hemicellulose and cellulose, which are obtained by subtraction calculation from the TG curves for NDF-ADF and ADF-SADF, are shown in Fig. 5.15. The pyrolysis behavior of lignin can be roughly represented by SADF pyrolysis, since SADF is mainly composed of lignin with a small addition of acid-insoluble ash.

The decomposition of hemicellulose and cellulose starts at 230 °C and 280 °C respectively. Both components undergo an obvious decomposition within a narrow temperature range with sharp DTG peaks and the weight loss rates reach their maxima at 310 °C and 350 °C, respectively. More volatiles are produced from cellulose pyrolysis

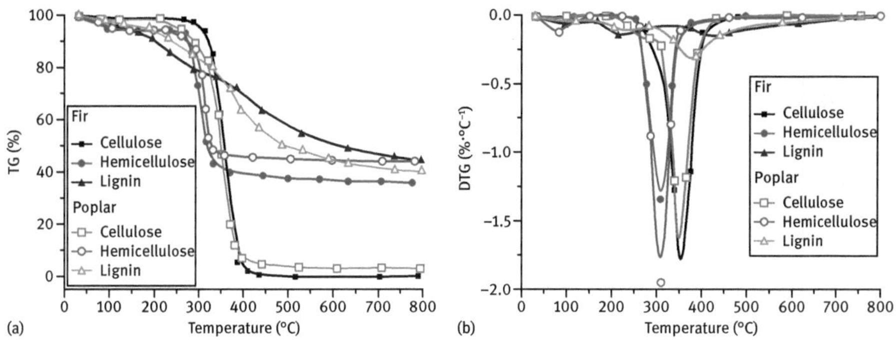

Fig. 5.15: TG/DTG curves for the main components in fir and poplar. Reprinted with permission from [17], © 2009 Springer.

and more char from hemicellulose. In comparison, lignin pyrolysis takes place in a broader temperature range with flatter DTG peaks and the highest yield of char. Although the decomposition of lignin starts at a relatively low temperature, its rate is very slow. This is in good agreement with the pyrolysis behavior of the three major components, suggesting that the most important factor for influencing the biomass pyrolysis process is the relative content of the component. Furthermore, compared to fir, the three major components of poplar and its detergent fibers undergo pyrolysis at a relatively low temperature and a high yield of char can be obtained, due to high acid-insoluble ash content found in poplar. This phenomenon confirms that the presence of acid-insoluble ash can contribute to the thermal weight loss of components.

5.3.2 Distribution of pyrolytic products for different detergent fibers

Fig. 5.16 shows the release of volatiles during the pyrolysis of poplar and its detergent fibers of at a heating rate of 20 °C/min. It is in good accordance with the DTG curves of the pyrolysis of the corresponding components. The pyrolysis of poplar SADF is considered to roughly reflect the pyrolysis of its lignin. Compared to raw poplar, SADF and ADF have much higher lignin contents, leading to the appearance of a peak in the early pyrolysis stage, which is characteristic of lignin pyrolysis. The products in the main stage of ADF pyrolysis are derived primarily from cellulose, then from lignin. The pyrolytic products of NDF are generated from the co-contributions of the three major components. Meanwhile, the formation of hemicellulose pyrolysis products can be determined by comparing the release spectra of NDF and ADF pyrolysis products.

According to the difference calculations of the TG curves of different detergent fibers, the maximum weight loss rates for hemicellulose and cellulose pyrolysis are at about 310 °C and 350 °C, respectively. If all components are pyrolyzed independently, the products of cellulose pyrolysis can be analyzed by comparing the IR spectrum of

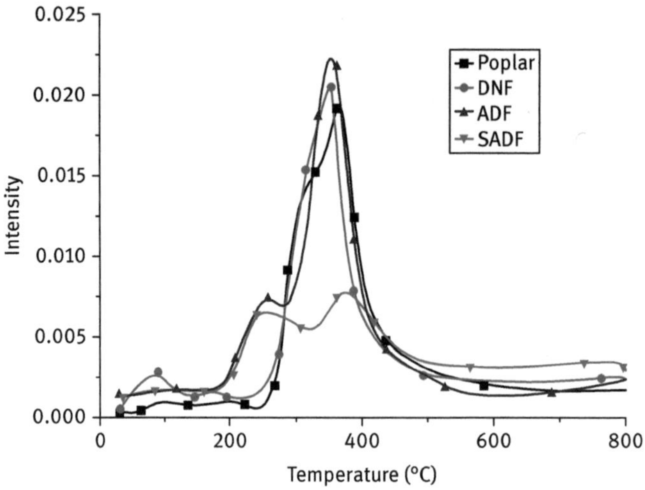

Fig. 5.16: Release of volatiles from the pyrolysis of poplar and its detergent fibers. Reprinted with permission from [17], © 2009 Springer.

SADF pyrolysis with that of ADF at 350 °C. Similarly, the products of hemicellulose pyrolysis can be analyzed by comparing the IR spectrum of NDF pyrolysis with that of ADF at 310 °C. The comparisons of the IR spectra are shown in Fig. 5.17 and Fig. 5.18 respectively. According to the IR spectra, the pyrolysis of all three detergent fibers produces small molecular gases like H_2O, CO_2, CO, etc. On the IR spectrum of SADF pyrolysis, the strong absorption peaks at 3,200–2,850 cm^{-1} can be assigned to hydrocarbon products, especially CH_4. The intense peaks at 1,085–9,60 cm^{-1} show the formation of abundant methanol. On the IR spectrum of ADF pyrolysis, the broad peaks at 3,200–2,700 cm^{-1} can be assigned to the formation of hydrocarbons, formaldehyde and acetaldehyde [10]. The intensive absorption peaks at 1,850–1,600 cm^{-1} represent the vibration of C=O, indicating the formation of various aldehydes and ketones. The characteristic peaks at 1,500–900 cm^{-1} correspond to the in-plane bending vibration of C–H and the skeleton vibration of C–O and C–C, suggesting the production of carboxylic acids and alcohols. In addition, the characteristic peaks at 1,183 cm^{-1} for LG and at 860 cm^{-1} for hydroxyacetaldehyde confirm that they are two typical products of cellulose pyrolysis.

The IR spectra for NDF and ADF are very similar, as shown in Fig. 5.18, which suggests the formation of acids, alcohols, ketones, aldehydes, and phenols. This can be ascribed to a similar polysaccharide structure in cellulose and hemicellulose, leading to the similarity in pyrolytic products. This also illustrates that hemicellulose has a weak influence on formation of volatiles in NDF pyrolysis. After excluding interference from the same band through spectrum subtraction, furfural (2,865–2,770 cm^{-1} and 1,435–965 cm^{-1}) is the major product of hemicellulose pyrolysis.

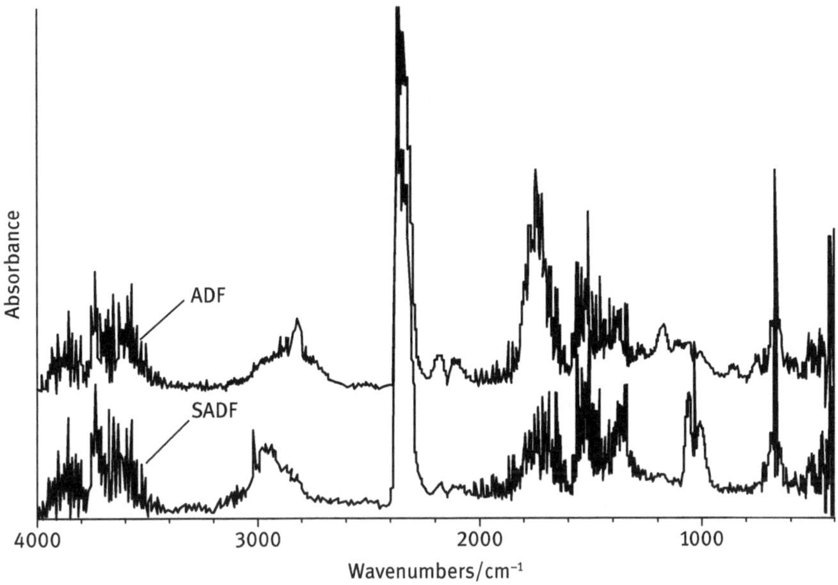

Fig. 5.17: FTIR spectra of products released from ADF and SADF pyrolysis at 350 °C. Reprinted with permission from [17], © 2009 Springer.

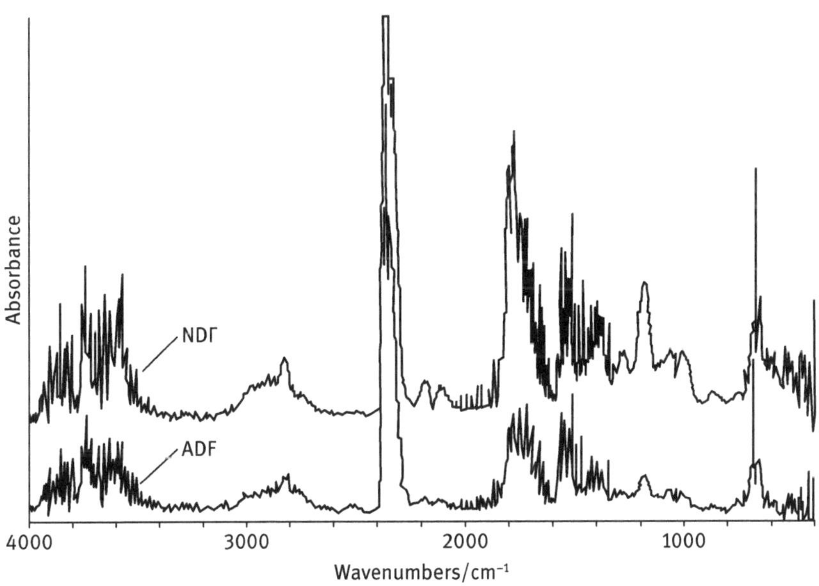

Fig. 5.18: FTIR spectra of products released from NDF and ADF pyrolysis at 310 °C. Reprinted with permission from [17], © 2009 Springer.

5.4 Influence of extractives on biomass pyrolysis

Two biomass materials, Mongolian Scots pine (MSP) and Manchurian ash (MA), were selected to obtain the extractives. The extractives were extracted by using a 2:1 (v/v) benzene-ethanol solution from biomass for 7 h. After extraction, the extracted residues were also obtained.

5.4.1 Pyrolysis behaviors of biomass extractives

The extractives are present in small amounts within biomass and are composed of free substances that do not constitute cell walls or intercellular layers. They can be extracted by organic solvents, steam, or water. They are nonstructural components, which include waxes, fats, resins, tannins, sugars, starches, pigments, etc. Due to their low abundance (less than 10 %), the influence of biomass extractives on biomass pyrolysis does not attract enough attention. However, their influence should not be ignored in biomass pyrolysis [22]. The extractives have a significant effect on biomass combustibility, since they have high volatility. The heating value of biomass increases with rising extractive content [23].

The pyrolysis of extractives occurs within a wide temperature range (130–550 °C), similar to raw biomass. As shown in Fig. 5.19, the TG curve of the raw biomass is in good accordance with that of the extracted residue, while the TG curve of the extractives seems obviously different. The extractives pyrolysis has a low weight loss rate and a high char yield. The apparent activation energies for the pyrolysis of MSP, its extracted residue and its extractives are 104.37 kJ/mol, 119.38 kJ/mol, and 58.78 kJ/mol, respectively. This shows that the extractives are most easily decomposed due to their low activation energy. The decomposition of raw biomass is easier than that of extracted residue, indicating that the existence of extractives possibly promotes the reactivity of biomass pyrolysis. Furthermore, Raveendran et al. [24] found that the pyrolysis behavior of biomass extractives is similar to that of lignin, except for a faster pyrolysis rate and a lower pyrolysis temperature. Thammasouk et al. [25] used ethanol to treat herbaceous biomass and found that the ethanol-soluble compounds were rich in the extractives. The water-ethanol extraction method removed almost 90 % of extractives, and also significantly reduced the content of ash, acid-insoluble lignin and soluble sugars in the biomass pyrolysis products [26]. In addition, the presence of extractives improves the bio-oil yield and suppresses the production of gases and biochar [27].

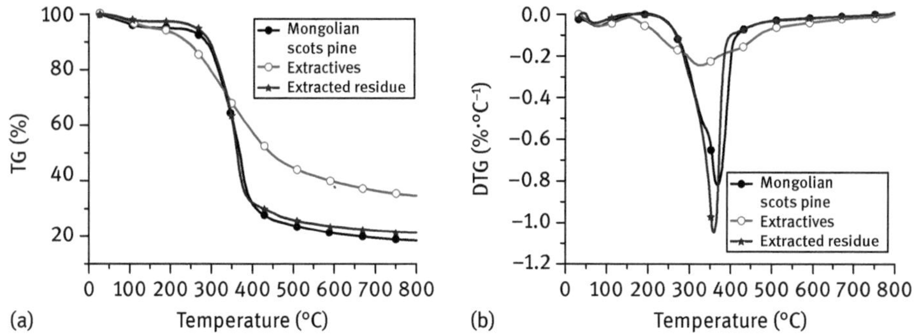

Fig. 5.19: TG/DTG curves for the pyrolysis of biomass, extractives, and extracted residue. Reprinted with permission from [28].

5.4.2 Influence mechanism of extractives on biomass pyrolysis

The evolution of pyrolytic products for the extracted residue is similar to that for raw biomass. The IR spectra for pyrolytic products at the maximum absorbance intensity are shown in Fig. 5.20. H_2O, CO, CO_2, acids, alcohols, and aldehydes can be found in the pyrolytic products of biomass as well as its extracted residue. Compared to raw biomass pyrolysis, the pyrolysis of extracted residue produces more CO_2 and less organics, and the release of products from extractives pyrolysis is much weaker.

Fig. 5.21 shows the comparison between the main pyrolytic products of raw MSP biomass and the main products of its extracted residue. In comparison with the pyrolysis of MSP, many of the products of the extracted residue pyrolysis are released earlier. Additionally, more H_2O, CO, and CO_2 are produced, while the yield of acids decreases sharply. The presence of extractives promotes the formation of acids and suppresses the production of CO and water.

With reference to the comparison between the pyrolytic products of the raw biomass, its extracted residue, and its extractives, we can investigate some effects on the composition and distribution of pyrolytic products. Compared to MSP, MA has a relatively high cellulose content, resulting in a higher bio-oil yield and a lower char yield from pyrolysis. This is in good agreement with the findings on the pyrolysis of individual components. The mixed biomass samples have a lower bio-oil yield and a higher small molecular gas yield compared to the extracted residue samples, while the char yields are almost the same (Fig. 5.22). One reason for this difference may be ascribed to the difference between the model compounds of the three major components and the natural components in biomass. Another reason for these differences may be the poor thermal stability of the mixed biomass samples due to the lack of natural crosslinked structures between components. As a result, the samples decompose completely and produce more small molecular gases. Couhert et al. [29] confirmed that the difference was primarily caused by the lack of natural crosslinked structures in the mixed sam-

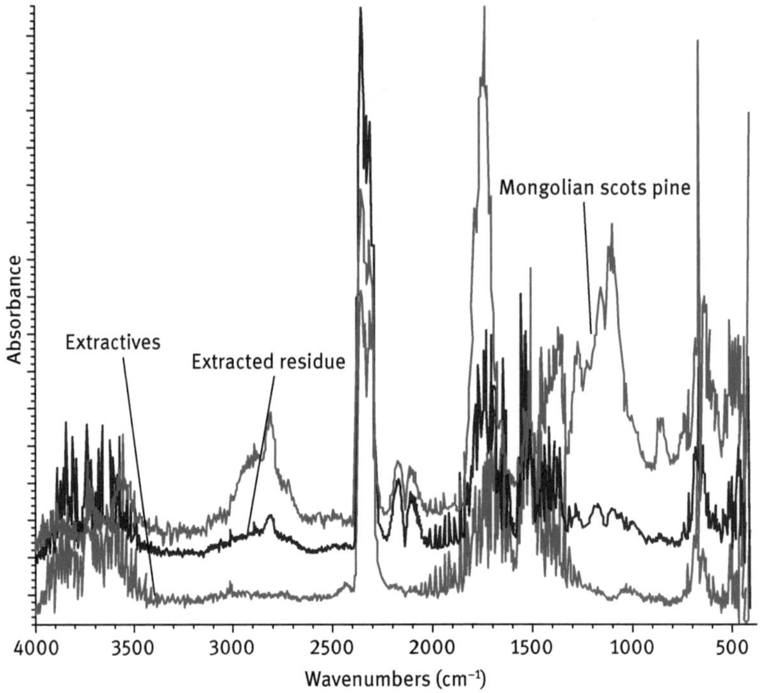

Fig. 5.20: FTIR spectra of the pyrolytic products of Mongolian Scots pine, extractives, and extracted residue at the maximum absorbance intensity. Reprinted with permission from [28].

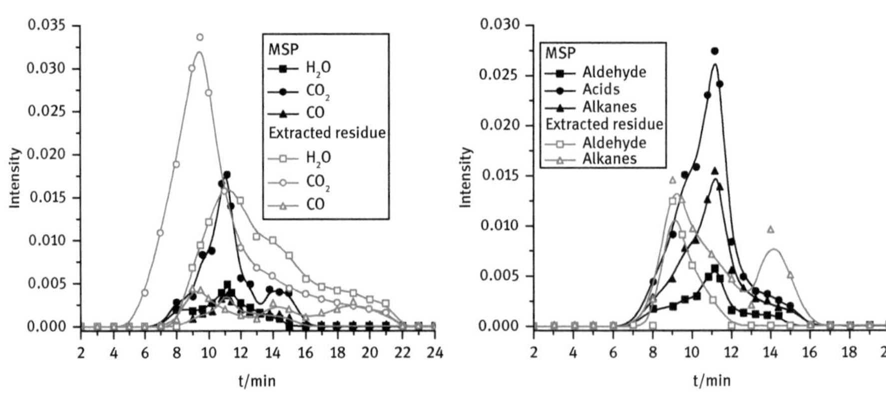

Fig. 5.21: Release of the main products of the pyrolysis of Mongolian Scots pine and its extracted residue. Reprinted with permission from [28].

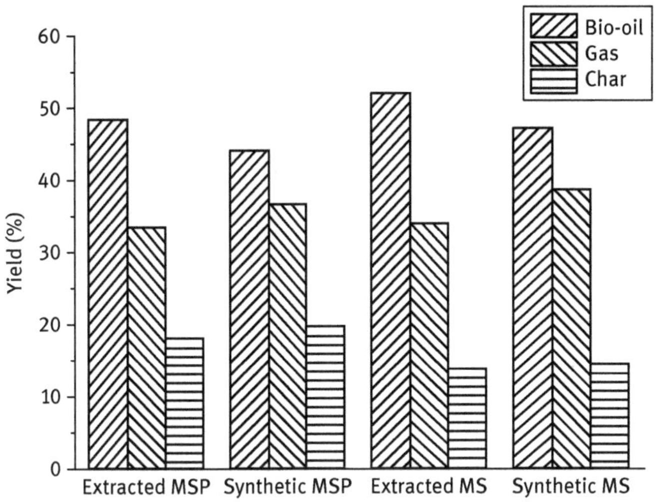

Fig. 5.22: Distributions of the products of the pyrolysis of extracted residues and mixed biomass samples.

ples. The compositions of bio-oils from the pyrolysis of MSP, the extracted residue and the mixed MSP sample are also different, especially for the typical organics. Compared to the extracted residue, the pyrolysis of the mixed sample produces more LG and hydroxyacetone, and less acetic acid, furfural, 2,5-diethoxytetrahydrofuran and phenols. As mentioned above, the presence of hemicellulose promotes the decomposition of cellulose to form 2,5-diethoxytetrahydrofuran and suppresses the formation of LG. Therefore, this influence of hemicellulose is more obvious in the extracted residue, while this effect is weakened in the mixed samples due to damage to the original structure. Acetic acid and furfural, mainly from hemicellulose pyrolysis, are less abundant in the pyrolytic product of the mixed samples than in the product of the extracted residue, suggesting that the natural crosslinked structures in the components have a significant influence on hemicellulose pyrolysis. Compared to the extracted residue, MSP has a higher LG content in its bio-oil, but less acetic acid, furfural, and hydroxyacetone, and only a small amount of 2,5-diethoxytetrahydrofuran. Consequently, the influence of extractives on bio-oil composition should not be ignored.

Biomass pyrolysis is an extremely complex process, which is influenced largely by its major components. In addition, the coupling interactions of components also influence biomass pyrolysis behaviors. The coupling interactions of the three major components, the inherent crosslinked structures between components, as well as extractives and acid non-soluble ash have some influence on the composition of biomass pyrolysis products. Therefore, the coupling interactions between components should not be ignored and biomass pyrolysis behaviors cannot simply be obtained by a superposition of the corresponding pyrolysis behavior of its three main components.

References

[1] Alén R, Kuoppala E, Oesch P. Formation of the main degradation compound groups from wood and its components during pyrolysis. Journal of Analytical and Applied Pyrolysis. 1996;36(2):137–148.

[2] Svenson J, Pettersson JB, Davidsson KO. Fast pyrolysis of the main components of birch wood. Combustion Science and Technology. 2004;176(5–6):977–990.

[3] Yang H, Yan R, Chen H, et al. In-depth investigation of biomass pyrolysis based on three major components: hemicellulose, cellulose and lignin. Energy & Fuels. 2006;20(1):388–393.

[4] Biagini E, Barontini F, Tognotti L. Devolatilization of biomass fuels and biomass components studied by TG/FTIR technique. Industrial & Engineering Chemistry Research. 2006;45(13):4486–4493.

[5] Caballero JA, Conesa JA, Font R, et al. Pyrolysis kinetics of almond shells and olive stones considering their organic fractions. Journal of Analytical and Applied Pyrolysis. 1997;42(2):159–175.

[6] Worasuwannarak N, Sonobe T, Tanthapanichakoon W. Pyrolysis behaviors of rice straw, rice husk, and corncob by TG-MS technique. Journal of Analytical and Applied Pyrolysis. 2007;78(2):265–271.

[7] Hosoya T, Kawamoto H, Saka S. Pyrolysis behaviors of wood and its constituent polymers at gasification temperature. Journal of Analytical and Applied Pyrolysis. 2007;78(2):328–336.

[8] Huber GW, Iborra S, Corma A. Synthesis of transportation fuels from biomass: chemistry, catalysts, and engineering. Chemical Reviews. 2006;106(9):4044–4098.

[9] Liu Q, Zhong Z, Wang S, et al. Interactions of biomass components during pyrolysis: A TG-FTIR study. Journal of Analytical and Applied Pyrolysis. 2011;90:213–218.

[10] Li S, Lyons-Hart J, Banyasz J, et al. Real-time evolved gas analysis by FTIR method: an experimental study of cellulose pyrolysis. Fuel. 2001;80(12):1809–1817.

[11] Yang C, Lu X, Lin W, et al. TG-FTIR study on corn straw pyrolysis-influence of minerals. Chemical Research in Chinese Universities. 2006;22(4):524–532.

[12] Piskorz J, Radlein D, Scott DS. On the mechanism of the rapid pyrolysis of cellulose. Journal of Analytical and Applied Pyrolysis. 1986;9(2):121–137.

[13] Hosoya T, Kawamoto H, Saka S. Solid/liquid-and vapor-phase interactions between cellulose- and lignin-derived pyrolysis products. Journal of Analytical and Applied Pyrolysis. 2009;85(1):237–246.

[14] Haensel T, Comouth A, Lorenz P, et al. Pyrolysis of cellulose and lignin. Applied Surface Science. 2009;255(18):8183–8189.

[15] Xie H, Yu Q, Qin Q, et al. Study on pyrolysis characteristics and kinetics of biomass and its components. Journal of Renewable and Sustainable Energy. 2013;5(1):13122.

[16] Zhang J, Choi YS, Brown RC. Cellulose-hemicellulose, cellulose-lignin interactions during fast pyrolysis: 2013 AIChE Annual Meeting, San Francisco, CA. 2013.

[17] Liu Q, Wang S, Wang K, et al. Pyrolysis of wood species based on the compositional analysis. Korean Journal of Chemical Engineering. 2009;26:548–553.

[18] Giudicianni P, Cardone G, Ragucci R. Cellulose, hemicellulose and lignin slow steam pyrolysis: Thermal decomposition of biomass components mixtures. Journal of Analytical and Applied Pyrolysis. 2013;100:213–222.

[19] Wang S, Guo X, Wang K, et al. Influence of the interaction of components on the pyrolysis behavior of biomass. Journal of Analytical and Applied Pyrolysis. 2011;91:183–189.

[20] Zhao K, Xiao J, Shen L, et al. Experimental study of biomass rapid pyrolysis based on three components. Acta Energiae Solaris Sinica. 2011;32(5):710–716.

[21] Caballero JA, Font R, Marcilla A. Comparative study of the pyrolysis of almond shells and their fractions, holocellulose and lignin. Product yields and kinetics. Thermochimica Acta. 1996;276:57–77.
[22] Liu L, Sun J, Cai C, et al. Corn stover pretreatment by inorganic salts and its effects on hemicellulose and cellulose degradation. Bioresource Technology. 2009;100(23):5865–5871.
[23] Demirbaş A. Relationships between lignin contents and heating values of biomass. Energy Conversion and Management. 2001;42(2):183–188.
[24] Raveendran K, Ganesh A, Khilar KC. Pyrolysis characteristics of biomass and biomass components. Fuel. 1996;75(8):987–998.
[25] Thammasouk K, Tandjo D, Penner MH. Influence of extractives on the analysis of herbaceous biomass. Journal of Agricultural and Food Chemistry. 1997;45(2):437–443.
[26] Tamaki Y, Mazza G. Measurement of structural carbohydrates, lignins, and micro-components of straw and shives: effects of extractives, particle size and crop species. Industrial Crops and Products. 2010;31(3):534–541.
[27] Wang Y, Wu L, Wang C, et al. Investigating the influence of extractives on the oil yield and alkane production obtained from three kinds of biomass via deoxy-liquefaction. Bioresource Technology. 2011;102(14):7190–7195.
[28] Guo X, Wang S, Wang K, et al. Influence mechanism of the extractives on biomass pyrolysis. Journal of Fuel Chemistry and Technology. 2010;38(1):42–46.
[29] Couhert C, Commandre J, Salvador S. Is it possible to predict gas yields of any biomass after rapid pyrolysis at high temperature from its composition in cellulose, hemicellulose and lignin? Fuel. 2009;88(3):408–417.

6 Catalytic pyrolysis of biomass components

Biomass resources, mainly wood chips and straw, can be converted to easily stored and transported bio-oil with higher energy density through fast pyrolysis technology. However, crude bio-oil has the disadvantages of high water and oxygen content, low heating and PH values, poor stability, etc., and is thus often used as a substitute fuel for combustion. To substitute for gasoline and diesel, further upgrading is required to improve bio-oil quality as a drop-in fuel. The concept of distributed biomass pyrolysis combined with centralized bio-oil upgrading seems more reasonable for the application of biomass pyrolysis technology. However, due to its complicated composition, the catalytic upgrading of bio-oil is difficult. The ready accumulation of coke deposits on the catalyst during the upgrading process leads to the rapid deactivation of the catalyst. If bio-oil quality can be improved in the pyrolysis process, the subsequent upgrading process could be simplified. In addition, it has been found that the distribution of products in biomass pyrolysis can be selectively controlled by adjusting the metal salts content or using some specific catalysts. Through catalytic pyrolysis, crude bio-oil quality can be improved by promoting the formation of target products, and suppressing the generation of undesired ones, especially acids.

This chapter mainly describes the influence of inorganic salts and the addition of catalysts on the pyrolysis of biomass components. Their influences on biomass pyrolysis will be presented in the next chapter. Cellulose has been chosen as the typical sample for its good consistency in structure across different biomass.

6.1 Influence of inorganic salts on the pyrolysis of biomass components

Biomass is mainly composed of carbon, hydrogen, and oxygen. It also contains a small amount of nitrogen, sulfur, and trace amounts of some metal elements, including K, Ca, Na, Mg, Al, Fe, Cu, etc. The most abundant inorganic salts in biomass are alkali salts and alkali earth metal salts, especially potassium and sodium salts, mainly in the forms of oxide, metasilicate, carbonate, chloride, phosphate, and so on. The types and amounts of inorganic salts significantly influence the pyrolysis behaviors of biomass and its components. By washing off the inherent inorganic salts in biomass or adding extra ones, the distribution of inorganic salts can be changed effectively, thus affecting the distribution of the products of biomass pyrolysis.

DOI 10.1515/9783110369632-008

6.1.1 Influence of inorganic salts on the kinetics of biomass components pyrolysis

Metal ion exchange and the soaking absorption method can be used to introduce metal ions into the materials. The former process is complex, costly, and therefore less feasible than the latter. Soaking materials in metal salt solutions with different concentrations can introduce metal salts to them. First, cellulose is soaked in the metal salt for 12 h, air dried, and then further dried at 60 °C. The amount of metal in the cellulose can be measured by atomic absorption spectroscopy.

To study the effect of metal salt addition on cellulose pyrolysis, KCl, $CaCl_2$, and $FeCl_2$ were selected as typical representatives of alkali salts, alkaline earth salts and transition metal salts, respectively. As shown in Fig. 6.1, after the addition of the three metal salts, the shoulder peaks on the TG curves of cellulose pyrolysis appeared at a lower temperature. In general, the initial steady weight loss stage can be attributed to the depolymerization of cellulose and the "glass transition" [1], followed by the breakage of intramolecular or intermolecular hydrogen bonds. Carbonyl groups were formed, and some small molecular gases and water were released. Active cellulose was also generated in this stage, resulting in the shoulder peak on the TG curves. The earlier shoulder peaks and the increase in the reaction rate indicate that the inorganic salts catalyzed the reaction, increasing the generation rate of active cellulose. At the main reaction stage, the reaction rate decreased slightly, while the char yield increased after the addition of inorganic salts. In comparison with the addition of K^+, the shoulder peaks on the TG curves appeared earlier with the addition of Ca^{2+}, and the weight loss range was wider. The DTG curves exhibited two weight loss peaks. This indicated that Ca^{2+} had a strong influence on the initial stage of cellulose pyrolysis, and promoted char formation. Shimada et al. [2] studied the effect of Na, K, Ca, and Mg on cellulose pyrolysis, and proved that the alkali earth metals had a stronger catalytic effect than the alkali metals. The catalytic effect of Fe^{2+} is milder compared to that of $CaCl_2$, and Fe^{2+} had little effect on the char yield. Khelfa et al. [3] also investigated the catalytic effect of transition metal ions (Zn^{2+} and Ni^{2+}) on cellulose pyrolysis, and found that they showed different effects. This was different from the effect of alkali/alkaline earth metals, which usually exhibited consistent performances.

By analyzing the TG curves of cellulose pyrolysis at different heating rates with the addition of inorganic salts, its apparent kinetic parameters can be obtained. As listed in Tab. 6.1, E_0 is the apparent activation energy for the formation of active cellulose from cellulose, and E_1 and E_2 are the apparent activation energies of tar and char formation from active cellulose, respectively. At a low pyrolysis temperature (< 330 °C), the weight loss rate was mainly controlled by active cellulose formation. When the pyrolysis temperature was higher than 330 °C, the active cellulose formation rate increased significantly, and was higher than its consumption rate. Thus, the weight loss process was controlled by the consumption rate of active cellulose leading to the production of tar and small molecular gases. The continuous reaction phenomenon was more obvious after $CaCl_2$ addition, where E_0 decreased from 266 to 135 kJ/mol. With

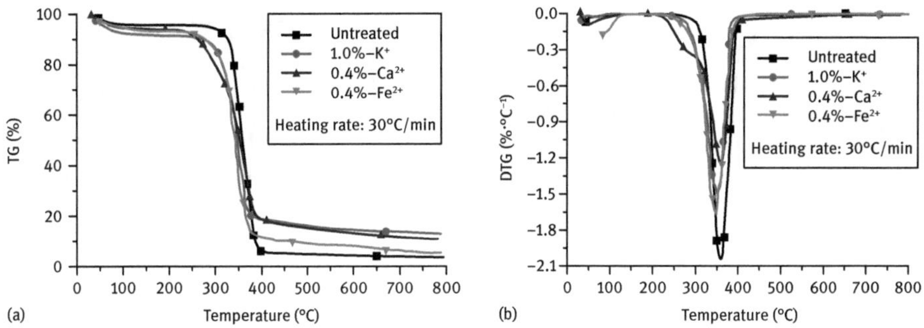

Fig. 6.1: The influence of inorganic salts on the TG and DTG curves of cellulose pyrolysis. Reprinted with permission from [4], © 2007 Springer.

Tab. 6.1: Apparent activation energy of each step in cellulose pyrolysis.

Reaction conditions	E_0 (kJ/mol)	E_1 (kJ/mol)	E_2 (kJ/mol)
Pure cellulose	266.0		186.6
1 % K^+ as catalyst		98.9	136.7
0.4 % Ca^{2+} as catalyst	135.0	85.5	154.1
0.4 % Fe^{2+} as catalyst			160.8
Bradbury et al. [1]	242.8	150.7	198.0

Note: The blank means the corresponding data could not be obtained through kinetic modeling.

the addition of KCl and $FeCl_2$, consecutive reaction features were not obvious from the calculation of apparent kinetic parameters. Furthermore, the shift of shoulder peaks to the lower temperature and the increase in weight loss revealed that the two inorganic salts catalyzed active cellulose formation, and consequently, active cellulose consumption became the dominant step. For the addition of alkali/alkaline earth metals, the apparent activation energy of the main weight loss stage constituted a turnaround phenomenon, presenting in a parallel reaction mode. For cellulose pyrolysis, the apparent activation energy of char formation was generally considered to be low, and the reaction usually took place at 130–260 °C [1]. However, the char formation rate did not decrease at a high temperature, it was just lower than the rate of volatiles formation, which competed with char formation. Therefore, char was mainly produced at a low temperature. After the addition of KCl and $CaCl_2$, char formation increased rapidly, leading to an increase of char yield at a lower reaction temperature. $FeCl_2$ had little influence on char formation, but enhanced volatiles formation during cellulose pyrolysis.

By comparing the different TG curves of cellulose pyrolysis after the addition of potassium salts with different concentrations (Fig. 6.2), it was found that as the K^+ concentration increased, the TG curves underwent an obvious change. It was also found

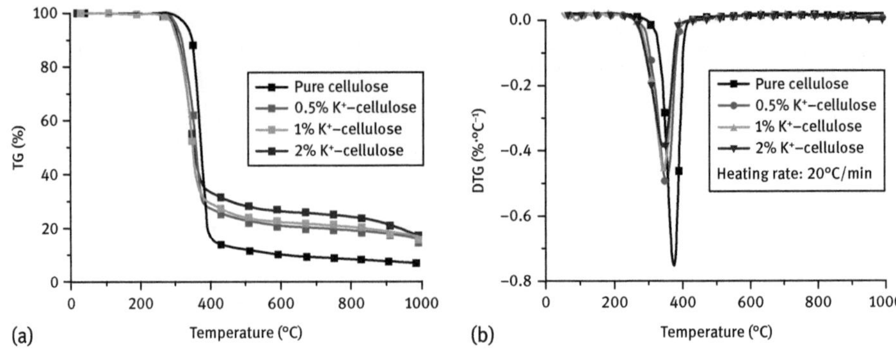

Fig. 6.2: The influence of potassium concentrations on the TG and DTG curves for cellulose pyrolysis. Reprinted with permission from [6], The Society of Chemical Engineers, Japan.

that the temperature corresponding to the initial weight loss and the maximum weight loss peaks decreased, as did the maximum weight loss rate, while the char yield obviously increased. Meanwhile, on-line FTIR analysis showed that the amounts of the small molecular gases (CO, CO_2, and H_2O) largely increased, and the small molecular gases were released at lower temperatures. However, some researchers also found that the composition of the gas products of cellulose pyrolysis did not change much with the addition of K^+ [5].

Similar to cellulose pyrolysis, K^+ lowered the temperature corresponding to the initial weight loss and the maximum weight loss peaks in the pyrolysis of xylan and lignin. The final char yield increased to a smaller extent compared with cellulose pyrolysis (Fig. 6.3). This is consistent with other results [5, 7]. K^+ had an intensive effect on xylan pyrolysis, leading to the presence of two weight loss peaks in the main reaction stage. The strength of the first peak was enhanced obviously, while that of the second peak remained almost unchanged. This indicates that K^+ had a strong catalytic effect on xylan decomposition at a low temperature.

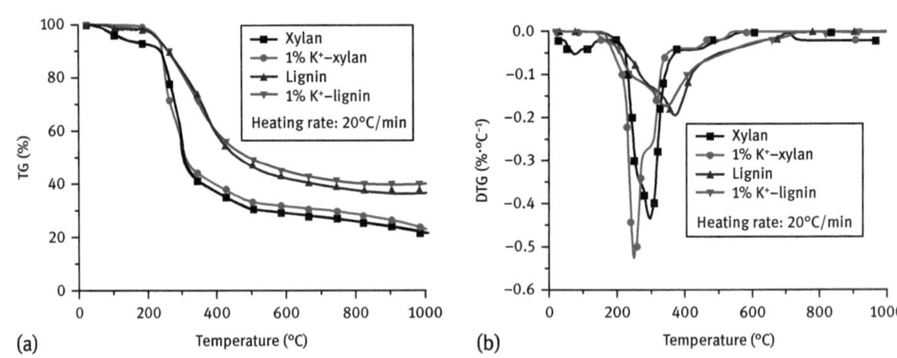

Fig. 6.3: Effects of potassium concentration on the TG and DTG curves for xylan and lignin pyrolysis.

Inorganic salts usually exist in the form of oxide, silicate, carbonate, chloride, and phosphate, and the effect of acid ions cannot be neglected. Fig. 6.4 displays the effects of the addition of different potassium salts on cellulose pyrolysis, and it was found that different types of acid ions had different influences on cellulose pyrolysis. For the different potassium salts, the strength of the promotion effect on char formation and the suppression effect on volatiles production decreased in the order of $K_2CO_3 > KCl > K_2SO_4$. Julien et al. [8] reported that the introduction of anions (SO_4^{2-} or Cl^-) to the cellulose surface increased the bio-oil yield, while decreasing the yields of char and small molecular gases. The introduction of SO_4^{2-} increased levoglucosan selectivity, and Cl^- increased glycolaldehyde selectivity. The results were similar to those obtained from the pyrolysis of biomass pretreated with H_2SO_4 and HCl by Piskorz et al. [9].

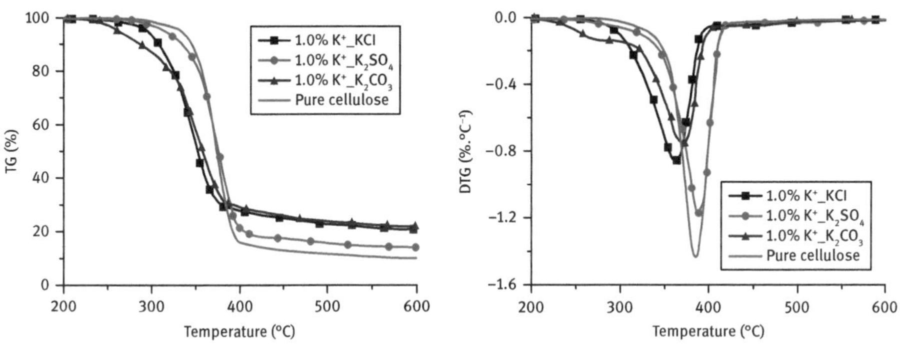

(a) (b)

Fig. 6.4: Effects of potassium salts with different acid ions on the TG and DTG curves for cellulose pyrolysis. Reprinted with permission from [6], The Society of Chemical Engineers, Japan.

6.1.2 Influence of inorganic salts on the distribution of cellulose pyrolysis products

The infrared radiation heating device mentioned above can also be used to study the effects of the addition of inorganic salts on the distribution of products of cellulose pyrolysis. As shown in Fig. 6.5, as the amount of K^+ rose, the bio-oil yield decreased and the char production increased, while the amount of small molecular gases remained almost unchanged. The quantity of oxygen-containing products in the bio-oil decreased and the water content increased, which indicated that K^+ promoted the dehydration reaction and the conversion of intermediates to char. This was identical to the reports of other researchers [10, 11]. In fact, it was not only K^+; other alkali/alkaline earth metals also promoted char formation to a certain extent [9, 12]. Potassium salt did not change the composition species, but did alter their relative content in bio-oil. K^+ greatly increased the acetone and glycolaldehyde content, and significantly decreased the levoglucosan content. The trend became a little stronger as more K^+ was added. This revealed that the addition of even small amounts of K^+ would affect the products' selectivity. K^+ also affected the formation of other products. After pyrolysis, most of the added K^+ was left on the char surface. To introduce K^+ to cellulose, the cellulose is soaked in potassium salt solution and a small amount of K^+ is adsorbed on the carboxyl and other functional groups, forming mobile organic K^+. During pyrolysis, this K^+ might enter into the volatile phase and could be detected in bio-oil [13]. Thus, the pyrolysis reaction was mainly catalyzed by K^+ on the surface of or inside the material, rather than in the vapor phase.

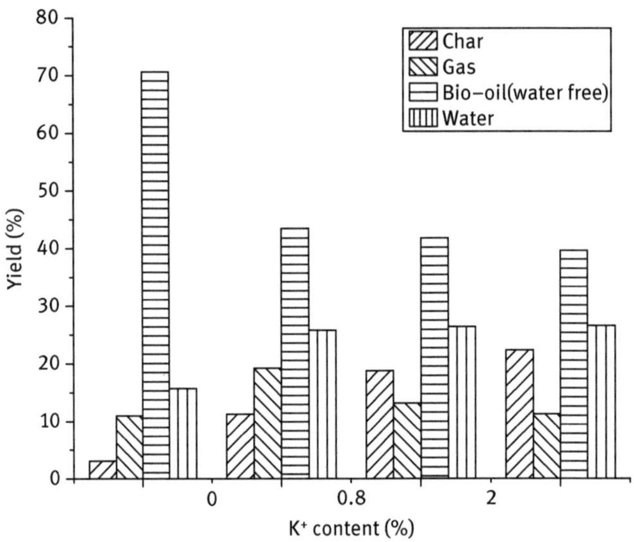

Fig. 6.5: Effect of K^+ content on the distribution of cellulose pyrolysis products.

The addition of Ca^{2+} has a similar effect on the distribution of products of cellulose pyrolysis to K^+ addition. After the introduction of Ca^{2+}, the production of char and small molecular gases increased, while bio-oil production decreased, as illustrated in Fig. 6.6. When added in the same quantity, Ca^{2+} had a greater promoting effect on char formation and the dehydration reaction than K^+, leading to a high water content of over 30 % in the bio-oil. The high water content of the bio-oil is not only attributed to the promoting effect of Ca^{2+} on the dehydration reaction, but is also probably due to the presence of some small amount of $CaCl_2$ in cellulose after pretreatment. $CaCl_2$ has strong moisture absorbability, and it is difficult to remove water efficiently during the drying process, leading to some water being left in the resulting bio-oil. Unlike K^+, Ca^{2+} suppressed the formation of glycolaldehyde and levoglucosan simultaneously, while promoting the formation of 5-hydroxymethylfurfural, acetaldehyde, acetone and so forth. This trend became stronger as more Ca^{2+} was added. It showed that Ca^{2+} had

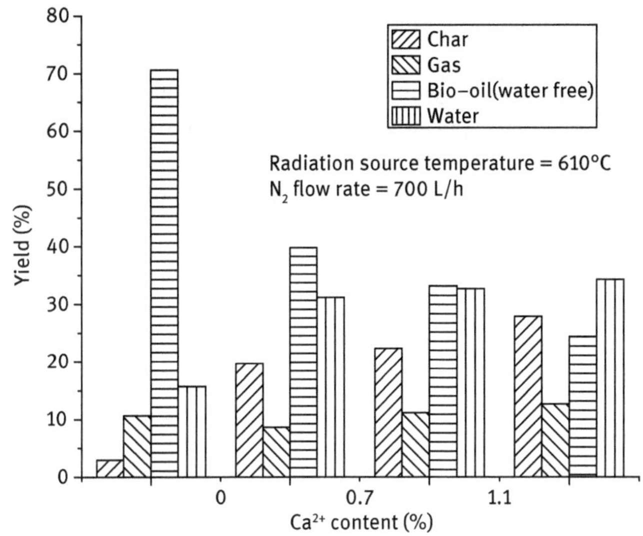

Fig. 6.6: Effect of Ca^{2+} content on the distribution of cellulose pyrolysis products.

a significant effect on cellulose pyrolysis. It promoted the ring-opening and dehydration of levoglucosan, followed by further cyclization to form a variety of furanic compounds.

Williams et al. [14] proposed that not only metal ions, but also the acidity or basicity of metal salts would affect cellulose pyrolysis. The acid catalysts facilitate the dehydration reaction to enhance the generation of levoglucosan, furans, and char. The basic catalysts are conducive to the dissociative and disproportionation reactions, promoting the production of glycolaldehyde, acetaldehyde, some small molecular carbonyl compounds, and char. However, our study revealed that the catalytic effect of inorganic salts cannot be explained on the basis of their acidity and basicity. The addition of alkali or alkaline earth metal salts promoted water and char formation, and suppressed levoglucosan formation in different ways. The addition of K^+ increased the amount of small molecular products in the bio-oil, while Ca^{2+} addition increased the amount of furans and pyrans.

In contrast, transition metal salts have a milder effect on cellulose pyrolysis. With the addition of $FeCl_2$, the levoglucosan yield decreased, the char yield increased slightly, and the small molecular volatiles yield increased. Among the metal ions, the effect of Fe^{3+} and Zn^{2+} on cellulose pyrolysis is relatively strong. They suppressed the formation of levoglucosan, glycolaldehyde, hydroxy-acetone, and other small molecular aldehydes and ketones, but facilitated the generation of furan compounds (such as furfural, 5-hydroxymethylfurfural, etc.), together with carboxylic acid compounds (such as formic acid, acetic acid, etc.) [15, 16]. Especially the addition of Zn^{2+} greatly increased furfural selectivity. The higher loading amount of Zn^{2+} benefited the cellulose dehydration reaction more significantly, resulting in a higher furfural yield [17].

Phosphorus is an essential element for plant growth, which also has an effect on the pyrolysis of biomass components, especially on cellulose pyrolysis, while its effect on lignin pyrolysis is weak. The addition of phosphate increased char production from biomass and cellulose pyrolysis, and improved the production of furfural and levoglucosan [18].

6.2 Catalytic effect of zeolite catalysts on the pyrolysis of biomass components

6.2.1 Classification and characteristics of zeolite catalysts

Zeolites are crystalline silicates or silica-alumina minerals, whose chemical compositions can be expressed as: $M_{2/n}O \bullet Al_2O_3 \bullet xSiO_2 \bullet yH_2O$, where M represents the metal ion, n represents the metal ion valence, and x and y represent the number of SiO_2 and H_2O, respectively. The SiO_2/Al_2O_3 molar ratio (x) is an important parameter for zeolite. The metal ions, including Na, K, Ca, etc., usually exist in the different types of molecular sieves synthetized by ion exchange method. Zeolites are composed of silicon-oxygen tetrahedrons or aluminum-oxygen tetrahedrons connected through oxygen-bridge bonds, and possess numerous equal-sized cavities connected by many pores of the same diameter, forming homogeneous microchannel structures. Zeolite has the ability to selectively sort molecules based on its channel sizes, hence it is called the "molecular sieve". The main differences between various kinds of molecular sieves are their chemical compositions and channel structures, and the former difference is reflected in the different molar ratio of Si to Al. By adjusting the molar ratio of Si to Al, the acidity of zeolites can be changed. Strong acid catalysts can accelerate the decarboxylation and dehydration of oxygen-containing products during biomass pyrolysis, forming small molecular hydrocarbons and unsaturated hydrocarbons, which can then undergo aromatization and polymerization to form aromatics and char, respectively. This results in a decrease in tar content and water, and an increase in the polycyclic aromatic hydrocarbon content in bio-oil [19, 20].

6.2.1.1 Catalytic performance of zeolite

Natural zeolites were originally used as adsorbents and desiccants, but then they were gradually developed as catalysts in industry and synthetized by hydrothermal synthetic method. As catalysts, zeolites have some unique advantages in promoting reactions [21].

The basic structure unit of zeolite is also called a cage and can be arranged to form different zeolite skeletons. The specific structure of zeolite enables it to adsorb many compounds, acting as a catalyst and a support [22]. Zeolites have a unique molecular shape selectivity, favoring specific reactants and products. In the petrochemical industry, zeolite is used as a catalyst to increase isoparaffin content in gasoline during its reforming process. For bio-oil production via biomass pyrolysis, zeolite promotes hydrocarbon content and improves bio-oil quality. Lu et al. [23] studied the influence of several zeolites, including HUSY, REY, HZSM-5, MLC (heavy oil catalytic cracking catalyst), and CIP (distillate bio-oil catalytic cracking catalyst), on the removal of oxygenated chemicals from wheat straw pyrolysis products and their selectivity for high octane components. The deoxygenation activity of the catalysts decreased in this order: REY = HUSY > HZSM-5, and the high octane component selectivity of the catalysts decreased in this order: REY > HUSY = HZSM-5. MLC and CIP were also highly active in deoxygenation.

In zeolites, the valence states of silicon and aluminum atoms are different, leading to an unbalanced charge, which can be utilized to prepare highly efficient composite catalysts to improve the reaction rate by the metal exchange method, such as introducing Ni^{2+}, Pt^{2+}, or Pd^{2+}, and forming highly dispersed metal species. Furthermore, acid and alkaline properties are the critical features of solid catalysts. Zeolite catalysts display excellent acid catalytic activity, which derives either from H proton exchange or from the hydrolysis of multivalent cations during dehydration. The number and the strength of the formed proton acid sites have a great influence on catalytic activity. Since zeolites have functional balance properties between acid properties, molecular shape selectivity, limited coking, and high stability, they have become important catalysts for biomass catalytic pyrolysis and bio-oil catalytic cracking [24].

6.2.1.2 Classification of zeolite catalysts

Zeolites are porous materials with ordered and uniform channel structures. According to channel diameter, zeolites can be classified as microporous zeolite, mesoporous zeolite, and macroporous zeolite (Tab. 6.2).

The pore diameter of microporous zeolite is less than 2 nm, similar to the diameters of common molecules. It has good molecular shape selectivity, and can be used as an excellent absorbing material in separation-adsorption technology, as a good catalyst in oil refining, and as an ion exchange material in waste liquid processing. Acid of a certain strength spread on zeolites facilitates the removal of aldehyde and ketone in

Tab. 6.2: Classification of zeolites.

Zeolites	Pore diameter (nm)	Typical zeolites
Microporous zeolite	< 2	ZSM-5, MOR, zeolite Y, zeolite β
Mesoporous zeolite	2–50	MCM-41, SBA-15
Macroporous zeolite	> 50	TiO_2 macroporous materials

biomass pyrolysis through decarbonylation and dehydration, forming small molecular hydrocarbons and unsaturated alkenes. Typical zeolites are ZSM-5, zeolite Y, and zeolite β.

The mesoporous zeolites are developed on the basis of microporous zeolites. Microporous zeolite confines the reactant diameter to the nanoscale, and it is difficult for large molecules to enter into its pore channel. However, this condition changes in the mesoporous zeolites. In biomass catalytic pyrolysis, mesoporous zeolites allow aromatics and some linear macromolecules produced by the primary biomass pyrolysis to get into the zeolite channel. These products combine with the strong acid sites, and undergo further pyrolysis, decreasing the amount of oxygen-containing products in the bio-oil and improving the bio-oil's quality. Furthermore, mesoporous and microporous zeolites have different channel types. The mesoporous zeolites are composed of amorphous walls, and have lower thermal stability and hydrothermal stability compared with microporous zeolites [25].

The pore diameter of macroporous zeolites is commonly greater than 50 nm, larger than the molecule size. Therefore, compared with micro- or mesoporous zeolites, it has almost no sieving ability, and is usually referred to as macroporous material. At present, research on ordered macroporous materials is rare, and only a small number of synthesis methods are reported. The common synthesis mechanism has not been summarized [26, 27]. Due to the large diameter and small specific surface area, macroporous zeolites are not recommended for use as catalysts [26, 27].

At present, microporous and mesoporous zeolites, used as solid acid catalysts or catalyst supports, are studied and applied in biorefinery [28], especially in the improvement of bio-fuel quality. Due to their acidity, these catalysts show potential catalytic effects on cracking, decarboxylation, hydrodeoxygenation, etc. [29, 30].

6.2.2 Catalytic effect of microporous zeolite on the pyrolysis of biomass components

6.2.2.1 Catalytic effect of zeolite on cellulose pyrolysis

In general, catalysts, such as zeolites and metal oxides, can be added to biomass or biomass components by simple blending, or the catalysts can be loaded onto the pyrolysis reactor which the volatiles produced during pyrolysis pass through. Three microporous zeolites, HZSM-5, H-β, and USY (attributed to the types of ZSM-5, zeolite β, and zeolite Y, respectively), were chosen and mixed mechanically with the raw material at a certain proportion to study the catalytic effect of zeolites on cellulose pyrolysis.

As shown in Fig. 6.7, the zeolite catalyst was simply blended in an equal ratio with cellulose. TG results showed that the initial dehydration was promoted and that weight loss increased after the addition of the zeolite catalyst. Weight loss increased especially, from 5 % to 20 %, after the addition of the USY catalyst. However, the tem-

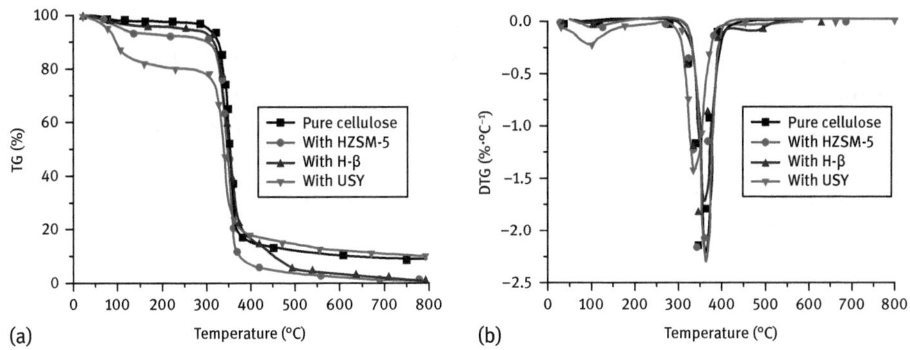

Fig. 6.7: TG and DTG curves of cellulose pyrolysis with added zeolites.

perature corresponding to the maximum weight loss peak remained at about 350 °C, and the maximum weight loss rate decreased. Moreover, the addition of HZSM-5 and H-β suppressed char formation, while promoting the formation of small molecular linear products. USY promoted dehydration at the initial stage, leading to lower weight loss at the main reaction stage.

Zeolites were not only shown to influence the thermal degradation process, but also to affect the distribution of products. As illustrated in Fig. 6.8, after the addition of microporous zeolites the amount of oxygen-containing products decreased obviously. By comparing the integrated peak areas, it was found that HZSM-5 decreased the peak intensity of aldehyde, acid, and esters by 49.7 %, 60.9 %, and 75.8 %, respectively. For USY, the three peak intensities decreased by 46.7 %, 25.6 %, and 54.0 %, respectively. With the addition of H-β, a different performance was displayed, with the peak in-

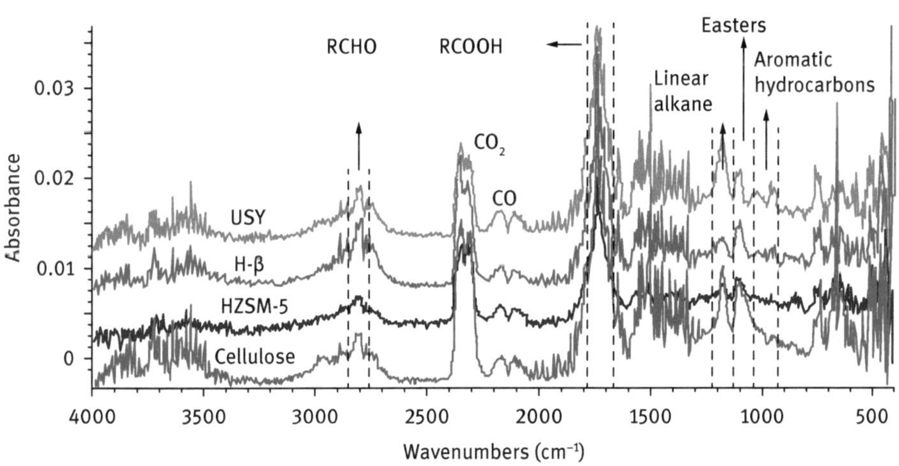

Fig. 6.8: Effect of zeolites on the release intensities of typical products of cellulose pyrolysis.

tensity of aldehyde and acid slightly increasing by 3.2 % and 5.9 %, and that of esters decreasing by 26.5 %. Furthermore, different microporous zeolites affected the linear alkane formation in different ways. USY promoted while HZSM-5 suppressed its formation. The presence of HZSM-5 decreased the amount of oxygen-containing products, while promoting the generation of aromatic hydrocarbons and alkenes, as has also been observed in other research [31]. The deoxygenation activity of the three microporous zeolites in cellulose pyrolysis decreased in the order of HZSM-5 > USY > H-β. At 350–450 °C, oxygen-containing products underwent a series of reactions including dehydration, decarboxylation, cracking, alkylation, condensation and polymerization, etc. Thus, macromolecular oxygen-containing products were decomposed into small molecular hydrcarbons, CO_2, CO, etc. [32, 33]. Fabbri et al. [34] found that the addition of NH_4-Y, H-Y, and NH_4-ZSM-5 decreased the anhydro-sugar content obviously. Du et al. [35] discovered that HZSM-5 promoted the production of aromatic hydrocarbons during cellulose pyrolysis, and that the yield increased remarkably with the ratio of catalyst to cellulose. Mihalcik et al. [36] studied the influence of HZSM-5 with different Si/Al ratios (23, 50, 280) on the formation of 15 major products of cellulose pyroloysis, and found that HZSM-5 with an Si/Al ratio of 23 was the most effective for producing aromatic hydrocarbons, and that HZSM-5 with an Si/Al ratio of 50 led to the lowest quantity of oxygen-containing componds in the bio-oil.

Zeolites have a large effect on cellulose pyrolysis. Experimental results on Py-GC/MS showed that HZM-5 greatly affected the distribution of products (see Fig. 6.9). With an increase in HZSM-5, the anhydro-sugars contents, including levoglucosan, levo-

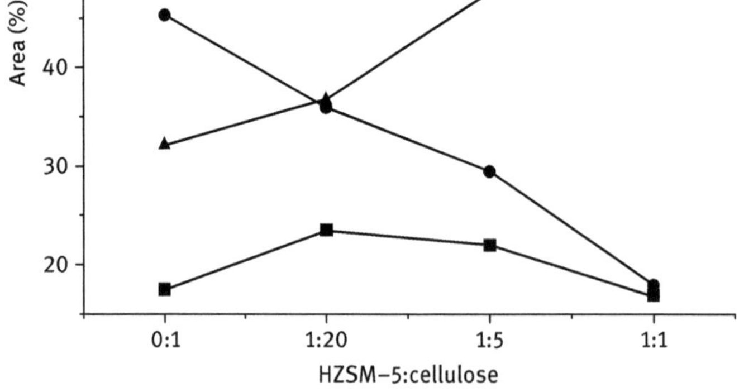

Fig. 6.9: Effect of zeolite addition on the distribution of cellulose pyrolysis products.

glucosenone and 1,4:3,6-Dianhydro-α-D-glucopyranose, as well as the furans content decreased, while the content of small molecular products, such as CO_2 and aromatic hydrocarbons, increased. CO_2 formation was mainly related to the secondary pyrolysis of anhydrocellulose and other volatiles, and the formation of aromatic hydrocarbons was related to the polymerization of alkenes or alkynes, or the polycondensation of C=O containing compounds [37, 38]. Thus, from the variation trends of small molecular products, it can be concluded that the strong acidity of HZSM-5 promoted the dehydration of cellulose and intermediates, forming different pyrolysis products.

The Si/Al ratio and channel structure are the main factors that influence the aromatic hydrocarbon content in cellulose pyrolysis products. Carlson et al. [39] compared the aromatic hydrocarbon content in bio-oils derived from the pyrolysis of cellulose, cellobiose and glucose over several zeolites. The highest aromatic hydrocarbon yield of 30 % was obtained from ZSM-5 (Si/Al = 60), followed by the yields from zeolite β and zeolite Y (Si/Al = 50), silica-zeolites, and the mixture of SiO_2 and Al_2O_3 (Si/Al = 8). The obtained aromatic hydrocarbons were mainly macromolecular compounds. Foster et al. [40] studied the catalytic pyrolysis of glucose, furan, and maple wood using semi-batch and fixed bed reactors, and optimized the product distribution of aromatic hydrocarbons across different types of ZSM-5. It was found that the acid sites had a decisive impact on product distribution. As a result of the selective removal of external acid sites, catalytic activity increased slightly, but the selectivity for aromatic hydrocarbons decreased. In view of the control of aromatic hydrocarbon and benzene content in gasoline, a breakthrough needs to be made in the alkylation of benzene and the hydrocracking of aromatic hydrocarbons to produce high quality liquid fuels from bio-oil [41].

6.2.2.2 Catalytic effect of zeolite on xylan pyrolysis

Similar to their performance in the catalytic pyrolysis of cellulose, some micropore zeolites also promote dehydration during xylan pyrolysis and increase weight loss during the initial stage (Fig. 6.10). The greatest weight loss was also obtained when using the USY catalyst. There was an obvious weight loss peak at 400–550 °C in the TG curve for pyrolysis when using the H-β catalyst. This may be because H-β promoted the catalytic cracking of xylan monomer. Some researchers found that 4-O-glucuronic acid with high thermal stability can be decomposed into small molecular products using certain catalysts [42]. In the final stage for char formation, char yields from the untreated xylan and the xylan with added USY were about 26 %. The char yields for xylan with added HZSM-5 and H-β were much lower, at about 16 %. Micropore zeolites have similar effects on xylan pyrolysis as they do on cellulose pyrolysis. USY promoted dehydration in the primary pyrolysis stage, and HZSM-5 and H-β influenced the char formation stage.

The influence of micropore zeolites on the distribution of products of xylan pyrolysis is displayed in Fig. 6.11. In comparison with the release intensities of pyrolytic

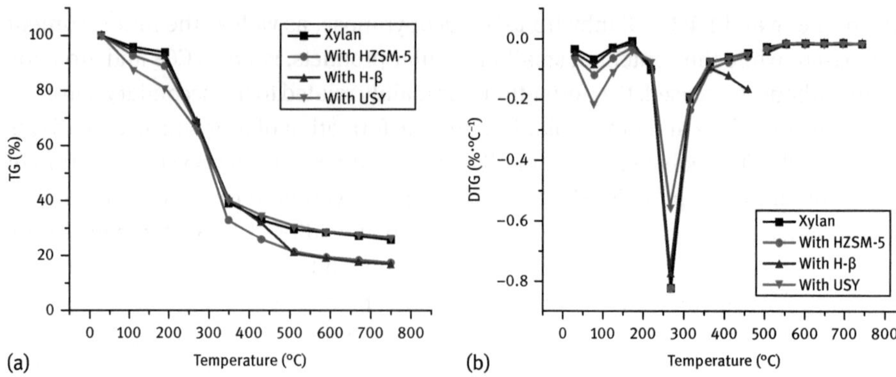

Fig. 6.10: Effect of zeolites on the TG and DTG curves for xylan pyrolysis.

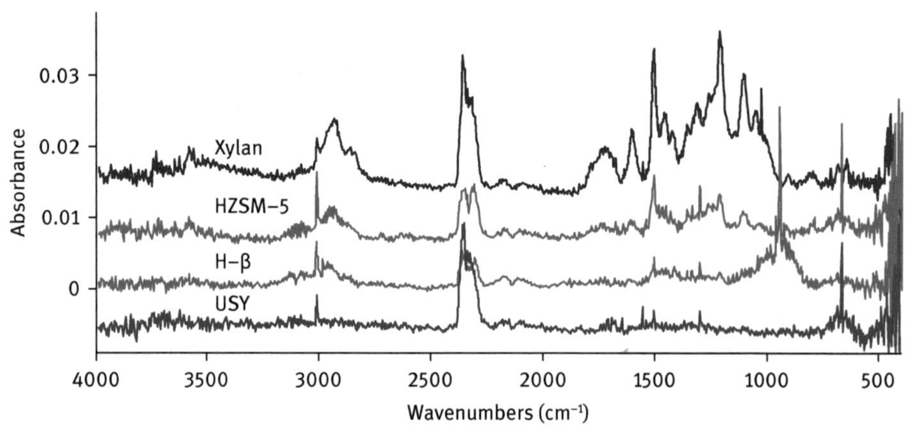

Fig. 6.11: Effect of zeolites on the release intensities for typical products of xylan pyrolysis.

products (oxygen-containing componds and small molecular linear products) from catalytic and non-catalytic pyrolysis, it can be inferred that H-β effectively promoted xylan decomposition into small molecular products, such as H_2O, CO_2, and alkanes. USY strongly promoted the deoxidization of oxygen-containing componds, and much more water was produced at the primary stage.

The influence of adding HZSM-5 on the distribution of the main products of xylan pyrolysis is displayed in Fig. 6.12. HZSM-5 promoted the decomposition of pyrans into aldehydes, and facilitated the deoxygenation of oxygen-containing componds, such as aldehydes and acids, to small molecular gases, such as CO_2. Compared to the catalytic performance of cellulose pyrolysis, the catalytic effect of HZSM-5 on xylan pyrolysis was weaker. This was related to the unstable chemical property of xylan, which started decomposing at a low temperature. Hence, the catalytic performance of HZSM-5 on xylan pyrolysis was not as obvious as on cellulose pyrolysis.

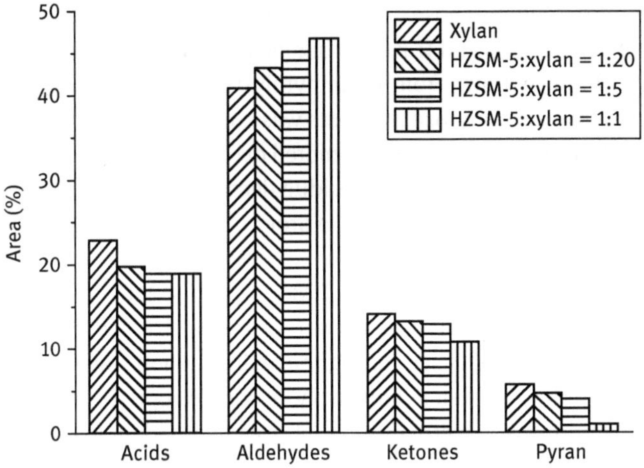

Fig. 6.12: Effect of HZSM-5 on the distribution of the main products of xylan pyrolysis.

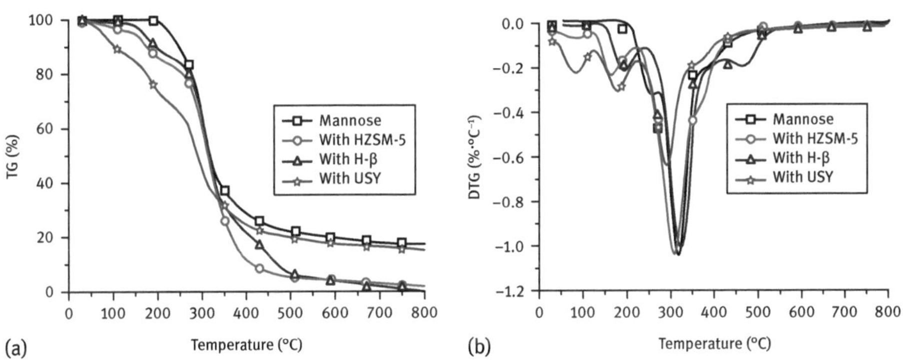

Fig. 6.13: Effect of zeolites on the TG and DTG curves for mannose pyrolysis. Reprinted with permission from [43], © 2013 Elsevier.

In addition to xylan, mannose was also selected as a model compound for hemicellulose to study the influence of catalysts on the pyrolysis behavior of hexose in hemicellulose. Fig. 6.13 displays the influence of zeolite on TG/DTG curves for mannose pyrolysis. Mannose was decomposed at 174–526 °C, and the highest weight loss rate was obtained at 302 °C. After zeolite addition, two dehydration peaks occurred at the early pyrolysis stage. The peak at 126 °C was ascribed to the evaporation of the physically absorbed water, and the peak in the range between 126 °C and 230 °C was attributed to the release of structural water [42]. The main reaction occurred at 230–342 °C, which was mainly attributed to mannose dehydration, along with the C–O and C–C breakage in the sugar ring, leading to the generation of CO, CO_2, and H_2O [44]. Similar to xylan pyrolysis, a weight loss peak at 406–510 °C appeared during H-β catalytic pyrolysis. In the char formation stage, untreated mannose showed a 17 % char yield, close to the char yield from USY (15 %). For HZSM-5 and H-β, the char yield decreased greatly. The conclusions agreed well with those drawn from xylan pyrolysis, and the influence of catalysts on TG/DTG curves for hemicellulose pyrolysis was consistent with their influence on the curves for cellulose pyrolysis. USY promoted deoxygenation and water formation, while HZSM-5 and H-β promoted solid residue decomposition. The dehydration activity of the three catalysts decreased in the order of USY > HZSM-5 > H-β. The char suppression activity was in the reverse order.

Three obvious weight loss peaks were observed at 228, 300, and 382 °C on the DTG curves for mannose pyrolysis. Large amounts of water were formed at 228 °C during the primary stage. The pyrolysis products released during the main reaction stage (about 300 °C) were similar to those obtained from xylan pyrolysis, including H_2O, CO, CO_2, formic acid, acetic acid, furfural, and 5-methylfuran. CO_2 was released at 382 °C, accompanied by the generation of a small amount of CO and acetic acid. Except for the generation of some ethylene from H-β at 350–500 °C, the products of catalytic and non-catalytic mannose pyrolysis were similar, albeit showing a certain difference in the intensity of the peaks.

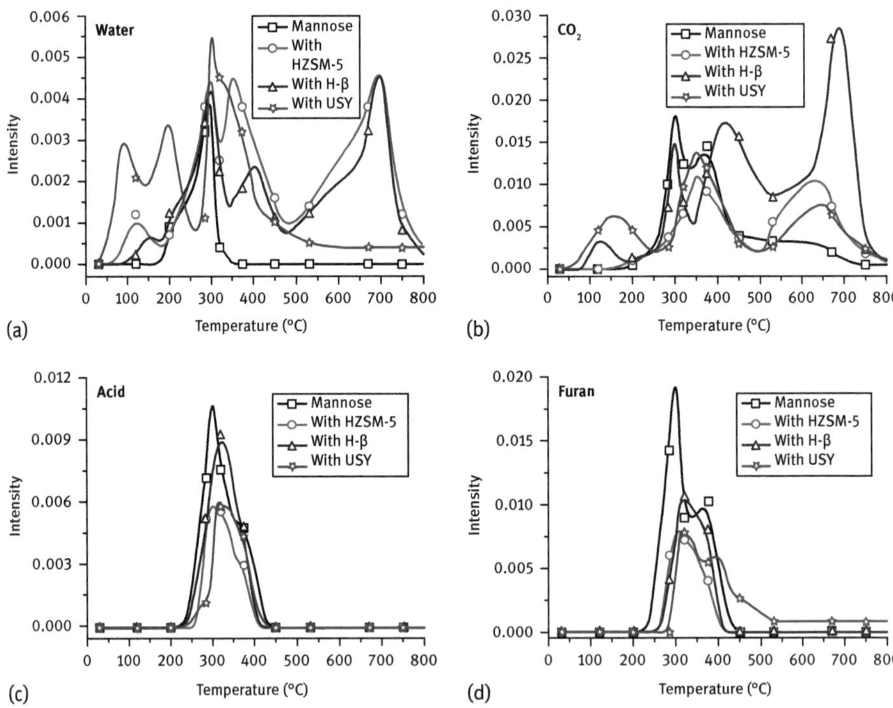

Fig. 6.14: The typical products of mannose catalytic pyrolysis. Reprinted with permission from [43], © 2013 Elsevier.

The typical products of mannose pyrolysis are shown in Fig. 6.14. The acid products include formic acid and acetic acid, and the furans include furfural and 5-methylfuran. The evolution of water and CO_2 from mannose pyrolysis was more complicated than from xylan pyrolysis. USY addition lowered the water release temperature, leading to a large amount of water forming at a lower temperature. HZSM-5 and H-β additions caused strong water release at 500–800 °C. In non-catalytic mannose pyrolysis, CO_2 release mainly occurred during the main reaction stage. With the addition of catalysts, the amount of CO_2 released during this stage decreased, while the amount released in the initial stage and the latter reaction stage obviously increased. The oxygen-containing products were mainly released at 200–450 °C, consistent with the results from xylan pyrolysis. With the addition of HZSM-5, H-β, and USY, the amount of oxygen-containing products decreased. This indicates that HZSM-5 effectively promoted mannose decomposition, and generated more water and less CO_2. Meanwhile, H-β led to a lower level of deoxygenation than HZSM-5, but promoted the decomposition of components at a high temperature, forming water, CO_2, and ethylene. The catalytic effect of USY was related to the deoxygenation activity, which mainly occurred at low temperature. For four selected hemicellulose monosaccharides, namely two pentoses (xylose and arabinose) and two hexoses (mannose and galactose), the effects of two different reaction processes by which the catalyst (HZSM-5) comes into contact with the substrate, namely, mixed with monosaccharide (in-bed) or layered above monosaccharide (in situ), were studied (Fig. 6.15) [45]. Under HZSM-5 catalytic conditions, the yield of aromatics was enhanced, and the corresponding volatilization temperature was lowered, especially for the in-bed process. Pentoses entered the zeolite pores more easily than hexoses because of their smaller molecular size; thus, the in-bed catalytic process drastically affected pentose pyrolysis.

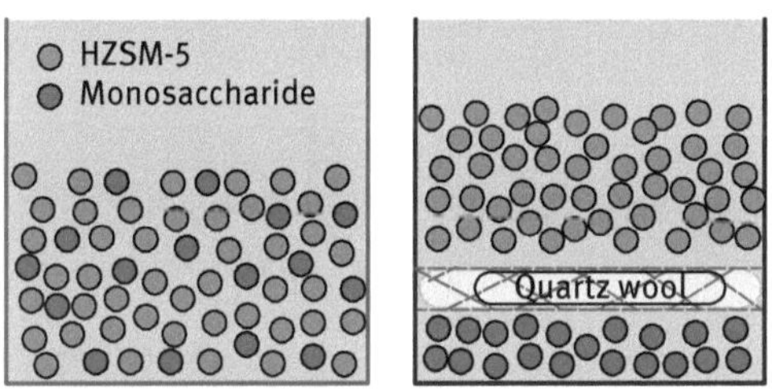

Fig. 6.15: Diagram of in-bed (A) and in situ (B) catalytic processes.

6.2.2.3 Catalytic effect of zeolite on lignin pyrolysis

Unlike the glycan structures of cellulose and hemicellulose, lignin is a complicated polymer connected by phenylpropanes through different branched chains. As shown in Fig. 6.16, micropore zeolites also promoted weight loss at the primary stage, and suppressed char formation. Among the three catalysts, USY influenced the primary stage of pyrolysis most obviously, and this corresponded with its influence on cellulose and xylan pyrolysis. HZSM-5 and H-β had little influence on lignin pyrolysis in

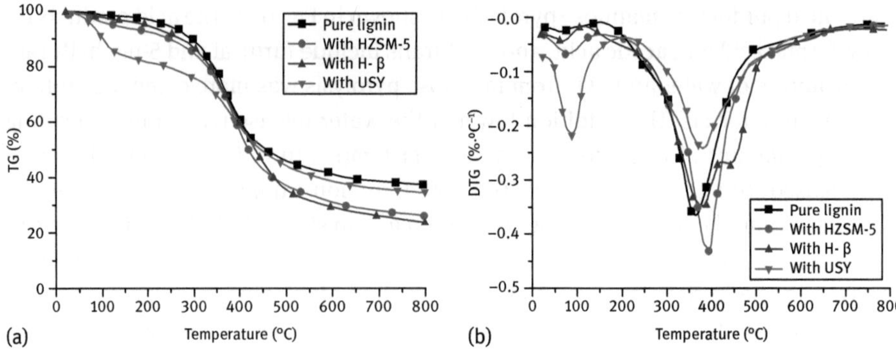

Fig. 6.16: The influence of zeolites on the TG and DTG curves for lignin pyrolysis. Reprinted with permission from [46], © 2012 Elsevier.

the primary stage, and they accelerated char decomposition, resulting in a char yield decrease from 40 % (for non-catalytic lignin pyrolysis) to 25 %.

The presence of micropore zeolites significantly suppressed the generation of phenols and methanol, and influenced the formation of small molecular gases. As displayed in Fig. 6.17, the addition of catalysts greatly changed product distribution. In lignin pyrolysis, the release of CO_2 occurred at three stages, namely the primary, the main, and the later reaction stage. USY addition made CO_2 release at lower temperatures during all the three stages, and promoted CO_2 release in the primary and the later stage, while suppressing it in the main reaction stage. H-β enhanced CO_2 release in the third stage, and made CO_2 release at higher temperatures in all the three stages. HZSM-5 promoted the release of CO_2 in the first two stages, but had little influence on the third stage. Thus, the influence of catalysts on CO_2 release occurred in differ-

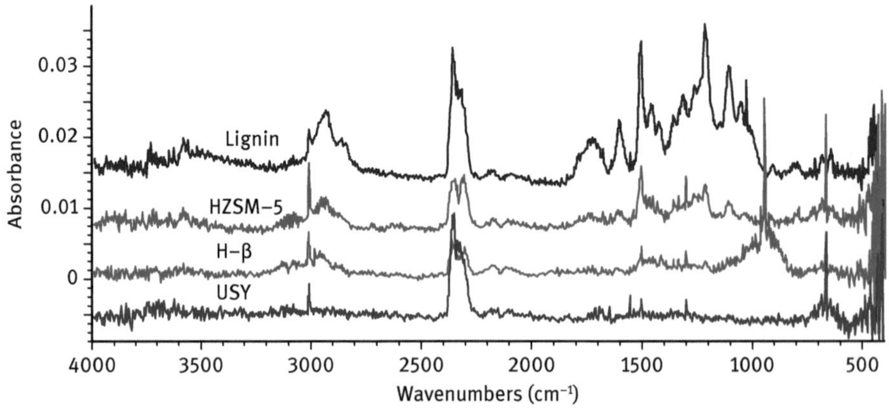

Fig. 6.17: Effect of zeolites on the release intensities of typical products of lignin pyrolysis Reprinted with permission from [46], © 2012 Elsevier.

ent stages, but the total amount of CO_2 released did not change much. Phenols are the main products of lignin pyrolysis, mainly including guaiacyl-type phenols and syringyl-type phenols. Micropore zeolites, especially H-β, suppressed phenol formation. When USY was used, no methanol was detected in the products, while HZSM-5 and H-β changed the methanol release temperature. After the addition of HZSM-5 and H-β, two release peaks were observed for methanol formation. Compared to cellulose and hemicellulose pyrolysis, lignin pyrolysis behavior was less influenced by zeolite addition. This may be attributed to the fact that the super-molecular structure of lignin is difficult to decompose, and the decomposed products are stable, which might block the catalyst channels and lead to a deactivation of catalysts.

Zeolites also have obvious effects on lignin pyrolysis products, especially the phenolic compounds and aromatic hydrocarbons. As shown in Fig. 6.18, Py-GC/MS results showed that the main pyrolysis products changed with the addition of HZSM-5. HZSM-5 promoted the generation of aromatic hydrocarbons, but suppressed the formation of phenolic compounds. This was consistent with results from previous research [47]. With an increase in HZSM-5, the aromatic hydrocarbon yield increased, while the yield of alkylphenol, guaiacyl, and syringyl phenols decreased. Li et al. [48] studied the catalytic pyrolysis of Kraft lignin, and found that the aromatic hydrocarbon content increased with the addition of HZSM-5. As the Si/Al ratio of HZSM-5 decreased from 200 to 25, the yield of phenolic compounds almost dropped to zero, while aromatic hydrocarbon production increased greatly. ZSM-5 also had a similar catalytic effect on Kraft lignin pyrolysis. ZSM-5 promoted the secondary reaction of volatiles, aliphatic hydroxyl, carboxyl, and methyl groups, decomposing them into

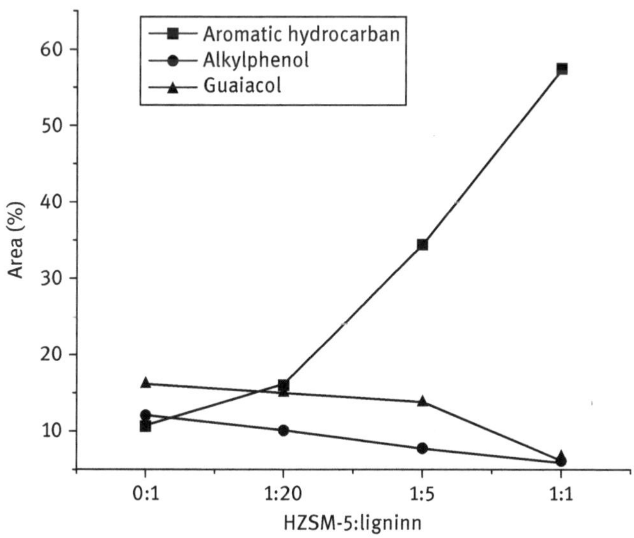

Fig. 6.18: Effect of HZSM-5 on the distribution of the main products of lignin pyrolysis.

small molecular products and gases, leading to a decrease in bio-oil yield and an increase in gas production [49].

Mihalcik et al. [36] studied the catalytic properties of zeolites (H-Mordenite, H-Ferrierite, HZSM-5, H-Y, and H-β) on the pyrolysis behaviors of three biomass components using Py-GC/MS, and analyzed the effect of catalysts on the yields of oxygen-containing compounds in bio-oil, gases, char and aromatic hydrocarbons. Structure and acidity were critical for catalyst activity. H-Mordenite and H-Ferrierite suppressed acetic acid formation, and had little influence on the production of aromatic hydrocarbons, whereas H-Y changed the product formation route, suppressing the formation of acetic acid and levoglucosan, promoting furan formation, and also having little influence on aromatic hydrocarbon generation. HZSM-5 and H-β both performed well at inhibiting the formation of oxygen-containing compounds and at increasing the aromatic hydrocarbon yield. However, catalyst deactivation was more serious due to an increased char yield.

6.2.3 Effect of mesoporous zeolites on the pyrolysis of biomass components

Since MCM-41 was first developed by the Mobil Corporation in 1992, mesoporous zeolites are becoming the new exploitation and application hotspot. In the field of catalytic pyrolysis of biomass, the channel size of the traditional micropore zeolites is small (< 2 nm), which decreases the bio-oil yield. The high surface acidity of micropore zeolites has a strong dehydration effect during biomass pyrolysis, and increases the water content in bio-oil. Mesoporous zeolites can be utilized to catalyze the pyrolysis of biomass or its components to improve bio-oil quality.

MCM-41 is the first synthesized zeolite with an ordered structure, existing in a regular arrangement of hexagonal mesopores with uniform channel sizes within a range of 2 to 10 nm. In the calcination process, the active aluminum species may be separated from the zeolite frameworks, forming pure silicon MCM-41 with low acidity and low activity. Many modification methods are usually utilized to increase its catalytic activity. The most common method is to introduce some heteroatoms into the MCM-41 framework, such as Al, Cu, Fe, Zn, etc. These atoms are not tetrahedrally coordinated in the framework, and can act as catalytic active sites [50]. To evaluate their catalytic activity, some catalysts (Al-MCM-41, Cu-Al-MCM-41, Fe-Al-MCM-44 and Zn-Al-MCM-41) with different Si/Al ratios were prepared to increase phenol content in bio-oil. It was found that these catalysts improved bio-oil quality, among which, Fe and Cu exhibited the best performance in increasing the phenol content in bio-oil, and that Al improved the bio-oil's stability [51]. Meanwhile, the introduction of Pt cations to mesopore zeolites increases their deoxygenation activity and the selectivity to aromatic hydrocarbons. This is because Pt take part in cracking and hydrogenolysis reactions, increasing the reaction rate of alkane dehydrogenation and promoting the conversion of alkanes to aromatic hydrocarbons. Introducing Ga ions increased the dehydrogenation rate, but it was unlikely to help improve bio-oil quality [52]. MCM-41 and some other mesopore zeolites, such as MCM-48 or MFI, show good catalytic effects on the deoxygenation of macromolecules to form phenols, and have a stronger catalytic effect than HZSM-5. However, since the aromatic hydrocarbons were mainly produced on the acid sites, the zeolites without acid sites, such as Al-MCM-41 and Al-MCM-48, do not have an effect on aromatic hydrocarbon formation. HZSM-5 and MFI obviously promote the generation of aromatic hydrocarbons due to the synergistic effect of the acid sites and their larger pore channels [53].

Zhang et al. [54] studied the catalytic pyrolysis of a mixture of pine and alcohol in the presence of ZSM-5 in a bubbling fluidized bed. For pine, below the temperature of 600 °C and WHSV of 0.35 h^{-1}, the yield of petrochemicals (aromatics, C_2–C_4 olefins, and C_5 compounds) reached a maximum of 23.7 %, while a maximum petrochemical yield from alcohol of 87.0 % was obtained at 400 °C. Thus the optimum reaction conditions for the two materials were different. Changing the ratio of pine to alcohol adjusted the effective hydrogen to carbon ratio, $(H/C)_{eff}$, favoring an increase in the petrochemical yield, especially at $(H/C)_{eff} = 1.25$. Torri et al. [55] studied cellulose pyrolysis in the presence of modified mesopore zeolite in a fixed bed at 500 °C. Mesopore zeolite addition increased the tar content of volatile components, while the char yield remained unchanged. For Sn-MCM-41, the mass fraction of tar in the products reached 47 %, higher than that yielded by the non-catalytic pyrolysis of cellulose (33 %). This indicates that mesopore zeolites can increase the liquid yield. Meanwhile, the proportions of the component in the tar changed. After the catalyst was added, the levoglucosan yield decreased greatly, while the levoglucosenone yield increased. The yield of other products, such as 2H-furanone, furfural, 5-methylfurfural, and dianhydro pyranose was hardly affected. Jackson et al. [56] studied lignin pyrolysis in the presence of the Al-MCM-41 catalyst at 600 °C, and found that the char yield decreased from 41 % to 36 % after catalyst addition. More phenols, dihydrobenzofuran and other oxygen-containing products in the bio-oil were converted into naphthalene, instead of simple aromatic hydrocarbons. The catalyst also promoted the generation of H_2, CH_4, and CO. Park et al. [57] synthesized a highly ordered mesopore zeolite called MM_{ZUSY}, based on the framework of the USY zeolite. The zeolite had high mass transfer performance, hydrothermal stability, and medium acidity, which increased the amount of small molecular gases and promoted the conversion of macromolecular oxygen-containing compounds into water or small molecular phenols during biomass pyrolysis.

In addition to MCM-41, SBA-15 is another commonly used mesopore material, with a hexagonal array of pores, thicker channel walls, adjustable pore sizes (5–30 nm) and high stability [58]. Jeon et al. [59] prepared a series of SBA-15 catalysts, including SBA-15, Pt/SBA-15, Al/SBA-15, and Pt/Al/SBA-15, and applied them to the catalytic pyrolysis of three biomass components. The introduction of Al increased the acid sites of SBA-15, which favored dehydration, cracking, decarbonylation, etc. Thus during the pyrolysis of cellulose and hemicellulose, a higher yield of high value-added compounds, such as furans and aromatics, was obtained from Al/SBA-15 and Al/Pt/SBA-15 catalysis, while the acetic acid yield was also enhanced. The catalytic pyrolysis of lignin promoted the production of phenols, especially those with fewer carbonyl groups. Moreover, Park et al. [52] found that the introduction of Pt increased the production of aromatics.

6.3 Catalytic effect of metal oxide on the pyrolysis of biomass components

6.3.1 Structural characteristics of metal oxide

Metal oxides have been widely used in industry as supports, promoters, and active components in the selective oxidation of hydrocarbons, and the disproportionation and polymerization of alkenes [60]. There are many kinds of metal oxides with different properties. The oxidation-reduction and acid-base properties of the metal oxides are closely related to their catalytic properties.

Metal oxides often present as crystals without a regular spatial structure. The electronegativity of the metals varies widely, but most of them have similar lattice imperfections, including electron holes, point defects, dislocations, and planar defects. The catalytic activity of metal oxides is closely related to their lattice imperfections.

Most of the metal oxides and their mixtures have acid, alkaline, or both properties. The solid acid contains Bronsted acid sites providing protons for the reactions and Lewis acid sites accepting electrons from the reactions, and the solid base provides electrons for the reactions and accepts protons from the reactions. For the metal oxides used in the catalytic pyrolysis of biomass, catalytic performance is closely related to the acidity and basicity of their surfaces. The common acidic oxides include Al_2O_3, Ti_2O_3, CeO_2, etc., and the usual base oxides include MgO and CaO. Metal oxides such as ZnO, having both acid sites and base sites, are called amphoteric oxides. Some processes can also be utilized to change the surface acid or base property of the metal oxides. For instance, the acidity of TiO_2 can be increased greatly through treatment with H_2SO_4, turning the obtained TiO_2 into a super solid acid.

In general, the surface acid sites of catalysts can promote the dehydration and decarboxylation of organic molecules. In this way, large organic molecules can be cracked into smaller molecules, increasing their bio-oil yield and water content. The base sites, on the other hand, promote the cleavage of the C–H bond, and accelerate the dehydrogenation reaction, increasing the yield of small molecular gases. The different acid and base sites of the metal oxides affect the distribution of bio-oil components, promoting or suppressing the formation of certain products.

6.3.2 Catalytic effect of metal oxides on the pyrolysis of biomass components

Metal oxide addition has a great influence on cellulose pyrolysis by increasing the yield of anhydro-sugars, such as levoglucosenone, levoglucosan, DGP (1,4:3,6-Dianhydro-β-D-glucopyranose), and LCA (δ-lactone of 3-hydroxy-5-hydroxymethyl-tetrahydrofuran-3-carboxylic acid) etc. Febbri et al. [34] studied the influence of nanopowder metal oxides on the distribution of products evolved from cellulose pyrolysis, and found that Al_2O_3, MgO, SiO_2, TiO_2, and Al_2O_3-Ti_2O_3 mainly affected the formation of anhydro-sugars. SiO_2 suppressed the generation of anhydro-sugars, whereas nanoscale Al_2O_3 and Al_2O_3-TiO_2 improved the levoglucosenone and LCA yields significantly, while showing less of an effect on levoglucosan and DGP yields. However, normal scale Al_2O_3-TiO_2 did not have such effects, which indicates that apart from the catalytic activity of the metal oxide itself, the surface properties of nanoscale catalysts also play an important role in catalytic activity. Moreover, the metal oxide catalysts treated with strong acid (SO_4^{2-}/TiO_2, SO_4^{2-}/ZrO_2, and SO_4^{2-}/SnO_2) also exhibited a strong influence on the pyrolytic behavior of cellulose [61, 62]. Due to their super acidity, the sugar products undergo further dehydration to form glycolaldehydes and furans (5-methylfurfural, and furfural etc.), and the different metal oxides showed different selectivity to furan compounds. For example, SO_4^{2-}/SnO_2 showed high selectivity to 5-methylfurfural, while SO_4^{2-}/ZrO_2 and SO_4^{2-}/TiO_2 displayed high selectivity to furan and furfural.

In addition to the inorganic salts, zeolites, and metal oxides, some other catalysts also exhibit certain effects on the pyrolytic behavior of biomass components. $Zr(SO_4)_2$ addition has been shown to increase the yield of liquid and solid products, the total of which can be higher than 95 % at 290–400 °C. The addition of this catalyst has also been found to lead to a more obvious promotion effect of temperature on the generation of levoglucosan, whose highest yield was obtained at 335 °C. However, catalyst activity was easily decreased due to SO_4^{2-} leaching, which had to be recovered through further impregnation with H_2SO_4 [63]. $ZnCl_2$ has also been shown to catalyze the pyrolysis of biomass components, changing the bio-oil state from a dark-brown water-oil emulsion to an orange liquid with well-separated water and oil phases. The results of the analysis showed that the oxygen-containing products content had decreased and that the hydrocarbon content had increased. Among the three major components, $ZnCl_2$ showed the most significant effect on cellulose pyrolysis [64]. Other studies have also revealed that biomass impregnated with $ZnCl_2$ favored furfural formation [65].

References

[1] Bradbury AG, Sakai Y, Shafizadeh F. A kinetic model for pyrolysis of cellulose. Journal of Applied Polymer Science. 1979;23(11):3271–3280.

[2] Shimada N, Kawamoto H, Saka S. Different action of alkali/alkaline earth metal chlorides on cellulose pyrolysis. Journal of Analytical and Applied Pyrolysis. 2008;81(1):80–87.

[3] Khelfa A, Finqueneisel G, Auber M, et al. Influence of some minerals on the cellulose thermal degradation mechanisms. Journal of Thermal Analysis and Calorimetry. 2008;92(3):795–799.
[4] Wang S, Liu Q, Liao Y, et al. A study on the mechanism research on cellulose pyrolysis under catalysis of metallic salts. Korean Journal of Chemical Engineering. 2007;24(2):336–340.
[5] Jensen A, Dam-Johansen K, Wójtowicz MA, et al. TG-FTIR study of the influence of potassium chloride on wheat straw pyrolysis. Energy & Fuels. 1998;12(5):929–938.
[6] Liu Q, Wang S, Luo Z, et al. Catalysis mechanism study of potassium salts on cellulose pyrolysis by using TGA-FTIR analysis. Journal of Chemical Engineering of Japan. 2008;41(12): 1133–1142.
[7] Nowakowski DJ, Jones JM, Brydson R, et al. Potassium catalysis in the pyrolysis behaviour of short rotation willow coppice. Fuel. 2007;86(15):2389–2402.
[8] Julien S, Chornet E, Overend RP. Influence of acid pretreatment H_2SO_4, HCl, HNO_3 on reaction selectivity in the vacuum pyrolysis of cellulose. Journal of Analytical and Applied Pyrolysis. 1993;27(1):25–43.
[9] Piskorz J, Radlein DSA, Scott DS, et al. Pretreatment of wood and cellulose for production of sugars by fast pyrolysis. Journal of Analytical and Applied Pyrolysis. 1989;16(2):127–142.
[10] Brown RC, Liu Q, Norton G. Catalytic effects observed during the co-gasification of coal and switchgrass. Biomass and Bioenergy. 2000;18(6):499–506.
[11] Sutton D, Kelleher B, Ross JR. Review of literature on catalysts for biomass gasification. Fuel Processing Technology. 2001;73(3):155–173.
[12] Richards GN, Zheng G. Influence of metal ions and of salts on products from pyrolysis of wood: applications to thermochemical processing of newsprint and biomass. Journal of Analytical and Applied Pyrolysis. 1991;21(1):133–146.
[13] Olsson JG, Jäglid U, Pettersson JB, et al. Alkali metal emission during pyrolysis of biomass. Energy & Fuels. 1997;11(4):779–784.
[14] Williams PT, Horne PA. The role of metal salts in the pyrolysis of biomass. Renewable Energy. 1994;4(1):1–13.
[15] Lu Q, Zhang D, Zhu XF. Catalytic effects of four metal chlorides on fast pyrolysis of cellulose (I) Py-GC/MS experiments. CIESC Journal. 2010;61(4):1018–1024.
[16] Lu Q, Zhang D, Zhu XF. Catalytic effects of four metal chlorides on fast pyrolysis of cellulose (II) Mechanism analysis. CIESC Journal. 2010;61(4):1025.
[17] Wang Z. Catalytic fast pyrolysis of biomass to prepare high-value chemicals. University of Science and Technology of China, Heifei; 2011.
[18] Nowakowski DJ, Woodbridge CR, Jones JM. Phosphorus catalysis in the pyrolysis behaviour of biomass. Journal of Analytical and Applied Pyrolysis. 2008;83(2):197–204.
[19] Aho A, Kumar N, Eränen K, et al. Catalytic pyrolysis of biomass in a fluidized bed reactor: influence of the acidity of H-beta zeolite. Process Safety and Environmental Protection. 2007;85(5):473–480.
[20] Aho A, Kumar N, Eränen K, et al. Catalytic pyrolysis of woody biomass in a fluidized bed reactor: influence of the zeolite structure. Fuel, 2008,87(12):2493–2501.
[21] Xu RR, Pang WQ, Tu KG. Zeolite molecular sieves structure and synthesis. Jilin University Press, Changchun; 1987.
[22] Liu DC. Heterogeneous catalysis principle. Fudan University Press, Shanghai; 1997.
[23] Lu CB. TG-FTIR research of biomass catalytic pyrolysis. Acta Energiae Solaris Sinica. 2007;28(6):638–643.
[24] Gayubo AG, Aguayo AT, Atutxa A, et al. Deactivation of a HZSM-5 zeolite catalyst in the transformation of the aqueous fraction of biomass pyrolysis oil into hydrocarbons. Energy & Fuels. 2004;18(6):1640–1647.
[25] Xu RR, Pang WQ. Molecular sieve and porous material chemistry. Science Press, Beijing; 2004.

[26] Qian K. Hydrothermal synthesis and characterization of germanosilicate zeolites. Jilin University, Changchun; 2012.
[27] Loiola AR, Da Silva LR, Cubillas P, et al. Synthesis and characterization of hierarchical porous materials incorporating a cubic mesoporous phase. Journal of Materials Chemistry. 2008;18(41):4985–4993.
[28] Perego C, Bosetti A. Biomass to fuels: The role of zeolite and mesoporous materials. Microporous and Mesoporous Materials. 2011;144(1):28–39.
[29] Stöcker M. Biofuels and biomass-to-liquid fuels in the biorefinery: Catalytic conversion of lignocellulosic biomass using porous materials. Angewandte Chemie International Edition. 2008;47(48):9200–9211.
[30] Corma A, Huber GW, Sauvanaud L, et al. Processing biomass-derived oxygenates in the oil refinery: catalytic cracking (FCC) reaction pathways and role of catalyst. Journal of Catalysis. 2007;247(2):307–327.
[31] French R, Czernik S. Catalytic pyrolysis of biomass for biofuels production. Fuel Processing Technology. 2010;91(1):25–32.
[32] Gayubo AG, Aguayo AT, Atutxa A, et al. Transformation of oxygenate components of biomass pyrolysis oil on a HZSM-5 zeolite. I. Alcohols and phenols. Industrial & Engineering Chemistry Research. 2004;43(11):2610–2618.
[33] Gayubo AG, Aguayo AT, Atutxa A, et al. Transformation of oxygenate components of biomass pyrolysis oil on a HZSM-5 zeolite. II. Aldehydes, ketones, and acids. Industrial & Engineering Chemistry Research. 2004;43(11):2619–2626.
[34] Fabbri D, Torri C, Baravelli V. Effect of zeolites and nanopowder metal oxides on the distribution of chiral anhydrosugars evolved from pyrolysis of cellulose: An analytical study. Journal of Analytical and Applied Pyrolysis. 2007;80(1):24–29.
[35] Du ZY, Hu B, Ma XC, et al. Catalytic pyrolysis of microalgae and their three major components: Carbohydrates, proteins, and lipids. Bioresource Technology. 2013;130:777–782.
[36] Mihalcik DJ, Mullen CA, Boateng AA. Screening acidic zeolites for catalytic fast pyrolysis of biomass and its components. Journal of Analytical and Applied Pyrolysis. 2011;92(1):224–232.
[37] Shin E, Nimlos MR, Evans RJ. Kinetic analysis of the gas-phase pyrolysis of carbohydrates. Fuel. 2001;80(12):1697–1709.
[38] Antal Jr MJ, Friedman HL, Rogers FE. Kinetics of cellulose pyrolysis in nitrogen and steam. Combustion Science and Technology. 1980;21(3-4):141–152.
[39] Carlson TR, Tompsett GA, Conner WC, et al. Aromatic production from catalytic fast pyrolysis of biomass-derived feedstocks. Topics in Catalysis. 2009;52(3):241–252.
[40] Foster AJ, Jae J, Cheng YT, et al. Optimizing the aromatic yield and distribution from catalytic fast pyrolysis of biomass over ZSM-5. Applied Catalysis A: General. 2012;423:154–161.
[41] Perego C, Ingallina P. Recent advances in the industrial alkylation of aromatics: new catalysts and new processes. Catalysis Today. 2002;73(1):3–22.
[42] Scheirs J, Camino G, Tumiatti W. Overview of water evolution during the thermal degradation of cellulose. European Polymer Journal. 2001;37(5):933–942.
[43] Wang S, Yan Z, Tao L, et al. Catalytic pyrolysis of mannose as a model compound of hemicellulose over zeolites. Biomass & Bioenergy. 2013;57(4):106–112.
[44] Zamora F, Gonzalez MC, Duenas MT, et al. Thermodegradation and thermal transitions of an exopolysaccharide produced by Pediococcus damnosus 2.6. Journal of Macromolecular Science, Part B. 2002;41(3):473–486.
[45] Wang S, Ru B, Lin H, et al. Pyrolysis mechanism of hemicellulose monosaccharides in different catalytic processes. Chemical Research in Chinese Universities. 2014;30(5):848–854.
[46] Luo Z, Wang S, Guo X. Selective pyrolysis of Organosolv lignin over zeolites with product analysis by TG-FTIR[J]. Journal of Analytical and Applied Pyrolysis. 2012;95:112–117.

[47] Mullen CA, Boateng AA. Catalytic pyrolysis-GC/MS of lignin from several sources. Fuel Processing Technology. 2010;91(11):1446–1458.
[48] Li X, Su L, Wang Y, et al. Catalytic fast pyrolysis of Kraft lignin with HZSM-5 zeolite for producing aromatic hydrocarbons. Frontiers of Environmental Science & Engineering. 2012;6(3):295–303.
[49] Ben H, Ragauskas AJ. Pyrolysis of Kraft lignin with additives. Energy & Fuels. 2011;25(10):4662–4668.
[50] O'Neil AS, Mokaya R, Poliakoff M. Supercritical fluid-mediated alumination of mesoporous silica and its beneficial effect on hydrothermal stability. Journal of the American Chemical Society. 2002;124(36):10636–10637.
[51] Antonakou E, Lappas A, Nilsen MH, et al. Evaluation of various types of Al-MCM-41 materials as catalysts in biomass pyrolysis for the production of bio-fuels and chemicals. Fuel. 2006;85(14):2202–2212.
[52] Park HJ, Park K, Jeon J, et al. Production of phenolics and aromatics by pyrolysis of miscanthus. Fuel. 2012;97:379–384.
[53] Lee HW, Jeon JK, Park SH, et al. Catalytic pyrolysis of Laminaria japonica over nanoporous catalysts using Py-GC/MS. Nanoscale Research Letters. 2011;6(33):459–472.
[54] Zhang H, Carlson TR, Xiao R, et al. Catalytic fast pyrolysis of wood and alcohol mixtures in a fluidized bed reactor. Green Chemistry. 2012;14(1):98–110.
[55] Torri C, Lesci IG, Fabbri D. Analytical study on the pyrolytic behaviour of cellulose in the presence of MCM-41 mesoporous materials. Journal of Analytical and Applied Pyrolysis. 2009;85(1):192–196.
[56] Jackson MA, Compton DL, Boateng AA. Screening heterogeneous catalysts for the pyrolysis of lignin. Journal of Analytical and Applied Pyrolysis. 2009;85(1):226–230.
[57] Park HJ, Jeon JK, Kim JM, et al. Synthesis of nanoporous material from zeolite USY and catalytic application to Bio-Oil conversion. Journal of Nanoscience & Nanotechnology. 2008;8(10):5439–5444.
[58] Lee HW, Cho HJ, Yim J, et al. Removal of Cu (II)-ion over amine-functionalized mesoporous silica materials. Journal of Industrial and Engineering Chemistry. 2011;17(3):504–509.
[59] Jeon M, Jeon J, Suh DJ, et al. Catalytic pyrolysis of biomass components over mesoporous catalysts using Py-GC/MS. Catalysis Today. 2013;204:170–178.
[60] Tang XD. Principles of industrial catalysis. Petroleum Industry Press, Beijing; 2003.
[61] Lu Q, Xiong WM, Li WZ, et al. Catalytic pyrolysis of cellulose with sulfated metal oxides: a promising method for obtaining high yield of light furan compounds. Bioresource Technology. 2009;100(20):4871–4876.
[62] Lu Q, Zhu XF. Production of levoglucosenone from pyrolysis of cellulose catalyzed by solid superacids. Journal of Fuel Chemistry and Technology. 2011;6(39):425–431.
[63] Wang Z, Lu Q, Zhu XF, et al. Catalytic fast pyrolysis of cellulose to prepare levoglucosenone using sulfated zirconia. ChemSusChem. 2011;4(1):79–84.
[64] Rutkowski P. Pyrolysis of cellulose, xylan and lignin with the K_2CO_3 and $ZnCl_2$ addition for bio-oil production. Fuel Processing Technology. 2011;92(3):517–522.
[65] Lu Q, Wang Z, Dong C, et al. Selective fast pyrolysis of biomass impregnated with $ZnCl_2$: Furfural production together with acetic acid and activated carbon as by-products. Journal of Analytical and Applied Pyrolysis. 2011;91(1):273–279.

7 Pyrolysis of biomass

7.1 Introduction to biomass pyrolysis

Based on the introduction of the distributions and characteristics of biomass components, the pyrolysis behaviors of cellulose, hemicellulose, and lignin have been presented in detail in the previous chapters. The pyrolysis mechanism of each component, which was based on the chemical reaction kinetics, pathways of product formation, and theoretical simulation at a molecular level, has also been discussed. We also investigated the influences of component cross-coupling, inner salts, and the addition of catalysts on the pyrolysis behaviors of biomass components. Therefore, in this chapter, the overall pyrolysis regulation of biomass is discussed. The pyrolysis behaviors of different biomass species and the corresponding evolution of pyrolysis products are studied on the basis of biomass component distributions. The operating conditions that affect the fast pyrolysis of biomass are summarized. In addition, this chapter proposes a graded upgrading scheme based on the highly efficient separation of bio-oil through molecular distillation, which provides an improved theoretical basis for the production of high-quality liquid fuels, as shown in Fig. 7.1.

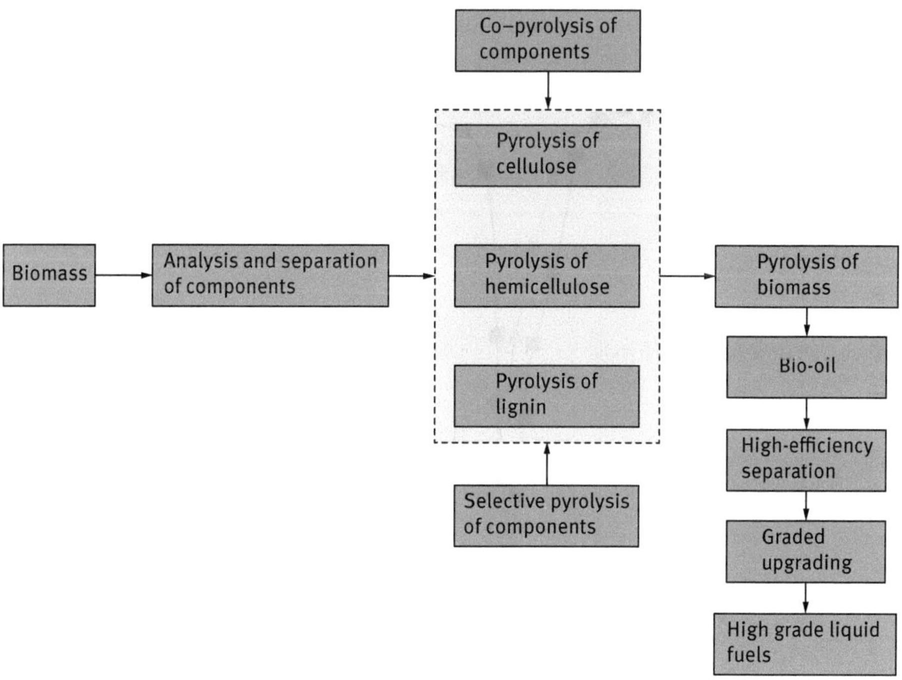

Fig. 7.1: Pyrolysis of biomass based on its components.

DOI 10.1515/9783110369632-009

In general, the basic process of biomass pyrolysis includes the following stages. The first stage is the drying stage (ambient temperature: 100 °C), also known as pre-dehydration stage. During this stage, biomass absorbs heat and releases external water with few chemical changes to its structure. This registers as a slight weight loss on the TG curve. The second stage is the pre-pyrolysis stage (100–250 °C). In this stage, the structure and composition of biomass begin to change. As the temperature increases, some unstable components, such as hemicellulose with the lowest initial pyrolysis temperature, start to decompose slightly and generate small molecules such as CO, CO_2, acetic acid, and so on. At the same time, lignin also begins to lose weight slightly and produce small molecules. The amorphous transition occurs in the crystalline regions of cellulose. This transition manifests as a platform in the TG curve, which is often called the "glass transition" process. The major pyrolysis stage (250–500 °C) is the main pyrolysis stage of biomass, where biomass undergoes significant decomposition. In this stage, the condensable volatile macromolecules are generated as well as some small molecules, such as CO_2, CO, CH_4, H_2, etc., which is shown as an obvious peak in the DTG curve. The last stage (> 500 °C) corresponds to the slow decomposition of residues. Some residues generated from incomplete pyrolysis undergo further decomposition in this stage, shown as a slow decline on the TG curve. Fig. 7.2 shows the weight loss stages of biomass pyrolysis. It is worth noting that the pyrolysis behaviors of different biomass species are quite different because the distribution of cellulose, hemicellulose, and lignin varies across different species

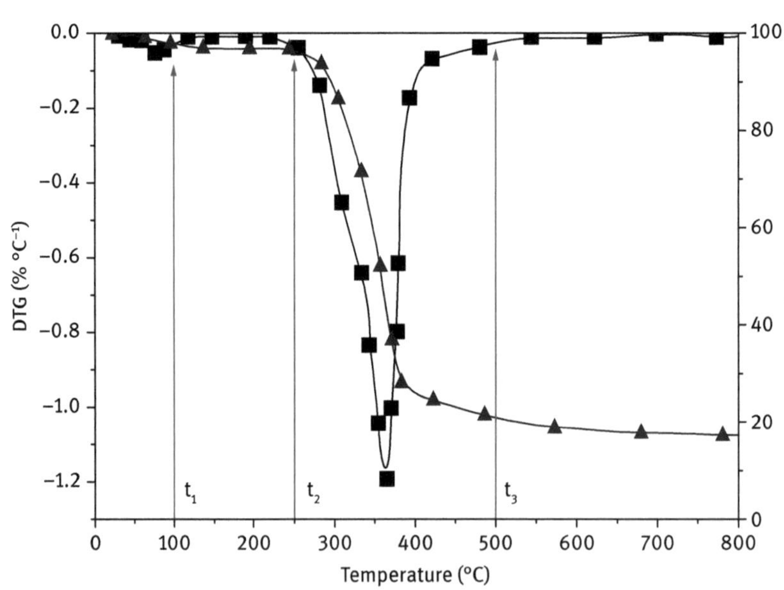

Fig. 7.2: Weight loss in biomass pyrolysis.

of biomass. Therefore, the classification of the above pyrolysis stages is rough, and it depends on the biomass species, which needs further investigation.

7.2 Pyrolysis of different biomass species

China has abundant lignocellulosic biomass resources. Taking into consideration the species of biomass resources, their geographical distribution and their economic feasibility, it has been found that the most suitable biomass feedstocks for pyrolysis include agricultural and forestry residues, rapidly growing herbaceous biomass and the abundant aquatic plants. As shown in Tab. 1.1 (in Chapter 1), biomass mainly consists of the elements carbon, hydrogen, and oxygen. The sulfur and nitrogen contents are low in most biomass species, but are slightly higher in aquatic plants because of their high protein content. The oxygen content of biomass can reach more than 30 %, thus for applications of biomass, deoxygenation is essential. Within the same biomass species, the amounts of fixed carbon and volatiles are similar, while some differences were found between different species of biomass. Rice husk and rice straw, the representatives of agricultural biomass, have the highest ash content across the four categories of biomass, followed by algae. Bamboo and grassiness also have high ash contents which reached 3.68 % and 2.44 %, respectively. Researchers from the Vienna University of Technology analyzed more than 800 species of biomass and found that the total carbon, hydrogen and oxygen contents could reach more than 90 %. The nitrogen content had a wider range than sulfur, especially for algae, where it reached more than 9 %. The ash content in biomass ranged from 1 % to 16 % [1]. Agricultural biomass has a low heating value due to its high ash content, while the low heating value of algae is due to the high ash content and the low fixed carbon content. Demirbas [2] pointed out that biomass with a high ash content is not an ideal fuel source, because ash content is one of the main factors that directly influence the heating value of biomass. The component distribution of different biomass species also showed great differences. As shown in Tab. 1.2 (in Chapter 1), the analysis of components in different species of biomass indicated that the total quantity of the three major components (cellulose, hemicellulose and lignin) was more than 70 % in lignocellulosic biomass. Algae are rich in protein, lipids, saccharides, vitamins, minerals, and trace elements, as well as sulfating polygalactose and provitamin: of these the total lipid and protein content was more than 50 %. Miao et al. [3] analyzed the components of microalgae and found that the protein content reached 52.64 % while the oil content was 14.57 %. Grassiness, an important fodder for livestock in southern China, has high quantities of protein and lipid. Because of the differences between the species of biomass, especially in terms of cellulose, hemicellulose, and lignin contents, different species of biomass showed different pyrolytic behaviors, eventually leading to a clear difference in the distribution of pyrolysis products.

7.2.1 Pyrolysis of forestry biomass

Forestry biomass mainly includes softwood and hardwood. Common softwood biomass includes spruce, Mongolian Scots pine, fir, and hemlock; Manchurian ash and rosewood are classified as hardwoods.

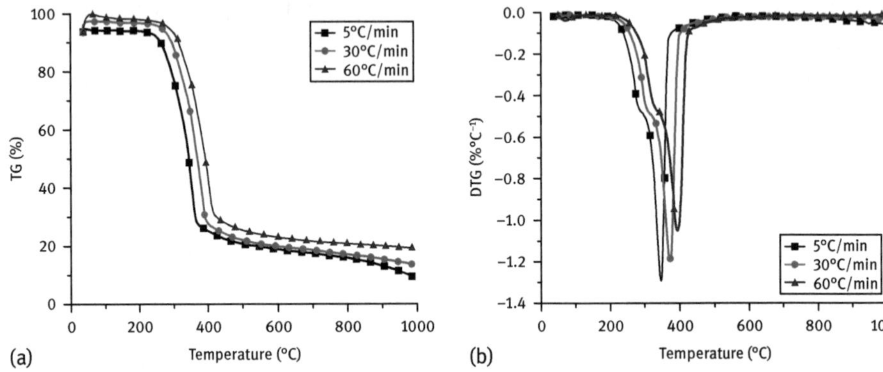

Fig. 7.3: TG/DTG curves of spruce pyrolysis.

Fig. 7.3 shows the TGA curve of spruce pyrolysis at different heating rates and it presents the typical pyrolysis characteristics of biomass. After the initial release of water, spruce only lost a tiny amount of weight in the temperature range from 110 to 210 °C, followed by an obvious weight loss due to the release of abundant volatiles. At about 380 °C, the maximum rate of weight loss was achieved, and it slowed down until around 430 °C. This region was the major pyrolysis stage of lignocellulose, and was also the main endothermic stage in the pyrolysis process. Finally, the residues were slowly decomposed until the end. Zhang et al. [4] found that the weight loss process of woody biomass was divided into three stages, including the dehydration stage before 200 °C, the major decomposition stage of the three components ranging from 200 °C to 430 °C, and the decomposition stage of lignin after 430 °C. At different heating rates, the TG and DTG curves displayed the same evolutionary trend. With an increase in heating rates, the initial and terminal temperatures of each stage moved slightly to the high temperature region, and the main reaction region also extended. This was due to the fact that when a higher heating rate is used, the required reaction time for reaching the same target temperature is shorter, and thus the weight loss is smaller. Zhao et al. [5] investigated the effects of heating rate on birch pyrolysis using the TG method, and found that the pyrolysis rate elevated linearly with the increase in heating rate. Meanwhile, the heating rate also had an impact on the temperature difference for heat transfer and the temperature gradient between the measuring point and the sample, as well as between the outer layer

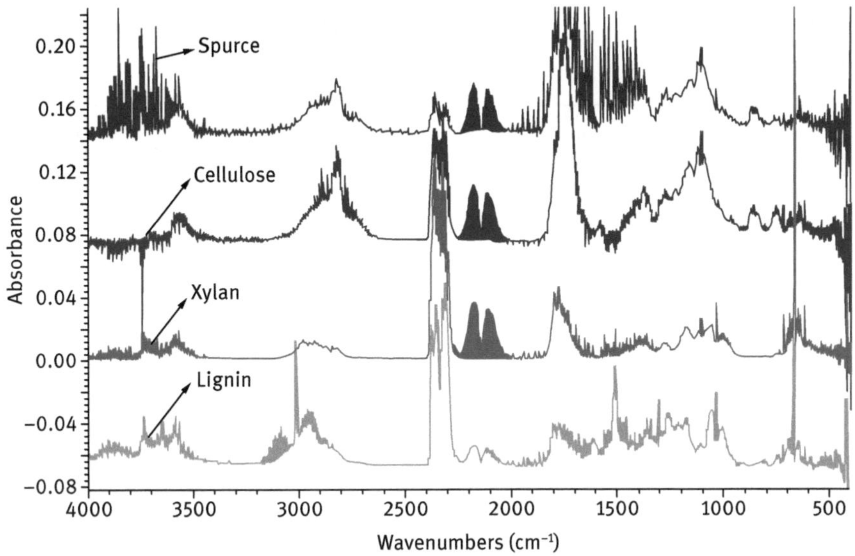

Fig. 7.4: Comparison of the pyrolytic products of spruce and its components. Reprinted with permission from [7].

and inner layer of the samples. Therefore, thermal hysteresis was aggravated and the curves moved to the higher temperature region.

Fig. 7.4 presents the comparison between the gases produced from the pyrolysis of spruce and its components including cellulose, xylan and lignin. These four raw materials all had the characteristic absorption peaks of H_2O, CO, CO_2, and methanol. The characteristic absorption peak of CH_4 of spruce corresponded to that of the pyrolysis of lignin, while another strong absorption peak (1,100–900 cm^{-1}) generated by the pyrolysis of spruce was consistent with the characteristic absorption peak of levoglucosan from the pyrolysis of cellulose. The two weak absorption peaks (1,620–1,450 cm^{-1} and 1,210 cm^{-1}) of spruce were assigned to the skeletal vibration of phenolic compounds from the pyrolysis of lignin. In addition, other absorption peaks (such as N_2O at 1,269 cm^{-1} and 1,314 cm^{-1}, NH_3 at 966 cm^{-1} and 932 cm^{-1}, NO at 1,915 cm^{-1}) could not be found in the pyrolysis of cellulose, xylan and lignin, indicating that these compounds were produced from the pyrolysis of the extractives such as pectin and protein rather than the three components. In spruce pyrolysis, a high yield of methanol was mainly from the pyrolysis of lignin, while acetic acid and carbon dioxide were generated mainly from the pyrolysis of hemicellulose.

Fig. 7.5 shows the TG/DTG curves of different species of forestry biomass. The DTG curve of fir is quite smooth, while that of Manchurian ash is slightly wavy and that of rosewood has an obvious shoulder peak. This might be caused by the differences in the component distribution of the three kinds of wood. Generally, the decomposition temperature ranges of hemicellulose and cellulose were narrow, and hemicellulose had a

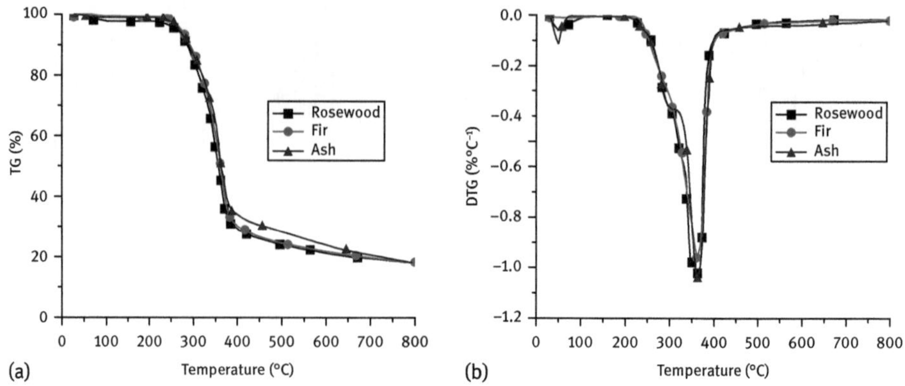

Fig. 7.5: TG/DTG curves from the pyrolysis of forestry biomass.

lower initial decomposition temperature than cellulose. The decomposition temperature of cellulose ranged from 315 °C to 400 °C, and that of hemicellulose ranged from 220 °C to 315 °C, while that of lignin was quite wide, from 160 °C to 900 °C [6]. Raveendran et al. [8] studied the pyrolysis behaviors of various species of biomass and the three components, and they reported that the pyrolysis of hemicellulose was the easiest followed by cellulose, while that of lignin was the most difficult. Consequently, under low heating rates, the DTG curve of biomass might present two separated peaks due to the pyrolysis of hemicellulose and cellulose. Whether the separated peaks occurred depended on the relative contents of hemicellulose and cellulose in the samples. The pyrolysis of samples with a higher content of hemicellulose tended to have an inflection point in this temperature range. Much research [8–10] has shown that the greater the amount of hemicellulose in the sample, the more apparent the occurrence of the shoulder peak in the DTG curve. As displayed in Tab. 1.2, rosewood had a higher content of hemicellulose compared with the other two species of biomass, and thus the shoulder peak on its DTG curve appeared more clearly. With regard to fir and Manchurian ash, the two separated peaks might be merged into one due to the lower hemicellulose content.

7.2.2 Pyrolysis of agricultural biomass

Rice straw and rice husk are the main agricultural biomass in China. The components analysis in Tab. 1.2 shows that they have a relatively high hemicellulose content. The peak of their DTG curve moved towards lower temperature compared with forestry biomass, as shown in Fig. 7.6. Yang et al. [12] selected four species of agricultural biomass and three species of forestry biomass to study their TG characteristics. The results showed that the initial decomposition temperature of forestry biomass was higher than that of agricultural biomass. The chemical reaction kinetic calculation revealed

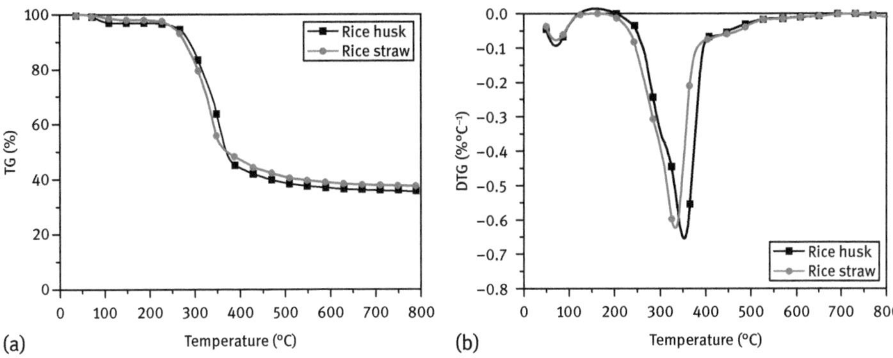

Fig. 7.6: TG/DTG curves from the pyrolysis of agricultural biomass.

that the activation energy of forestry biomass ranged from 74–96 kJ/mol, while that of agricultural biomass was in the range of 55–78 kJ/mol, indicating that a higher cellulose content in forestry biomass led to a better stability compared with agricultural biomass. Zhu et al. [13] found that the DTG curve of corn stalk with a high hemicellulose content had a shoulder peak, and the DTG curve of corncob even displayed two obvious peaks. However, the DTG curve of wheat stalk with a low hemicellulose content had only one peak. It should be noted that the shoulder peak did not definitely appear when the hemicellulose content of agricultural biomass was higher than that of forestry biomass. It could also be due to the different linkages between the components in various species of biomass. When the distributions of components in biomass were similar, the DTG curve from the pyrolysis of biomass with a high hemicellulose content tended to have an inflexion, or even have multiple peaks. In addition, the distribution of pyrolysis products also displayed distinct characteristics. Fu et al. [14] studied the release behavior of pyrolysis products from rice straw, and they reported that besides CO_2, CO, and small molecular weight gases, relatively high quantities of acids, aldehydes, ketones, alcohols, and ethers were also observed in pyrolysis products. These chemical families corresponded to the pyrolysis products from hemicellulose. Sun et al. [15] analyzed the chromatography figures of pyrolysis products from rice straw and rice husk under different photon energy, and pointed out that the kinds of pyrolysis products from these two feedstocks were roughly the same, but the abundance of typical products from lignin pyrolysis like methyl-guaiacols in rice husk was higher than that in rice straw, owing to the higher lignin content in rice husk.

7.2.3 Pyrolysis of herbaceous biomass

Herbaceous biomass is currently one of the most promising resources. Grassiness, the main biomass resource in south China, has the advantages of fast growth, rapid propagation and renewability. The components of herbaceous biomass are similar to those of agricultural biomass. For example, the TG curves of grassiness and bamboo display certain similarities. In comparison, herbaceous biomass contains more protein and aliphatics, belonging to the extractives. The extractive contents of grassiness and bamboo were 27.69 % and 20.79 %, respectively. The pyrolysis temperature of extractives was located in a wide range of 130–550 °C. Li et al. [16] reported that the main pyrolysis temperature of grassiness occurred in the range of 180–380 °C, and pyrolysis was nearly completed at 380 °C, with the final char yield of 26 % at 630 °C. Fu et al. [17] selected reed, rice straw, pennisetum alopecuroides, and miscanthus as the representatives of herbaceous biomass, and the TG results showed that their weight losses were concentrated in the range of 190–400 °C.

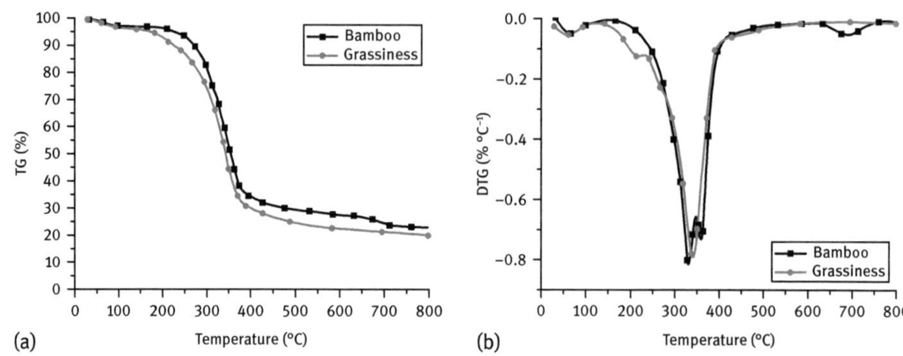

Fig. 7.7: TG/DTG curves from the pyrolysis of herbaceous biomass.

7.2.4 Pyrolysis of aquatic biomass

Algae from aquatic plants are abundant, grow fast, and occupy a small area of land. The TG curve of algae is shown in Fig. 7.8, and obvious differences were found between its pyrolysis behavior and that of lignocellulosic biomass. Its initial pyrolysis temperature was lower and multiple peaks in the DTG curve appeared, which might be due to the large amount of extractives and a small portion of cellulose. Algae biomass contained protein and polysaccharides, as well as a few insoluble polysaccharides like cell wall cellulose. The pyrolysis of protein and polysaccharides was faster than insoluble polysaccharides and also had lower initial temperature and a wider decomposition region, with more final solid residues.

The pyrolytic characteristics of three kinds of red algae were studied by Li et al. [18]. Three stages appeared during pyrolysis and the initial pyrolysis temperature was low, indicating its complexity compared to other species of biomass. Wang et al. [19, 20] compared the pyrolytic behaviors of typical seaweeds and land plants and the results showed that the thermal stability of land plants was higher and the low activation energy of seaweed pyrolysis resulted in its easy decomposition. Zhao et al.

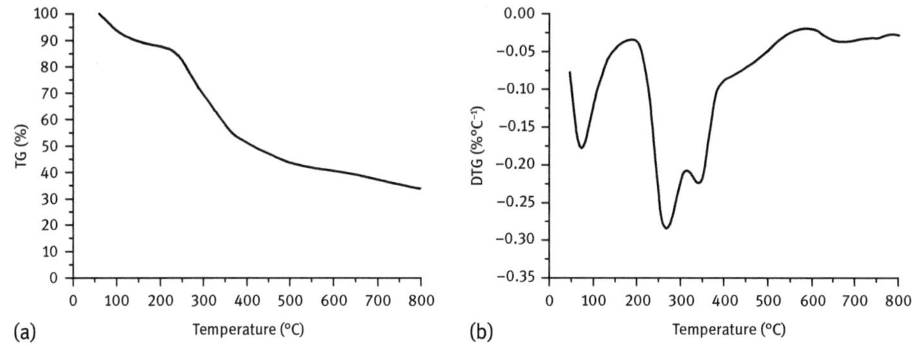

Fig. 7.8: TG/DTG curve from the pyrolysis of algae.

[21, 22] also carried out an investigation on the pyrolytic characteristics and kinetics of marine macroalgae, and the results also showed that the activation energy of macroalgae pyrolysis was lower than that of cornstalk and sawdust, illustrating that they were more easily decomposed. Wang et al. [23] analyzed the pyrolytic characteristics and mechanism of typical macroalgae by thermogravimetric analysis, differential thermal analysis and mass spectrometry. The major pyrolysis stage of algae biomass was in the range of 180–580 °C, and the initial pyrolysis temperature was lower than that of land plants. The composition of algae resulted in the complexity of its pyrolysis products, including H_2, CH_4, CO, CO_2, H_2S, SO_2, NO_x, etc., of which H_2S and SO_2 were related to the acidophobe in polysaccharides while NO_x was related to proteins.

In general, forestry biomass has relatively low hemicellulose content, while its cellulose content is high, leading to its DTG curve being similar to that of cellulose. Agricultural and herbaceous biomass have relatively low lignin contents but high hemicellulose and cellulose contents. As a result, the corresponding DTG curve has a notable shoulder peak corresponding to hemicellulose pyrolysis and a main peak corresponding to cellulose pyrolysis. The DTG curve from aquatic algae biomass displays multiple peaks due to the high extractive content and low cellulose content.

7.2.5 Comparison of pyrolysis products from different biomass species

TG-FTIR analysis showed that at the maximal weight loss in the pyrolysis of different species of biomass, the corresponding FTIR figures of pyrolysis products were different. The comparison of pyrolytic products from Mongolian Scots pine, rice husk, bamboo and algae is displayed in Fig. 7.9. The pyrolysis of forestry biomass has the strongest absorbance at the wavenumbers of condensable volatiles such as hydrocarbons, ketones, aldehydes, and acids, while the absorbance of non-condensable small molecular gases like CO and CO_2 is weak. Therefore, the pyrolytic products of forestry biomass have a high condensable volatiles content, indicating that a high bio-

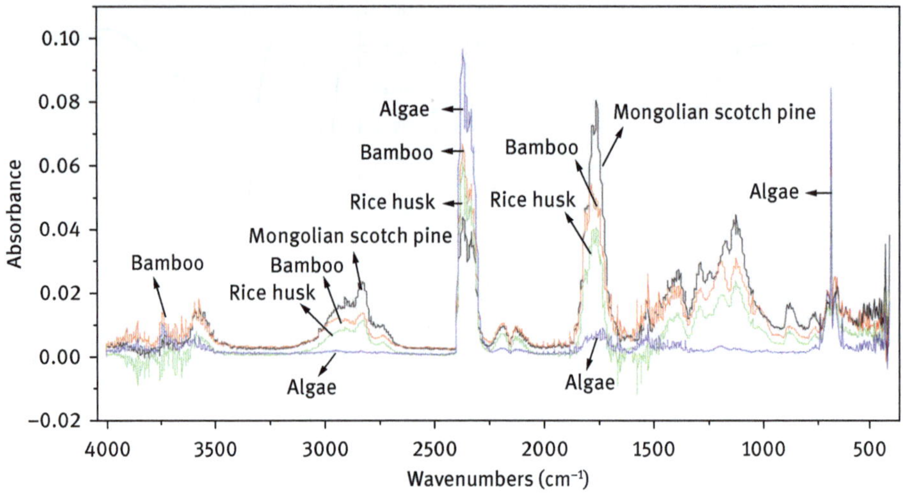

Fig. 7.9: Comparison of pyrolytic product distribution from the pyrolysis of different species of biomass.

oil yield could be achieved. The distribution of pyrolytic products from agricultural biomass is similar to that from forestry biomass, but the absorbance is weaker, implying a lower bio-oil yield. The pyrolytic products of bamboo have strong absorbance at the wavenumbers of CO_2 and water. Especially water shows the strongest absorbance in all the biomass materials, indicating the high water content in its bio-oil. The peak intensity of CO_2 in pyrolytic products of algae was the strongest compared to other functional groups, resulting in a low bio-oil yield and a simple composition of bio-oil.

7.3 Fast pyrolysis of biomass for bio-oil production

Fast pyrolysis of biomass is one of the advanced technologies applied in renewable energy development. The main difference between various pyrolysis technologies lies in the pyrolysis reactors, as shown in Tab. 7.1. Currently, Ensyn, Dynamotive, ROI, and BTG are the main companies engaged in research and promotion of biomass fast pyrolysis. Fast pyrolysis technology in China is also developing rapidly. Since the introduction of the rotating cone flash pyrolysis reactor from the Netherlands by Shenyang Agricultural University, many universities and scientific research institutes in China have also been carrying out studies on biomass fast pyrolysis, mainly focusing on experimental research and pilot scale tests.

Tab. 7.1: Typical fast pyrolysis reactors for biomass.

Type	Reactor	Institute	Fundamental principles
Contact mechanical reactor	Vortex	National Renewable Energy Laboratory	The biomass particles enter tangentially to the reactor with the injection of high speed nitrogen and superheated steam and are then decomposed. Recycling of biomass particles achieves the high efficiency decomposition.
	Rotating cone	University of Twente	The biomass particles and inert carriers are fed into the bottom of the reactor and move upward along with the cone under the centrifugal force to realize sufficient blending.
	Ablative	University of Aston	The biomass particles are sent the into the reactor bottom by the rotation of the vane and then decompose due to the high speed motion on the hot reactor surface.
Hybrid reactor	Rapid ejection	Georgia Institutes of Technology	The mixture of biomass particles and hot gas flows from the bottom mixing chamber upward to the reactor and is then decomposed.
	Fluidized bed reactor	University of Waterloo	The fluidized biomass particles are rapidly pyrolyzed under the inert carrier gas in the reactor.
	Vacuum	Laval University	The biomass particles are heated and decompose between two horizontal homothermal metal plates heated by molten salts in the vacuum reactor.

7.3.1 Reaction process of biomass fast pyrolysis

When the biomass particle passes through the hot area, heat is transferred to its surface through convection and radiation. Heat conduction is the main means of heat transfer from its surface to the inside, resulting in increased temperature of the particle. After the removal of water, the increase in temperature leads to the pyrolysis of biomass, the release of volatiles and the formation of coke. During this process, the reaction thermal effect and the flow of volatiles inside the particles influence the inner heat transfer of biomass particles. Since the orientation of heat transfer in biomass particles is from their external surface to the inside during pyrolysis, the biomass particles can be divided into three regions: the external surface where the reaction is finished, the middle region in pyrolysis, and the inner region in which pyrolysis does not occur. The primary pyrolysis of biomass produces bio-oil, non-condensable gases, and coke. The primary bio-oil in the inner of porous biomass particles is further pyrolyzed to produce non-condensable gases, secondary bio-oil, and coke. Meanwhile, the volatiles also react with ambient gaseous chemicals after they escape from the biomass particle. The secondary reactions of volatiles inside and outside the biomass particle change the distribution of final products. In order to improve the yield of bio-oil, the volatiles produced from primary pyrolysis should be condensed rapidly to inhibit secondary pyrolysis. Fast pyrolysis of biomass aims at the maximum yield of bio-oil, which is affected by the reaction temperature, the heating rate, the residence time and the species of feedstock.

7.3.2 Effect of reaction parameters on biomass fast pyrolysis

7.3.2.1 Effect of reaction temperature

The pyrolysis of biomass is influenced by various factors, among which temperature is the key factor for the pyrolysis process of biomass and the ultimate composition of products, while other factors include the control of the heating rate to the reaction temperature or the residence time of volatiles in the reaction zone. Therefore, the effect of temperature on pyrolysis characteristics should be analyzed first. For the different pyrolysis technologies, the distribution of all products varies with the reaction temperatures. In general, the pyrolysis of biomass proceeds slowly and the products mainly consist of char and non-condensable gas when the temperature is below 400 °C. In the range of 450–600 °C, the bio-oil yield first increases with elevating temperature and then declines with the continuous rise of temperature after reaching its maximum value. The gas yield rises with increasing temperature and it becomes dominant when the temperature is higher than 650 °C. The distribution of pyrolysis products from rosewood and spruce according to pyrolysis temperature is shown in Fig. 7.10. The char yield gradually decreases with increasing temperature and then tends to a certain value. The yield of non-condensable gas increases uniformly with temperature. The bio-oil yield first increases slightly with rising temperature, but then decreases with the continuing rise in temperature after reaching its highest value at a certain temperature. This is because bio-oil undergoes secondary decomposition to produce non-condensable gas, char and secondary bio-oil when the temperature

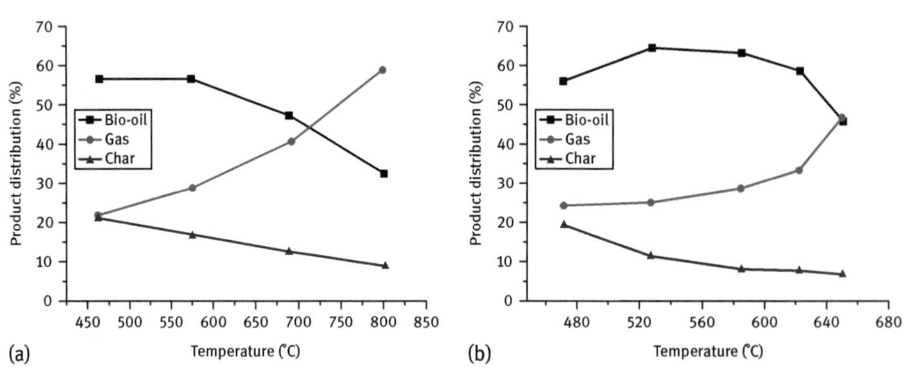

Fig. 7.10: Distribution of pyrolysis products according to pyrolysis temperature.

gets too high, which results in the decrease of bio-oil yield. The optimal temperature for the maximum bio-oil yield was in the range of 500–550 °C, depending on various experimental conditions. However, the distribution of ultimate products was similar. Horne et al. [24] used mixed wood waste as the feedstock to study its pyrolysis characteristics in a fluidized bed reactor at different temperatures. The results showed that when the temperature increased, the char yield decreased obviously whereas the yields of liquid and gas products increased. However, gasification, instead of liquefaction, was the main reaction when the temperature continuously elevated, which led to a reduction of liquid yield and an increased gas yield. Liu et al [25] studied the pyrolysis behavior of sawdust in a self-design fluidized bed reactor and obtained a high bio-oil yield of 58.74 % at 500 °C. The results in another fluidized bed reactor carried out by Scott et al. [26, 27] also showed that the yield of pyrolysis oil from aspen-popular sawdust increased with the increasing temperature, and reached the highest value at 500–520 °C, whereas it declined with a further rise in temperature. Conti et al. [28] fed bagasse into a fluidized bed pyrolyzer in the temperature range of 400–750 °C and obtained a maximum oil yield of 44 % at 500 °C. The pyrolysis of agricultural residues under different conditions was carried out by Encinar et al. [29]. They also found that increasing temperature could lower the char yield and raise the gas yield, and in most cases, a temperature between 600 °C to 700 °C led to a maximum liquid production.

The composition of bio-oil is also affected by reaction temperature. The H/C and O/C ratios decreased with increasing temperature, which indicates that the secondary or condensation reactions of primary bio-oil occur to form stable organics with low oxygen content such as benzene and naphthalene. Comparison of the composition of bio-oils from spruce pyrolysis at different temperatures illustrated that the content of unstable organics like carboxylic acids declined sharply owing to the fact that carboxylic acids could be decomposed into gas in the form of alkanes, CO, CO_2, and water at a high temperature. The quantities of stable chemicals like phenols increased slightly with increasing temperature and meanwhile polycyclic aromatic hydrocarbons like naphthalene and acenaphthene started to be generated and their quantities also increased. Zhang et al. [30] also reported that the content of oxygenated compounds including aldehydes, ketones and esters declined with a further temperature increase.

The reaction temperature also has some effects on the composition of gaseous products. The main gaseous products during biomass pyrolysis are CO, CO_2, H_2, and CH_4. CO_2 is mainly derived from the primary cleavage of the uronic acid structures in hemicellulose at low temperatures, and the cleavage of carboxyls in lignin at high temperatures also releases a small amount of CO_2. As a result, the CO_2 content decreases with increasing temperature. CO, formed from the breakage of unstable carbonyls in volatiles, is derived from the secondary decomposition of primary volatiles at high temperatures. Therefore, the yield of CO increases with increasing temperature. Zhang et al. [30] carried out an investigation of the pyrolysis of soybean stalk, rice husk, and sawdust and found that the CO content was low at 400 °C, while it increased to 40 %

at 800 °C. The decomposition of methoxyl groups rich in lignin favors the formation of CH_4, and the CH_4 yield is increased by the deep decomposition of lignin with increasing temperature.

7.3.2.2 Effect of heating rate

The heating rate, an important characteristic for distinguishing the type of reactors, is determined by the type of reactor, the reaction temperature and the particle size of biomass. Scott et al. [26] pointed out that the heating rate depends on the heat transfer, particle size, etc. Maschio et al. [31] reported that a high heating rate required higher temperature, shorter residence time and smaller particle size. The yields of gas, liquid, and char products are almost equal under a lower temperature and a longer residence time. Generally, a low heating rate will lead to extending the residence time in the low temperature region, leading to the delay of internal particle temperature reaching the set pyrolysis temperature, which results in increasing char yield due to the dehydration and charring reactions of cellulose and lignin. A high heating rate, often adopted in biomass pyrolysis, is beneficial to increasing bio-oil yield and reducing char formation. This is due to the fact that the residence time of biomass particles in the low temperature region is shortened, resulting in the reduction of dehydration and polycondensation of cellulose and lignin, as well as the increasing difficulty in the formation of carbon skeleton. Broido et al. [32] reported that a low heating rate favored the formation of char rather than tar. Demirbas et al. [33] investigated the effects of heating rate on beech bark pyrolysis, and found that the yield of bio-oil and its heating value increased uniformly with rising heating rate. Onay [34] carried out fast pyrolysis of safflower seed, and obtained a bio-oil yield of 54 % under a heating rate of 300 $°C\,min^{-1}$, which was about 7 wt% higher than that under 100 $°C\,min^{-1}$. The heating rate is affected by the structure of the reactor and the particle size. Therefore, fluidized bed, entrained bed, rotating cone and spouted bed reactors are the most common reactors to realize fast pyrolysis.

7.3.2.3 Effect of residence time

The residence time can refer to that of solids or gaseous products. Drummond et al. [35] pointed out that biomass feedstock should be in the pyrolyzer for a certain time, otherwise the pyrolysis of biomass was incomplete. For fast pyrolysis, the residence time generally refers to the residence time of gaseous products, which roughly equals the ratio of reactor volume to gas volumetric flow. The residence time significantly affects the yield of bio-oil. The condensable volatiles, which escape from the particle in the initial stage of biomass fast pyrolysis, can undergo secondary cracking to form char, secondary oil, and non-condensable gas in the gaseous phase space, resulting in the decline of bio-oil yield. A longer residence time can lead to deeper secondary decomposition. As shown in Fig. 7.11, a study of spruce pyrolysis in infrared radiation

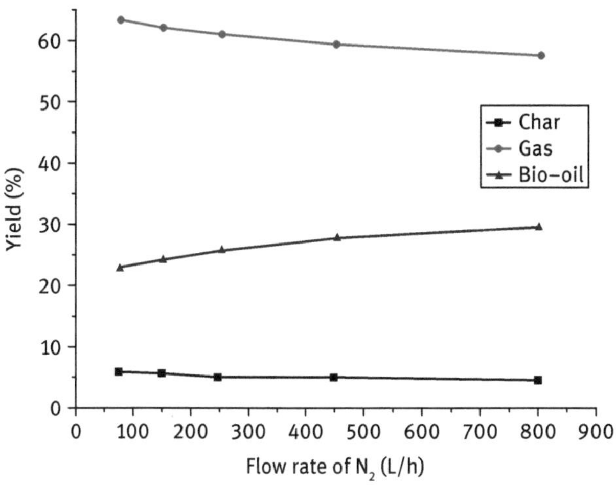

Fig. 7.11: Effect of residence time on spruce pyrolysis products. Reprinted with permission from [36].

apparatus showed that the yield of spruce bio-oil declined slightly with decreasing carrier gas flow rate, which corresponded to increasing residence time. Meanwhile, the non-condensable gas yield also increased. These observations also match with other results obtained by other researchers [37, 38].

7.3.2.4 Effect of inorganic salts and catalysts

Biomass mainly contains carbon, hydrogen and oxygen, as well as a certain content of inorganic elements like K, Na, Ca, Mg, Fe, and Cu. Although the quantities of these inorganic elements are very low, they have significant effects on biomass pyrolysis and have been attracting more and more attention in recent years.

Fig. 7.12 shows that with the addition of potassium, the initial and final temperatures of the pyrolysis process, as well as the temperature corresponding to the maximum weight loss, all moved towards to the low temperature region, while the release of volatiles was reduced and the char yield increased. The addition of 7 % of potassium increased the char yield from 20 % to 35 % at 530 °C, and this tendency became more obvious with the addition of high amounts of potassium. This phenomenon was in accordance with that of potassium on cellulose pyrolysis. It is worth noting that the original shoulder peak, which occurred in the DTG curve of spruce pyrolysis at around 300 °C, shifted closer to the low temperature region after the addition of potassium. A similar regulation was also observed by Müller-Hagedorn et al. in their study of effects of potassium on pine weight loss [39]. The reason for this phenomenon might be due to the fact that potassium accelerated the initial pyrolysis of hemicellulose.

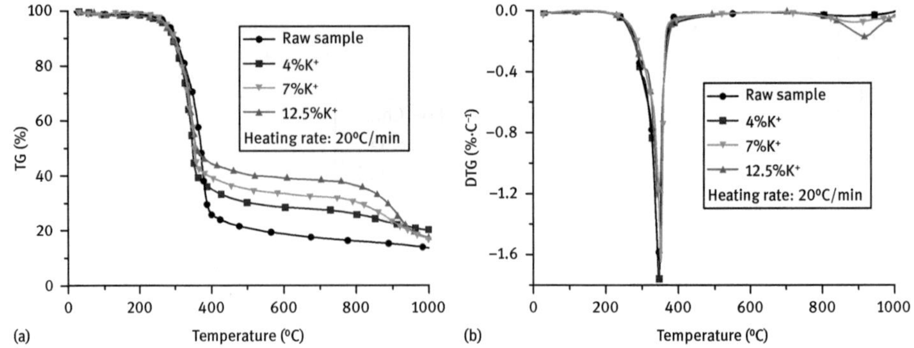

Fig. 7.12: TG/DTG curves of spruce pyrolysis with the addition of potassium.

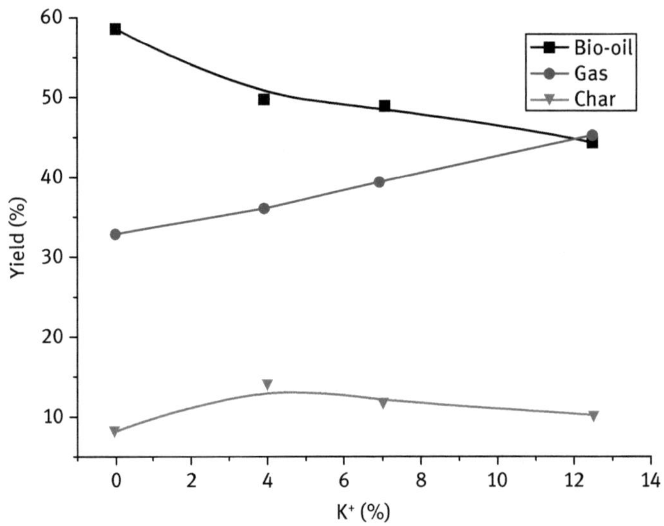

Fig. 7.13: Effect of potassium on the distribution of pyrolysis products. Reprinted with permission from [40].

Fig. 7.13 shows the effect of potassium on the distribution of pyrolysis products. The yield of bio-oil reduced gradually, while that of gaseous products increased gradually. The char yield elevated noticeably at a potassium content of 4 % and then remained stable with further addition of potassium. By the addition of 12.7 % potassium, water production was enhanced obviously from 13.18 % to 20.03 %. Jensen et al. [41] employed TG-FTIR to study the effect of KCl on wheat straw pyrolysis. The release of water increased with the addition of KCl, illustrating that KCl promoted the dehydration. A similar variational tendency was also observed by Wang et al. in their study of the effect of K_2CO_3 on pine pyrolysis [42]. The analysis of bio-oil distribution suggested that the acid and furan contents declined obviously while those of alkanes and phenols increased slightly when potassium was added. With regard to gaseous products, the amounts of H_2, CO, and CO_2 increased.

Calcium, another significant element widely present in biomass, also has an important effect on biomass pyrolysis. The TG/DTG curves of spruce pyrolysis with the addition of calcium are shown in Fig. 7.14. The temperature region of spruce pyrolysis moved toward the low temperature region with the addition of calcium, which was similar to the effect of potassium. Furthermore, the char yield increased obviously after 530 °C. A higher char yield was achieved with the addition of calcium compared to potassium, indicating that calcium's catalytic effect on char formation was more intensive than that of potassium.

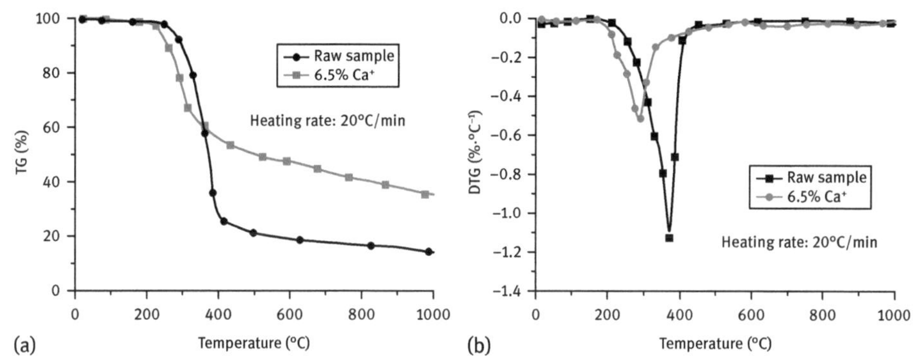

Fig. 7.14: Effect of calcium on the TG/DTG curves of spruce pyrolysis.

Fig. 7.15 displays the effect of calcium on the distribution of pyrolysis products. The yield of bio-oil decreased while that of char increased as the concentration of calcium addition increased. The gaseous products yield did not show much variation. However, Wang et al. [42] found that with the addition of $Ca(OH)_2$, the yields of bio-oil and gaseous products increased slightly with a decrease in the char yield, which might be due to the different anions. Meanwhile, acids and aldehydes in pyrolysis oil disappeared and the yields of sugars, furans, and guaiacols declined, while the yield of alcohols increased remarkably. Richards et al. [43] used TG-FTIR to study the effects of metal salts on the release of volatiles and observed that potassium significantly promoted the formation of CO, CO_2, formic acid, acetic acid, and ethanol in contrast to calcium.

In addition to the inorganic salts in biomass, adding catalysts can also change pyrolysis behaviors. Fig. 7.16 displays the effect of zeolite catalysts on the pyrolysis of Manchurian ash. All types of zeolites promoted dehydration at the initial stage, resulting in a large weight loss at this period and the movement of the main reaction temperature to the high temperature region. Char yields increased, especially under

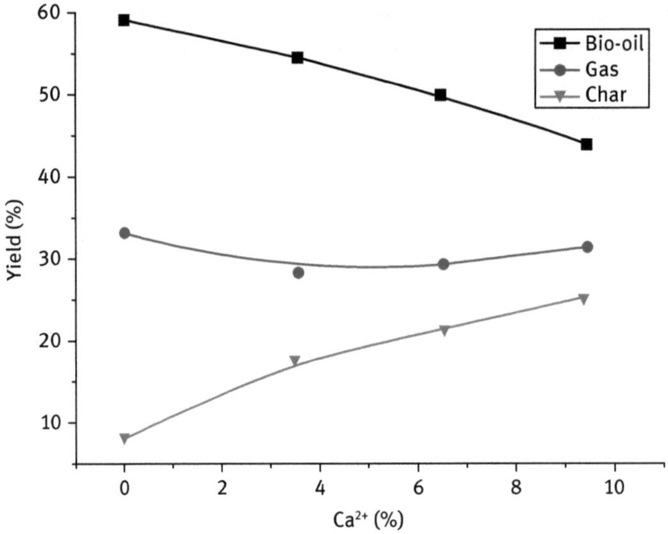

Fig. 7.15: Effect of calcium on the distribution of pyrolysis products. Reprinted with permission from [40].

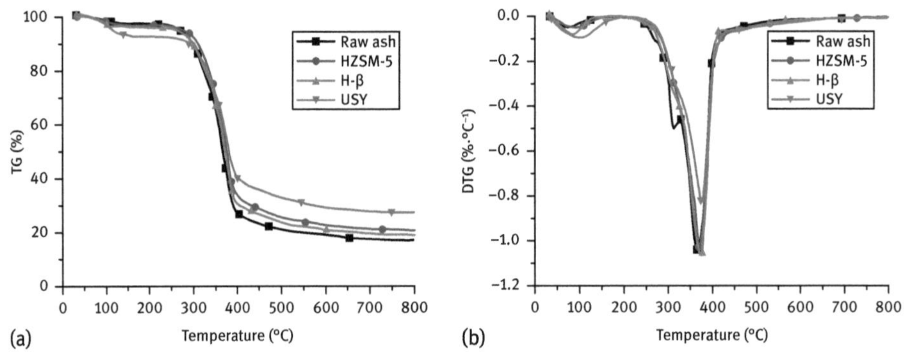

Fig. 7.16: Effect of the addition of zeolite catalysts on the pyrolysis of Manchurian ash. Reprinted with permission form [44], © 2010 Taylor & Francis.

the addition of USY. These phenomena were different from the effect of zeolites on the three individual components in biomass, indicating that the interaction of the three components, the small amounts of extractives and inorganic salts also had significant effects on char formation and the pyrolysis routes.

The addition of zeolite catalysts also affects the distribution of pyrolysis products. Among these catalysts, the effect of USY is the most outstanding, as shown in Fig. 7.17. The quantities of oxygenated compounds such as acids, aldehydes and esters decreased significantly, whereas the contents of alkanes and aromatics products

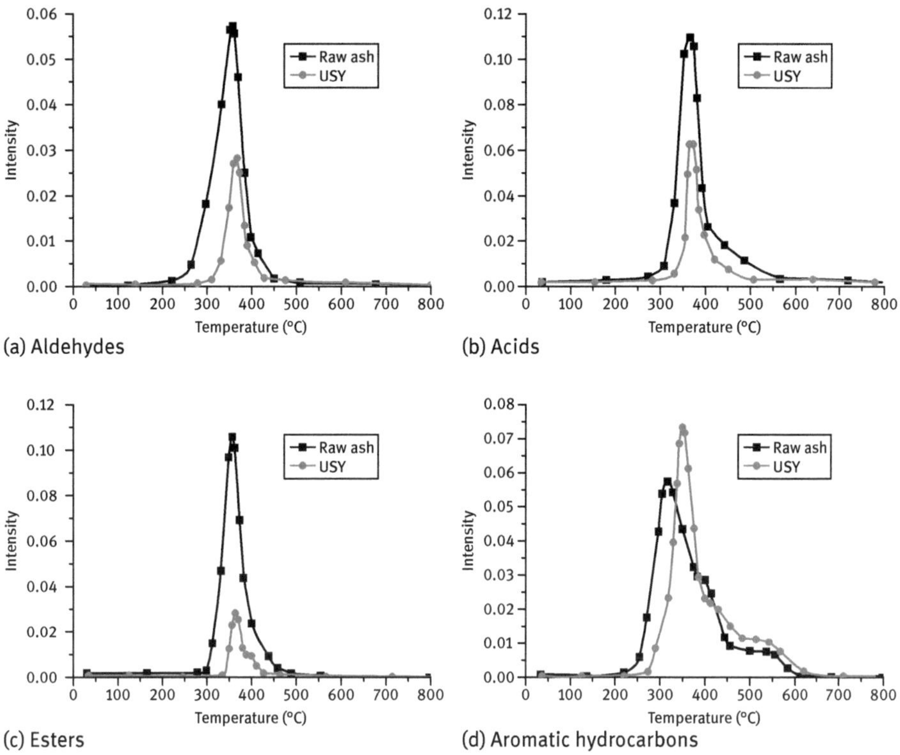

Fig. 7.17: Effect of the addition of USY on the pyrolysis of Manchurian ash. Reprinted with permission form [44], © 2010 Taylor & Francis.

increased. After the addition of USY, the contents of acids, aldehydes and esters decreased by 65.86 %, 49.63 %, and 80.56 % respectively, while the content of aromatics increased by 5.15 %. The increase in hydrocarbon products improved the quality of bio-oil, but resulted in a decrease in bio-oil yield. Therefore, the influence of USY on removing the oxygenated compounds was obvious, which was in agreement with its effect on the pyrolysis of cellulose.

Aho [45] selected H-β, H-Y, HZSM-5, and Mordenite as the catalysts to investigate their effects on the pyrolysis of pine. The addition of catalysts significantly promoted the dehydration of pine pyrolysis, and the water content increased from the initial 5.4 % without catalysts to 13.9 %, 16.7 %, 13.0 %, and 14.4 %, while the bio-oil yield decreased from 27.3 % to 15.1 %, 9.0 %, 20.7 %, and 17.6 %, respectively. The catalytic behaviors of β-zeolites with different silica to alumina ratios were also studied by Aho et al. [46]. By analyzing the relative contents of acids, aldehydes, alcohols, phenols, ketones and polyaromatic hydrocarbons in bio-oil, it was found that the yields of several chemical families were clearly influenced by the zeolite's acidity. Zeolites with stronger acidity produced more water and polyaromatic hydrocarbons rather than oxygenated organics. The effects of ZSM-5, H-Y, and USY on fast pyrolysis of corn stalks were investigated by Uzun et al. [47]. These catalysts led to a reduction in bio-oil yield and char, but a promotion of gas production. Meanwhile, a high deoxygenation was achieved. The oxygen content in bio-oil declined from the initial 24.50 % to 20.23 %, 15.70 %, and 19.98 %, respectively. Among the catalysts used, H-Y enhanced the formation of aliphatics, and USY gave a high amount of aromatics, whereas ZSM-5 had a neutral performance. Williams and Nugranad [48] used ZSM-5 as the catalyst in the pyrolysis of rice husks in a fluidized bed reactor. The addition of ZSM-5 enhanced dehydration and bio-oil yield and its oxygen content decreased remarkably, while an obvious increase in the quantities of monocyclic and polycyclic aromatic hydrocarbons was observed. Qi et al. [49] studied the effect of the NaY zeolite on bamboo pyrolysis, and they reported that the addition of NaY led to a significant increase in the yield of liquid products. Therefore, different types of zeolites result in various effects on the distribution of pyrolysis products, which is due to the intrinsic properties of the catalysts themselves, as well as the distribution of components in biomass.

7.3.2.5 Effect of biomass feedstocks

In different species of biomass, the contents of the three components vary and the structures of hemicellulose and lignin also differ due to the various primary units and linkages in them. In addition, the interactions between the biomass components, the inorganic salts and the extractives also influence the overall pyrolysis behavior of biomass. As a result, the pyrolysis behaviors for different species of biomass are different, indicating that it is necessary to study the product distributions and properties during the pyrolysis of various species of biomass.

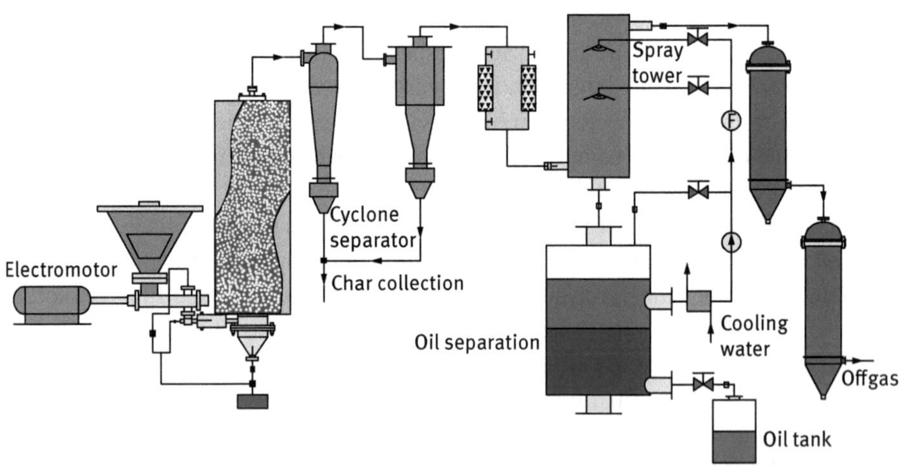

Fig. 7.18: Schematic diagram of fluidized bed reactor for biomass fast pyrolysis.

The pyrolysis properties of different biomass species were studied in a self-design fast pyrolysis reactor with a spray condenser, as shown in Fig. 7.18. Seven different biomass species, including Mongolian Scots pine, rosewood, bamboo, grassiness, rich husk, rice straw, and algae, were selected as the typical representatives of forestry wastes, agricultural wastes, and herbaceous and aquatic plants.

The bio-oil and char yields are shown in Figs. 7.19 and 7.20, respectively. Both woody and agricultural biomass feedstocks gave high bio-oil yields. Although both bamboo and grassiness belong to herbaceous plants, the bio-oil yields from their pyrolysis differed greatly. The bio-oil yield from bamboo pyrolysis was similar to that from forestry wastes, while that from grassiness pyrolysis was only 17.65 %, which was the lowest among the selected feedstocks. Furthermore, the pyrolysis of algae also achieved a low bio-oil yield of 20.54 %.

The difference in bio-oil yields is related to the different intrinsic properties of biomass feedstocks, because the contents of the three main components and ash in biomass significantly influence the distribution of pyrolysis products. As mentioned in the above sections, the pyrolysis of cellulose gives the highest bio-oil yield, and the pyrolysis of hemicellulose favors the formation of gaseous products, while lignin pyrolysis contributes remarkably to char formation. Similar phenomena have also been observed by other researchers [50–54]. The pyrolysis of cellulose and hemicellulose mainly produces the volatiles, an especially small amount of residue is obtained from cellulose pyrolysis. Char is one of the major products from lignin pyrolysis, so that lignin is regarded as the main source for pyrolytic char formation. In addition, as

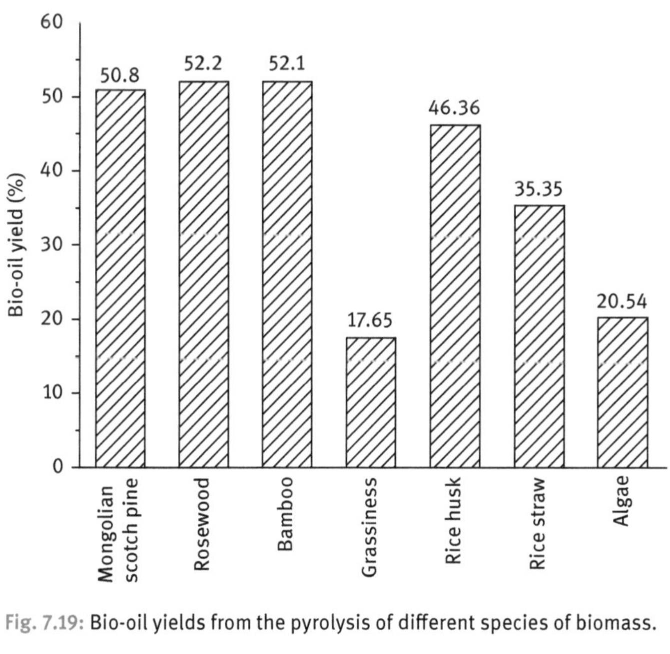

Fig. 7.19: Bio-oil yields from the pyrolysis of different species of biomass.

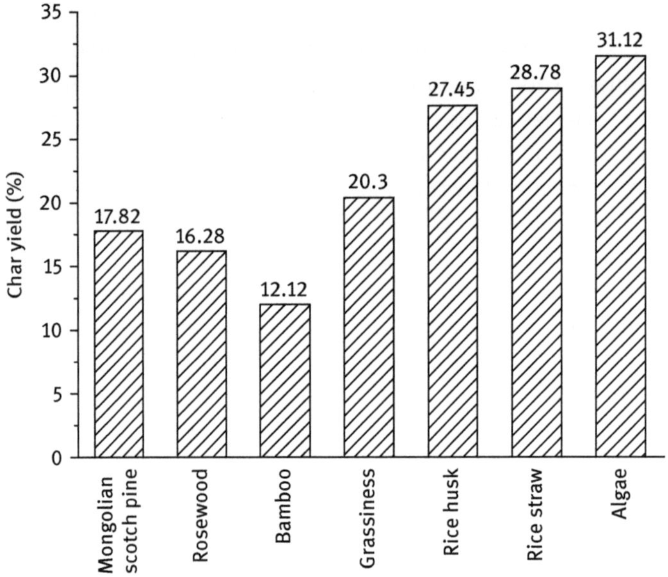

Fig. 7.20: Char yields from the pyrolysis of different species of biomass.

mentioned before, the metal salts in biomass promote the formation of char and gas through secondary condensation and cracking reactions of large molecular weight compounds in bio-oil, which leads to the decrease in bio-oil yield and increasing char and gas yields.

Based on the distribution of pyrolysis products from the three main components in biomass, combined with the proximate analysis of seven species of biomass as well as the measurement of the three components' contents, the distribution of pyrolysis products of real biomass was analyzed. The cellulose content of Mongolian Scots pine and rosewood belonging to forest biomass were close to 50 %, and they had relatively low hemicellulose content and the lowest ash content, therefore favouring the improvement of bio-oil yield. The pyrolysis of real forest biomass also had the highest bio-oil yield and a relatively low char yield. The cellulose and hemicellulose content in rice husk and rice straw were high, while the lignin content was below 10 %. In addition, the ash content in rice husk and rice straw were high. Based on these results, we could conclude that the pyrolysis of agricultural biomass produced more char but less bio-oil compared with forest biomass. The results of Liu et al. [53] also showed that forestry biomass tended to produce a bio-oil yield up to 60–65 % higher than agricultural and herbaceous biomass. A big difference was found in the distribution of pyrolysis products between bamboo and grassiness. The amounts of the three components in bamboo were similar to that in agricultural biomass. Consequently, the distribution of its pyrolysis products was similar to that of rice husk and rice straw, achieving a high bio-oil yield. The of extractives content and acid-insoluble ash content in grassiness were up to 27.69 % and 5.99 % respectively, while the cellulose content was only 33.21 %, leading to a low bio-oil yield and a high char yield. Algae had the highest char yield of these seven species of biomass at 31.12 %, while its bio-oil yield was the lowest, which was due to its high extractives content of up to 50 %. The chemical distribution of seven bio-oils from different species of biomass is shown in Fig. 7.21. These bio-oils had similar chemical families covering the main products from the three components' pyrolysis, namely acids, alcohols, ketones, aldehydes, sugars, and phenols. Especially acetic acid, furfural, hydroxyacetone, guaiacol, and levoglucosan accounted for high percentages. A large amount of lipids and proteins in algae resulted in more nitrogenous compounds in its bio-oil.

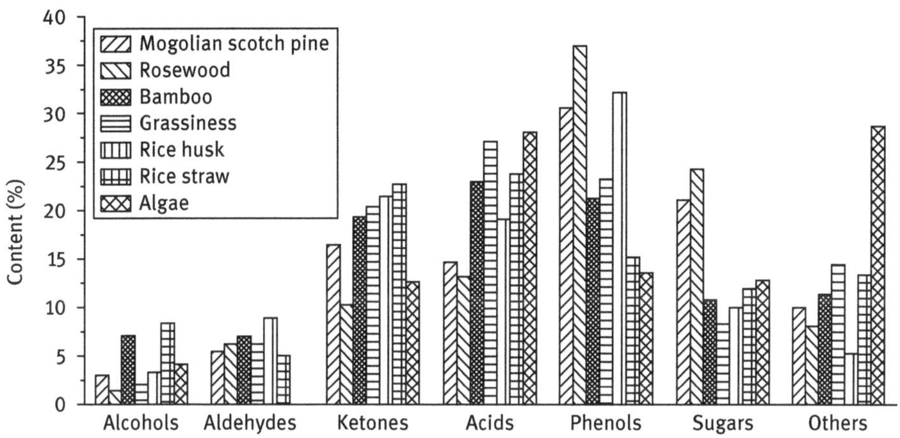

Fig. 7.21: Distribution of bio-oil compounds obtained from different species of biomass.

As mentioned in the individual pyrolysis study of the three components in biomass, sugars (such as 3,4-anhydro-altrose and levoglucosan) and acetic acid are typical products from cellulose and hemicellulose pyrolysis respectively, whereas the typical products of lignin pyrolysis are phenol, guaiacol, syringol, 2,6-dimethoxy-4-(2-propenyl)phenol. Additionally, furfural and ketones (such as hydroxyacetone, 2-2-hydroxy-cyclopenten-1-one and 3-methyl-2-2-hydroxy-cyclopenten-1-one) are produced from the pyrolysis of cellulose and hemicellulose. Forestry biomass has a high cellulose content, leading to the high sugars content in its bio-oil, of which levoglucosan and 3,4-anhydro-altrose are the main products. Furthermore, typical pyrolysis products from cellulose and hemicellulose, including aldehydes (like furfural) and ketones (like hydroxyacetone), can also be found in forestry biomass bio-oil, especially in bio-oil from Mongolian Scots pine. In addition, due to the high lignin content in forestry biomass, its bio-oils contain the highest quantities of phenolic compounds, including phenol, guaiacol, syringol, etc. Agricultural biomass is abundant in hemicellulose and cellulose, resulting in the high quantities of acids, ketones, and aldehydes, of which the typical compounds are acetic acid, hydroxyacetone, furfural, etc. Bamboo, belonging to herbaceous biomass, also has a high hemicellulose content, leading to high quantities of acids, aldehydes, and ketones in its bio-oil. The bio-oil yields of grassiness and algae are far lower than those of other species, so that the real yields of chemical compounds in bio-oil are lower, which is in accordance with their high extractives contents and low contents of the three components. Although the distribution of compounds in bio-oil has a certain relationship with the distribution of the three components in biomass, it is also affected by inorganic salts and the interaction between the three components. For example, the lignin content in Mongolian Scots pine is higher than that in rosewood, but its phenols content is lower than rosewood's. The cellulose content in these two species of biomass is similar, but the contents of acids, aldehydes, and ketones in their corresponding bio-oils are different.

7.3.2.6 Effect of torrefaction

Torrefaction, also known as mild pyrolysis, is usually carried out as a pretreatment technology for biomass at low temperature (200–300 °C). It can remove water and preserve about 90 % of energy in raw biomass [55]. In addition, this technology also improves the grindability and hydrophobicity of biomass. Hence, torrefaction is considered one of the promising pretreatment technologies before biomass gasification, pyrolysis or combustion.

In general, biomass undergoes structure changes and releases some small molecular gases during torrefaction. Fig. 7.22 shows the variations in the composition of fast growing poplar and torrefaction yields at different torrefaction temperatures. Cellulose content remains nearly unchanged below 250 °C, and decreases slightly when the temperature is further elevated. Lignin only loses about 20 % of its original weight, and its content increased to about half of the torrefied biomass. Hemicellulose changes significantly when the torrefaction temperature exceeds 250 °C. The typical hemicellulose representative, xylan, only occupies about 36 % of the original weight after torrefaction at 300 °C (Fig. 7.22 (b)). This might be due to the fact that during torrefaction, O-acetyl branches in hemicellulose could be dissociated because of their weak thermal stability. It indicates that small molecules released during torrefaction are mainly from hemicellulose. The ratios of H/C and O/C decrease with increasing torrefaction temperature. This might be due to that some volatiles, such as water and acids, are released by dehydration and deacetylation. Prins et al. [56] also found acetic acid and water were the main products during biomass torrefaction. It should be noted that the elemental composition also showed remarkable changes at torrefaction temperatures above 250 °C. These variations all showed that 250 °C was a turning point, at which a significant torrefaction influence on biomass started to appear.

The evolution of some typical functional groups during torrefaction was investigated using the perturbation-based two-dimensional correlation method (2D-PCIS). The original hydrogen bond network was broken, resulting in increasing vibration in-

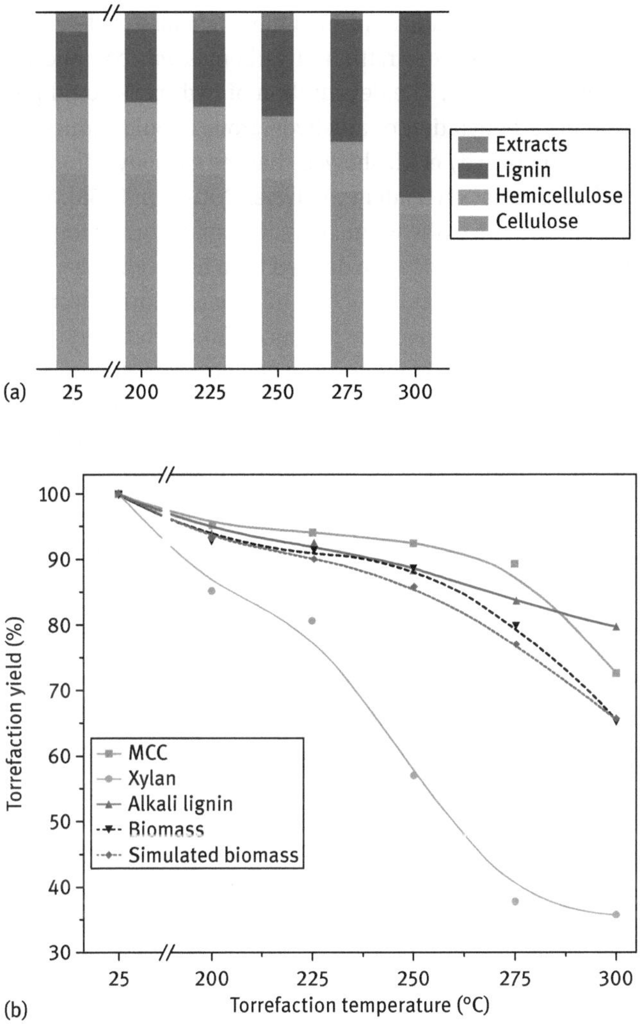

Fig. 7.22: Variations in biomass component distributions (a) and yields (b) during torrefaction. Reprinted with permission from [57], © 2015 American Chemical Society.

tensity for free hydroxyls and the decrease in vibration intensity for associated hydroxyls. Alkane groups were transformed into alkene groups, such as the dehydration of hydroxyls at the side chains of lignin or the dehydration of sugar units. Some linkages among biomass units, including glycosidic bonds, O-acetyl branches, and aryl ether linkages, could be broken easily at low temperature. These reactions might be more vigorous above 250 °C. The dissociation of O-acetyl branches was the main reason for the formation of acetic acid. The most abundant aryl ether linkages, β-O-4, could be cleaved within the torrefaction temperature, resulting in the depolymerization of lignin. The methoxyls attached to the aromatic units in lignin might also undergo dissociation. The torrefaction influence on the variation of carbonyls was complicated and might be affected by biomass species. The dehydration of hydroxyls could produce carbonyls, whereas the decarbonylation of aldehyde groups could reduce the carbonyls. The results reported by Harvey et al. showed that the carbonyls from different biomass exhibited different variation tendency between 200–300 °C [58]. Since hemicellulose made a huge contribution to the weight loss of biomass during torrefaction, the main variations of functional groups were derived from hemicellulose.

Apparently, the variations in the chemical structure of biomass during torrefaction influence its ultimate distribution of pyrolytic products. Since some aryl ester linkages and side chains can break at low temperature, some liner small compounds, especially acetic acid, exhibit low yields in bio-oil from the further pyrolysis of torrefied biomass. As mentioned in Chapter 3, acetic acid was mainly from hemicellulose which was easily cleaved, while furans were mainly produced by stepwise dehydration of pentosyl residues in hemicellulose. However, hemicellulose decomposed remarkably during torrefaction pretreatment. As a result, the yields of acetic acid and furans decrease in the bio-oil from torrefied biomass. When the torrefaction temperature increases, the yield of levoglucosan, a typical compound from cellulose pyrolysis, also decreases. Pelaez-Samaniego et al. [59] suggested that this phenomenon can be ascribed to the crosslinking of pyrolytic intermediates from cellulose degradation during torrefaction. Since the decomposition of lignin is difficult, phenolic compounds are the most abundant compounds in the pyrolytic products from torrefied biomass. The side chains of lignin might undergo degradation at high torrefaction temperatures, leading to an increasing yield of phenolic compounds without side chains. Neupane et al. [60] performed the torrefaction of biomass at 225–275 °C for 15–45 min, and also found that the bio-oil obtained from torrefied biomass contained more phenolic compounds. They propose that torrefied biomass has the potential to produce hydrocarbons.

7.4 Bio-oil graded catalytic upgrading

The bio-oil produced from fast pyrolysis is a complex organic mixture, in which there are many oxygenated components belonging to different chemical families, as well as some sugars and phenolic oligomers with high molecular weight, leading to the inferior fuel quality of bio-oil [54, 61]. The carbolic acids in bio-oil are responsible for the corrosiveness, and the high water content not only decreases its heating value, but also leads to instability during combustion. Some conventional indexes of fuel quality of bio-oil are listed in Tab. 7.2. Compared to gasoline and diesel, the overall fuel quality of bio-oil is low. Its water content reaches 15–50 %, the LHV ranges from 13–18 MJ/kg, and its pH value varies from 2 to 3. Consequently, crude bio-oil is difficult to be used directly as a high-grade liquid fuel. Instead, it is usually used as a furnace fuel.

Tab. 7.2: Fuel quality index of bio-oil.

Index	Bio-oil	Diesel	Gasoline
Water content (wt%)	15–50	Trace	N/A
Solid (wt%)	< 1	N/A	N/A
Ash (wt%)	< 1	< 0.01	/
Stability	Unstable	Stable	Stable
Viscosity (cst)	15–35 (40 °C)	1.8–8 (20 °C)	0.55–0.7 (30 °C)
Density (15 °C/kg.m^{-3})	1100–1300	190–850 (20 °C)	700–800 (20 °C)
Flash point (close cup/°C)	40–110	≥ 55	/
Char yield (wt%)	17–23	≤ 0.3	/
Lower heating value (MJ/kg)	13–18	46	43
pH	2–3	/	/

Bio-oil needs to be upgraded for it to realize its high-grade utilization. Several upgrading technologies have been developed to improve the quality of bio-oil, including catalytic esterification, catalytic cracking, hydrodeoxygenation, steam reforming, and emulsification [62, 63]. However, a single upgrading technology cannot realize the efficient conversion of all of the components due to the complexity of bio-oil, but causes problems such as catalyst deactivation and reactor blockage [64, 65]. Consequently, to achieve high upgrading efficiency for the whole bio-oil, it is necessary to adopt separation pretreatment of bio-oil to enrich the compounds suitable for the corresponding upgrading technologies.

The common separation and purification technologies for bio-oil are column chromatography, extraction, and distillation. Column chromatography is based on the different adsorption capacity of individual components retained in the stationary phase. By using different eluents, bio-oil can be separated into aliphatics, aromatics, and other polar chemicals [66, 67]. However, the fractions obtained by this method are not suitable for the subsequent graded upgrading of bio-oil. Solvent extraction, a conventional separation technology based on the principle of similarity and intermiscibility, also provides a fraction rich in the components with similar polarity. Therefore, similar to column chromatography, it cannot achieve the enrichment of chemicals appropriate for a certain upgrading technology either [68, 69]. Traditional distillation separates the chemicals by the difference in their boiling points. Since bio-oil has a large portion of thermal-sensitive chemicals and oligomers with large molecular weight, the distillation yield of traditional distillation is often low, furthermore, coking and deterioration of the residual fraction often occur simultaneously [54].

Molecular distillation is a separation technology which can separate the chemicals at a temperature far below their boiling points, making it more appropriate for the separation of thermal-sensitive bio-oil. Through molecular distillation, bio-oil can be divided into distilled fraction (light fraction and middle fraction) with high reactiv-

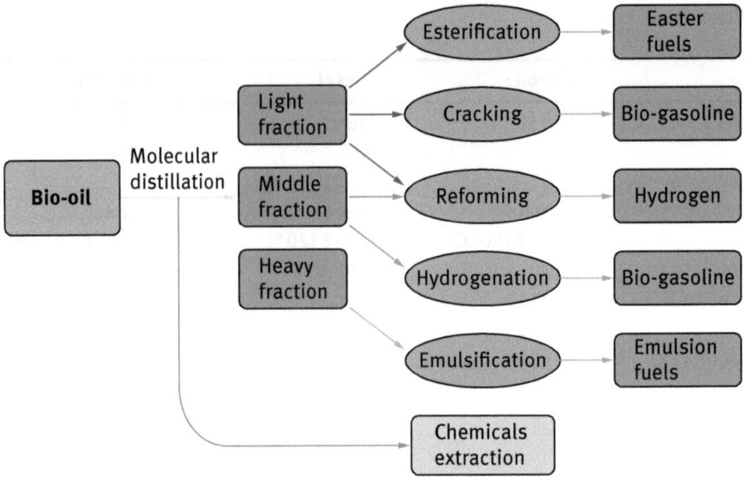

Fig. 7.23: A scheme of bio-oil graded upgrading based on molecular distillation.

ity, and residual fraction (heavy fraction) with a relatively low reactivity. The graded upgrading technology based on molecular distillation is exhibited in Fig. 7.23. The light fraction can be used to produce ester fuels by catalytic esterification, hydrocarbon fuels by catalytic cracking and hydrogen by steam reforming. Hydrogen and hydrocarbon fuels can also be obtained from the middle fraction by the corresponding upgrading technologies. The emulsification of the residual fraction and diesel can obtain emulsion fuels. In addition, valuable chemicals can be also extracted by molecular distillation.

7.4.1 High-efficiency separation of bio-oil based on molecular distillation

7.4.1.1 Introduction to molecular distillation

Molecular distillation is a special liquid-liquid separation technology, which relies on the various mean free paths of different substances (Fig. 7.24 (a)). The mean free path of a molecule is the mean distance of free motion before it collides with another molecule. Chemicals with low molecular weight in the distilled fraction have shorter mean free paths than those with high molecular weight in the residual fraction. The distance between evaporation and cooling surfaces is shorter than the mean free paths of chemicals in the light fraction but longer than those in the residual fraction. This special design makes the volatile compounds condense once they escape from the evaporation surface, which breaks the balance of gas-liquid around the evaporation surface, and then leads to the continuous distillation of light fraction. It is difficult for the chemicals in the residual fraction to reach the cooling surface and therefore high efficiency separation of chemicals can be achieved. Furthermore, the mean free path of a molecule is proportional to the temperature and inversely proportional to the pressure. Consequently, both high temperature and low pressure favor obtaining a high yield of distilled fraction.

The molecular distillation apparatus consists of four units: an evaporation unit, a heating and cooling unit, a feeding unit, and a vacuum unit. A typical molecular distillation apparatus of KDL-5 type is shown in Fig. 7.24 (b). During the molecular distillation process, an oil film is formed through the oriented distribution plate and rollers. The volatiles in the oil film escape from the surface of the evaporator and are cooled by the internal condenser, while the residual components flow down along the evaporator surface.

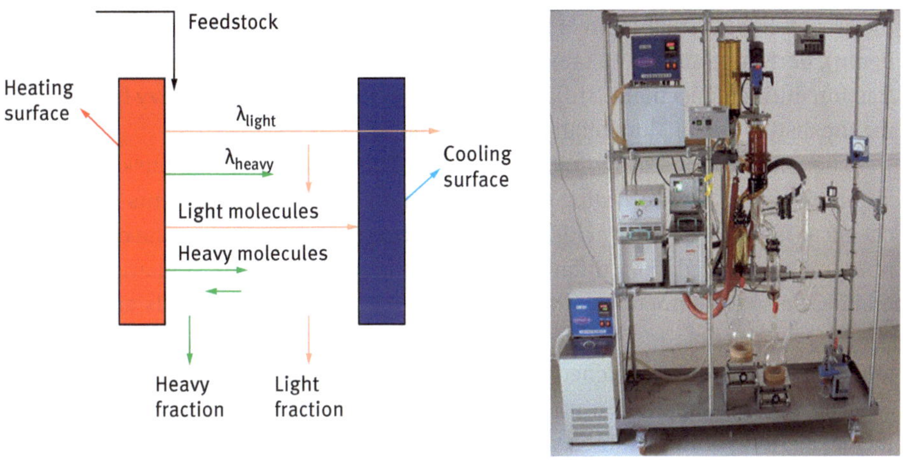

(a) (b)

Fig. 7.24: Scheme of molecular distillation; (a) schematic diagram, (b) KDL-5 molecular distillation apparatus.

7.4.1.2 Separation of bio-oil through molecular distillation

The separation characteristics of molecular distillation for bio-oil vary for different bio-oils due to the differences in the distribution of components. In general, a bio-oil with high water content and many light molecular weight compounds has better evaporation properties. The molecular distillation characteristics of bio-oils from pyrolysis of three types of biomass feedstock (Mongolian Scots pine, rice husk, algae) are displayed in Tab. 7.3. The evaporation temperature was 100 °C and the operation pressure was 100 Pa. Small molecular weight chemicals were evaporated and reached the second condenser and were collected as the light fraction (LF). The chemicals with middle molecular weights were cooled at the internal condenser surface and named the middle fraction (MF). The residual chemicals were called the heavy fraction (HF). Algae bio-oil had the highest distillation yield of 90.35 % among three bio-oils, of which the light fraction reached 87.83 %. This was attributed to the high water content in algae

Tab. 7.3: Separation characteristics of bio-oil from different species of biomass by molecular distillation.

Bio-oil	Recovery (wt%)	Total distillation (wt%)	LF (wt%)	MF (wt%)	HF (wt%)
Mongolian Scots pine	97.22	72.48	65.26	7.22	24.74
Rice husk	96.33	76.79	58.89	17.90	19.54
Algae	98.56	90.35	87.83	2.52	8.21

bio-oil, which was enriched in the light fraction. The distilled fraction yield of rich husk bio-oil was slightly higher than that of Mongolian Scots pine bio-oil, owing to the greater lignin structure in Mongolian Scots pine. Phenolic oligomers (or pyrolytic lignin), derived from the uncompleted pyrolysis of lignin, were enriched in the heavy fraction during molecular distillation, leading to the high yield of heavy fraction during the distillation of Mongolian Scots pine bio-oil.

The compositions of bio-oils from Mongolian Scots pine, rice husk, and algae are shown in Fig. 7.25 according to a classification of bio-oil components into several chemical families, such as acids, aldehydes, ketones, phenols, etc. The composition of bio-oils produced from different species of biomass exhibits similarity for light, middle, and heavy fractions. The light fraction mainly contains acids and ketones, whereas the phenols and sugars contents are relatively higher in the middle fraction, and they are obviously enriched in the heavy fraction.

The mean free path of a molecule is proportional to the temperature, therefore increasing temperature can elevate the mean free paths of all molecules, and more chemicals with higher molecular weight will reach the cooling surface, leading to an increasing overall distillation yield. However, the composition of the distilled fraction is correspondingly more complex. Considering the severe coking of bio-oil at high temperature due to condensation, the operation temperature of molecular distillation is kept lower than 150 °C. Bio-oil was separated into three fractions under 60 Pa and at 70–130 °C, as shown in Tab. 7.4. The yield of the distilled fraction (light and middle fractions) reached 56.75 %, and it increased to 82.6 % without any coking occurrence as temperature increased up to 130 °C. This result illustrates that increasing temperature was beneficial to the distillation of components and thus raising the distilled fraction yield. However, high distillation yield also means inferior enrichment of volatiles in the distilled fraction. For example, the water content in the light fraction decreased from 70 % to 57 %. The chemical distributions in the light and middle fractions show that the acid contents start to decline at 100 °C and the enrichment of chemicals with higher molecular weight like catechol was improved obviously. As water and most acids in bio-oil were enriched in the light fraction at a relatively low temperature, only little or no water remained in the middle and heavy fractions. The heating values of these two fractions achieved levels above 20 MJ/kg, whereas the pH values remained almost unchanged with temperature.

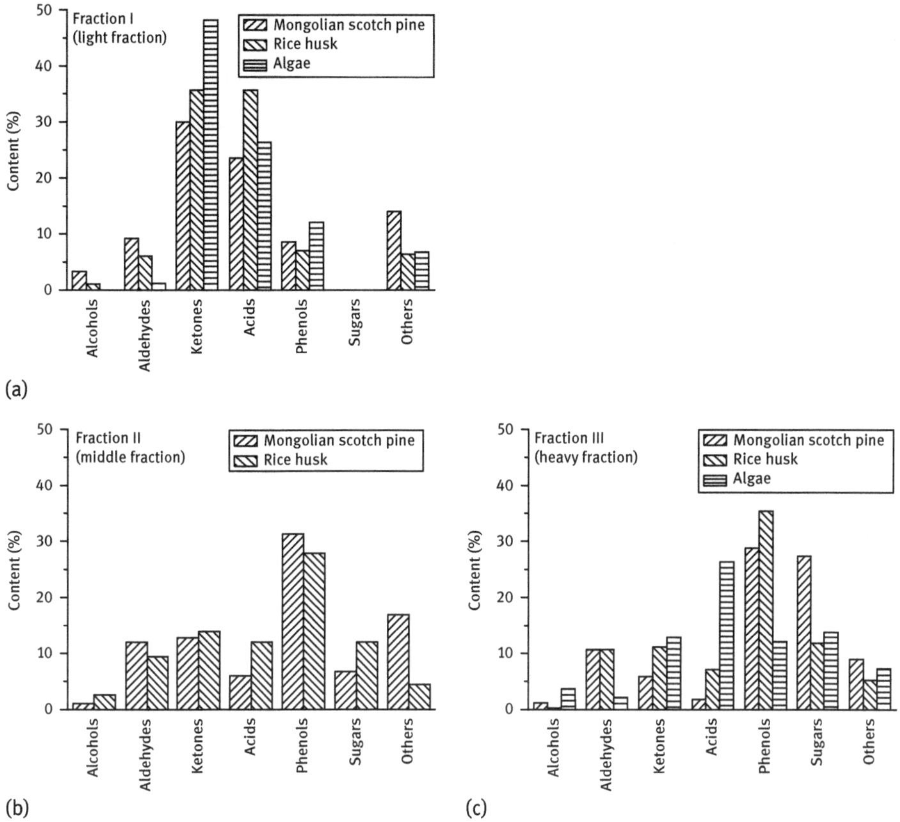

Fig. 7.25: Distribution characteristics of chemical families in bio-oil fractions.

Tab. 7.4: Separation characteristics of bio-oil from Mongolian Scots pine at different distillation temperatures.

	70 °C			100 °C			130 °C		
	LF	MF	HF	LF	MF	HF	LF	MF	HF
Yield (wt%)	50.0	6.75	40.9	50.6	11.75	35.2	63.2	19.4	15.6
Water content (wt%)	70	2	/	71	1	/	57	1	/
pH value	2.13	4.98	5.2	2.17	4.78	5.1	2.14	4.8	5.3
Heating value (MJ/Kg)	/	20.6	21.4	/	22.6	23.9	/	21.9	24.2

The operation pressure during molecular distillation has a more significant influence on the separation results than the temperature. This is due to the fact that the range of pressure is more flexible (60 Pa – atmospheric pressure), while the temperature adjustment is limited by the thermal-sensitivity of bio-oil. As mentioned before, a portion of components with low molecular weights can reach the second cold trap

Tab. 7.5: Physical characteristics of bio-oil fractions obtained from molecular distillation.

Oil type		Yield (wt%)	Water content (wt%)	pH value	Heating value (MJ/kg)
Crude bio-oil		/	23.30	3.18	16.40
3000 Pa	DF-1	37.37	49.51	2.59	8.44
	RF-1	62.63	7.17	3.29	18.86
2000 Pa	DF-2	39.78	47.35	3.01	9.55
	RF-2	60.22	7.98	3.27	18.89
700 Pa	DF-3	56.50	35.57	2.90	12.62
	RF-3	43.50	4.20	3.25	20.13

and are condensed as the light fraction. However, under high pressures (> 700 Pa), even the components with low molecular weight like water and acetic acid can only arrive at the internal condenser but not the second cold trap. As a result, two fractions are obtained under relatively high pressures, namely the distilled fraction (DF) collected from the internal condenser and residual fraction (RF). The physical properties of the obtained fractions under 90 °C and 700–3,000 Pa are shown in Tab. 7.5. The yield of the distilled fraction is significantly improved with declining pressure, increasing from 37.37 wt% at 3,000 Pa to 56.50 wt% at 700 Pa. Because of the increase in distilled fraction yield, the water content in the corresponding distilled fraction decreases from 49.51 % to 35.57 %. The water content in the residual fraction exhibits a downward trend with declining pressure. The water content in the residual fraction under 700 Pa was only 4.2 %, leading to a heating value of residual fraction up to 20.13 MJ/kg. However, this water content is relatively higher than that in the heavy fraction obtained from 60 Pa, indicating that the high vacuum favored the distillation of volatile components. The comparison of the distribution of chemical families in the fractions shows that the relative content of low molecular weight carboxylic acids like acetic acid in the distilled fraction decreases with reducing pressure, which was attributed to the enrichment of other oxygenated compounds in the distilled fraction under a low pressure. The ketones performed a partial enrichment in the distilled fraction, which was not as obvious as acids. This is due to the fact that they have larger molecular weight and weaker volatilities. With regard to phenols, phenol and its derivatives have better volatility, while the distillation of diphenols is more difficult, but can be improved by lowering pressure. Sugars (levoglucosan) with large molecular weight and non-volatile property are enriched in the residual fraction.

In order to obtain more fractions suitable for the subsequent upgrading technologies, a multiple molecular distillation study was carried out. A typical two-step molecular distillation was performed as follows: first the bio-oil was distilled under 1,600 Pa and 80 °C to obtain the distilled fraction 1 and residual fraction 1. After that, the residual fraction 1 was further distilled under 340 Pa and 80 °C to obtain distilled fraction 2

Tab. 7.6: Physical characteristics of bio-oil fractions from molecular distillation.

Oil type	Appearance	Heating value (MJ/Kg)	Water content (wt%)	pH value
Crude bio-oil	Black	18.49	8.27	2.32
DF-1	Dark yellow	12.30	30.40	1.65
RF-1	Dark black	21.29	1.49	3.20
DF-2	Dark red	16.62	6.46	1.61
RF-2	Dark black	22.34	/	3.55

and residual fraction 2. The yields of distilled fractions for the two distillation processes were 26.36 % and 22.58 % respectively. The main physical properties of the obtained fractions are shown in Tab. 7.6. Carboxylic acids, ketones, furans, and ethers were more effectively enriched in the distilled fraction, whereas aldehydes, phenols and sugars with non-volatility remained in the residual fraction even after the second distillation process.

7.4.1.3 Multiple separation of bio-oil

Apart from molecular distillation, extraction with pH regulation provides another pathway for the subsequent upgrading of bio-oil fractions, through which monophenols with high purity and pyrolytic lignins with different activity were obtained. Water was used to separate the bio-oil into aqueous phase and oil phase (water-insoluble phase). The aqueous phase, which had high reactivity and volatility, contained polar carbohydrates and small molecular weight compounds like acids, ketones, aldehydes and a small portion of monophenols. In contrast, the composition of the viscous oil phase was complex. It mainly contained the pyrolytic products of lignin such as phenol, guaiacol, syringol and phenolic oligomers. The strong alkaline solution could break the network structure covering the whole bio-oil volume, and the phenolic fraction with high purity was obtained by pH control and solvent extraction, as shown in Fig. 7.26. The overall relative content of phenolic compounds obtained around pH 6 reached 94.35 %, of which the guaiacols content was as high as 48.72 %, while the syringols content with double methoxyls was only 1/3 of the guaiacols content. The primary structural units of high and low molecular weight pyrolytic lignins were of the guaiacyl and syringyl types, and more active hydroxyls could be found in the low molecular weight pyrolytic lignin.

7.4.2 Upgrading of bio-oil fractions from molecular distillation

7.4.2.1 Catalytic esterification of bio-oil fractions

Bio-oil contains abundant carboxylic acids like acetic acid, leading to its low pH value and strong corrosiveness. Catalytic esterification means the reaction of car-

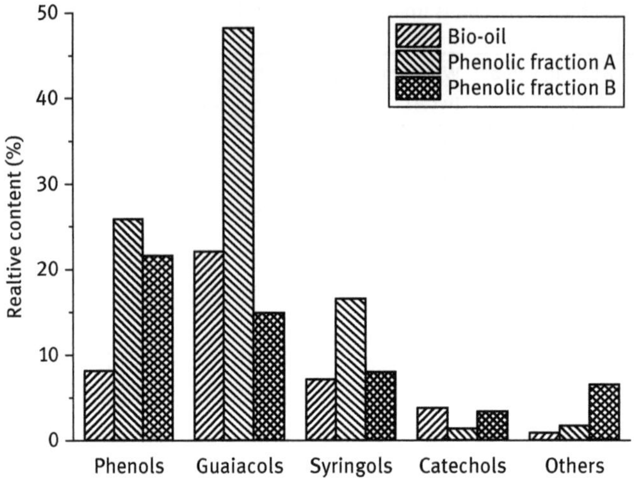

Fig. 7.26: Distribution of phenolic compounds in bio-oil and its fractions. Reprinted with permission from [70], © 2014 Elsevier.

boxylic acids and alcohols into neutral esters under acidic or base catalysts, which can effectively decrease the carboxylic acids content in bio-oil. Considering the separation and collection of products, the selected catalysts are usually solid acid or base catalysts. During the catalytic esterification process, solid acid acids show high catalytic efficiency. The study of bio-oil model compounds (acetic acid and propionic acid) with ethanol showed that the conversion of acids was only 20 % without catalyst, while it reached about 80 % after the addition of ion-exchange resin 732.

Catalytic esterification is widely investigated because the operation is relatively simple. The corrosiveness of upgraded bio-oil is reduced, and the heating value is obviously increased. Zhang et al. [71] found that the catalytic activity of solid acid $40SiO_2/TiO_2{-}SO_4^{2-}$ was better than solid base $30K_2CO_3/Al_2O_3{-}NaOH$, and the fluidity of upgraded bio-oil was improved and the heating value was increased markedly. Xiong et al. [72] used ionic liquid as the catalyst for bio-oil esterification with ethanol. The properties of upgraded bio-oil were significantly improved. The higher heating value reached 24.6 MJ/kg, the pH value increased from 2.9 to 5.1, and the water content decreased from 29.8 to 8.2 wt%. Wang et al. [73] selected 732 and NKC-9 type ion-exchange resin as the esterification catalysts to upgrade bio-oil. They found that the acidities of upgraded bio-oil were lowered by 88.54 % and 85.95 %, while the heating values were elevated by 32.26 % and 31.64 %, respectively. Xiong et al. [74] investigated the esterification of bio-oil and its distillation fraction on sulfo-acid ion-exchange resin. The corresponding heating values were improved from 16.80 MJ/kg and 12.76 MJ/kg to 20.08 MJ/kg and 18.33 MJ/kg, respectively.

The reactivity of carboxylic acids in bio-oil during the esterification process is high, but some other chemical families like phenolic fractions mainly remain stable. In addition, some complex side reactions may also occur. Hu et al. [75] found that cyclic ethers and terpenoids in mallee bio-oil could be converted into aromatic hydrocarbons through the removal of oxygen on acidic catalysts. The side reactions of ethanol with aldehydes during esterification were observed by Lohitharn et al. [76]. The results showed that acetalization occurred between one acetaldehyde molecule and two ethanol molecules, which affected acetic acid conversion. Gunawan et al. [77] reported the hydrolysis and glycosidation of sugars during the esterification of bio-oil. Levoglucosan in bio-oil could be hydrolyzed to D-glucose, which further reacted with methanol to form methyl-α-d-glucopyranoside under acid catalysts. Consequently, the composition of upgraded bio-oil remains complex, which limits its high-grade application. The aforementioned investigation of molecular distillation illustrates that the light distillation fraction mainly containing carboxylic acids and water can be obtained under low pressures. Therefore, studying catalytic esterification of acid-enriched fractions is more effective, and high-quality liquid fuels rich in esters can be obtained. A study on the esterification of acid-enriched fractions obtained from molecular distillation under La-promoted solid acid catalyst (La^{3+}–SO_4^{2-}–TiO_2–SiO_2) showed that the composition of the upgraded fraction differed greatly from that of the bio-oil fraction (Fig. 7.27).

The compounds in the bio-oil fraction and its upgraded fraction were classified and are shown in Fig. 7.28. The bio-oil fraction only contained volatile components such as carboxic acids, esters, and ketones, of which the carboxylic acids content reached 18.39 %. After upgrading the carboxylic acids content decreased from 18.39 % to 2.70 %. In contrast, the esters content increased sharply from 0.72 % to 31.17 %. These phenomena indicate that the carboxylic acids were efficiently converted to the corresponding esters.

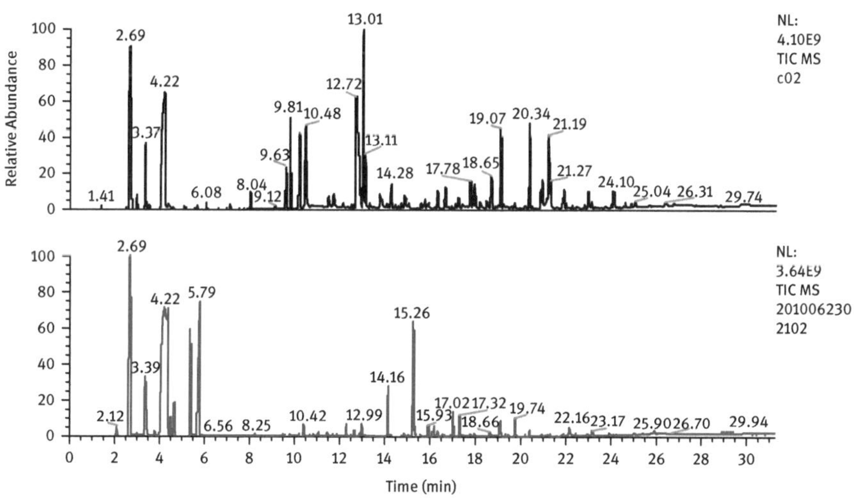

Fig. 7.27: Spectra of bio-oil fraction (before esterification; after esterification).

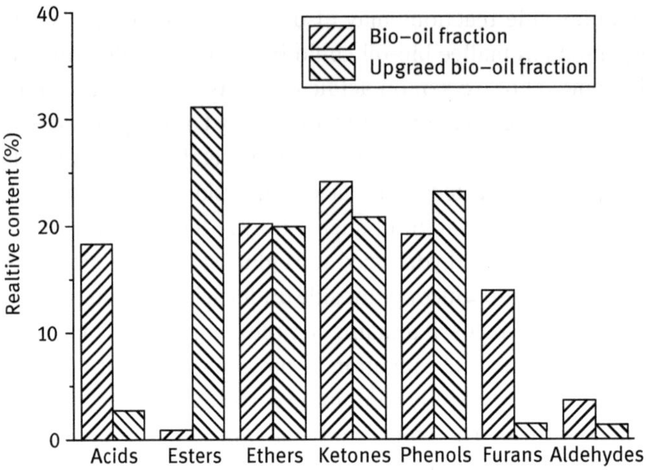

Fig. 7.28: Comparison between the original and the esterified bio-oil fraction.

7.4.2.2 Catalytic cracking of bio-oil fractions

The high oxygen content in bio-oil is the main reason leading to its low fuel quality. Among the oxygenated compounds in bio-oil, the presence of carboxylic acids leads to its strong corrosiveness; ketones and aldehydes with carbonyl and aldehyde groups are responsible for its storage instability; while sugars and phenolic oligomers increase its viscosity. Therefore, the deoxygenation of bio-oil to produce hydrocarbons is the main pathway to realize the high-grade utilization of bio-oil. Catalytic cracking is one of the potential deoxygenation technologies. Catalytic cracking of bio-oil refers to the reaction whereby oxygen is removed in the forms of CO, CO_2, and H_2O, yielding a hydrocarbon-rich high-grade liquid fuel. The main reaction pathways are shown in Fig. 7.29. The intermediate products (such as olefins) are formed after the deoxygenation reactions of oxygenated compounds, including decarbonylation, decarboxylation, and dehydration. Thereafter, the targeted products hydrocarbons are generated through secondary reactions, which include aromatization, polymerization, alkylation, and isomerization.

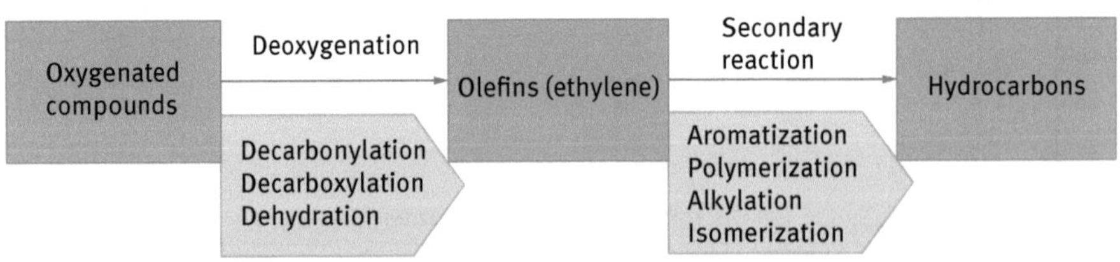

Fig. 7.29: Main reaction pathway of catalytic cracking of oxygenated compounds in bio-oil.

Catalytic cracking of crude bio-oil has been studied for a long time. Although a proportion of bio-oil is converted into hydrocarbons, the yield of coke reached above 20 %, resulting in severe deactivation of the catalyst [78, 79]. Gayubo et al. [80] investigated the coke deposit formed at 400–500 °C during the bio-oil catalytic cracking process in a fixed bed reactor by TPO analysis. The TPO curves identified two types of carbon, namely thermal carbon and catalytic carbon. Thermal carbon was located on the outside of the catalyst particles, as well as the macropores. In contrast, the catalytic carbon was located inside the micropores of catalysts and its removal by combustion required a higher temperature (520–550 °C) than thermal carbon (450–480 °C). The major coke component leading to catalyst deactivation was catalytic carbon, which could directly cover the active sites. Furthermore, catalyst deactivation was also accelerated by the thermal carbon deposit. Guo et al. [81] used HZSM-5 catalyst for bio-oil upgrading and analyzed the precursor of coke on the catalyst by means of TGA, FTIR, and ^{13}C NMR. Precursors of coke deposited on the outer surface of the catalyst were identified as saturated aliphatic hydrocarbons with a boiling point below 200 °C. Those in the pores of the molecular sieve were mainly aromatic hydrocarbons with a boiling point ranging from 350 °C to 650 °C. They believed that deactivation started from the inside of the catalyst and pores were blocked by high molecular compounds, which led to catalyst deactivation. An investigation of reactivity of different chemical families in bio-oil was carried out by Gayubo et al. [82, 83]. Alcohols, ketones, and carboxylic acids could be efficiently converted into aromatic hydrocarbons, while phenols showed low reactivity and tended to form carbon deposits. Sugars and phenolic oligomers with high molecular weight might undergo condensation to form coke deposits, which would block the reactor and lead to catalyst deactivation [84]. In addition, because bio-oil has a low hydrogen content and a high oxygen content, and the dehydration during cracking was accompanied by the loss of hydrogen, the H/C ratio of the final cracking product would be low, and carbon deposits with large aromatic structures tended to be formed, which might influence the activity of the catalyst and lead to by-product formation [64, 85]. Catalytic cracking of acids (acetic and propionic acids) showed that the organic phase contained aromatic hydrocarbons and phenolic by-products. Ketone by-products were also observed accompanied by hydrocarbons during the direct cracking of ketones (cyclopentanone and hydroxyacetone). Therefore, there are two pathways to achieve the efficient cracking of bio-oil fractions, namely the enrichment of acids and ketones suitable for cracking in one specific fraction and the improvement of the feedstock H/C ratio to suppress the formation of coke.

Carboxylic acids with low molecular weight can be enriched in the distilled fraction by molecular distillation, so that the distilled fraction has higher reactivity than the crude bio-oil. Since the intrinsic hydrogen content of crude bio-oil is insufficient, hydrotreatment, hydrogenation-cracking (hydrocracking), or co-reactants with high H/C ratios can be employed to overcome this drawback. Vispute et al. [86] carried out a low-temperature hydrogenation pretreatment of bio-oil to convert the components with C=C and C=O bonds into the corresponding thermally stable alcohols (monohydric alcohols and diols), which increased the integral hydrogen content of the feedstock. The subsequent cracking results showed that the coke yield was only 12.6 %, much lower than that of direct cracking of bio-oil. At the same time, the selectivities of aromatics and olefins were 18.3 % and 43 %, respectively. Ausavasukhi et al. [87] investigated the deoxygenation of benzaldehyde over HZSM-5 and gallium-modified ZSM-5 catalysts in a fixed bed reactor. On the HZSM-5 catalyst, the presence of H_2 did not significantly affect the conversion of benzaldehyde. However, Ga became an active center of H_2 on the Ga/HZSM-5 catalyst, and hydrogenation occurred to enhance the deoxygenation of benzaldehyde. The deoxygenation of benzaldehyde was also investigated by Peralta et al. [88] on CsNaX, and NaX zeolite catalysts at 475 °C. It was observed that CsNaX displayed a high activity and the initial conversion was 100 % at the start and only decreased by 10 % after 5 h. However, a higher deactivation rate of CsNaX was observed under He atmosphere, and the conversion decreased by 90 % after the same time, indicating that H_2 could participate in reactions of hydrogen transfer to improve the reaction stability. Graça et al. [89] introduced a standard gasoil with high $(H/C)_{eff}$ as a co-cracking reactant into each key model bio-oil oxygenated compound (acetic acid, hydroxyacetone and phenol). This approach reduced the coke yield and increased the overall conversion of model bio-oil compound to fuel gas, liquefied petroleum gas (LPG), and gasoline. Alcohols are another kind of effective co-cracking reactant and have high reactivity in similar reactions. Mentzel et al. [85] investigated the co-conversion of various model compounds with methanol, which had a distinct positive effect on catalyst lifetime. The conversion capacity of the catalyst was up to 10 times that without methanol addition. Valle et al. [84] pretreated the bio-oil with a thermal step, followed by co-cracking with methanol. A feed containing 60 wt% methanol can obtain 90 wt% conversion of the bio-oil with a selectivity of aromatics of 40 %, and the selectivity to the total benzene, toluene, xylenes (BTX) of 25 %. Hence, the co-cracking of bio-oil distilled fraction with alcohols (methanol or ethanol) can suppress coke formation and improve the selectivity of liquid hydrocarbons.

The distilled fraction of bio-oil is rich in acids and ketones. Therefore, acetic acid, hydroxyacetone, and cyclopentanone were selected as the representative model compounds and their co-cracking characteristics with alcohols were studied and are shown in Tab. 7.7. Due to the promoting effect of the alcohols the three model compound displayed good performances. The conversion of reactants achieved 100 %, and the selectivities to oil phases were above 30 %. More than 97 % of the oil phase was liquid hydrocarbons, mainly aromatic hydrocarbons.

The influences of temperature and pressure on the co-cracking of ethanol and distilled fraction of bio-oil over HZSM-5 are shown in Tab. 7.8. The ratio of distilled fraction to ethanol was 1 : 2. At low temperatures, the deoxygenation efficiency was low and by-products like alcohols, esters and ethers were produced. The oil and aqueous phases were even difficult to separate at 340 °C. The yield of oil phase was only 5.5 % under atmospheric pressure, and oxygenated compounds were also found in the oil phase. When the temperature was 400 °C and the pressure was 2 MPa, the

Tab. 7.7: Catalytic co-cracking of bio-oil model compounds and ethanol.

Model compound	Alcohol	Conversion (wt%)	Selectivity (wt%)		Composition (%)		
			Oil phase	Water phase	Aromatics	Aliphatics	Others
Cyclopentanone	Methanol	100	31.6	42.6	96.1	1.0	2.9
Hydroxyacetone	Ethanol	100	31.9	36.2	94.6	5.4	/
Acetic acid	Ethanol	100	39.0	48.4	87.7	12.3	/

Note: Reaction temperature: 400 °C; Reaction pressure: 2 MPa; Weight hourly space velocity: 3 h^{-1}; Alcohol concentration: 70 %.

yield of oil phase reached 27.7 %, in which the hydrocarbons content was 98.2 %. The appearance of bio-gasoline produced from the co-cracking of distilled fraction and ethanol was similar to commercial gasoline, as shown in Fig. 7.30. The total ion chromatography of the bio-gasoline phase is displayed in Fig. 7.31. The main chemicals were C_7–C_9 hydrocarbons, including toluene, xylene, trimethylbenzene, and methyl-ethyl-benzene. A further study over a modified Ga_2O_3/HZSM-5 further promoted the formation of C_7–C_9 monoaromatic hydrocarbons with an elevated oil phase by 7.7 %.

Tab. 7.8: Catalytic co-cracking performance of distilled fraction and ethanol.

Temperature (°C)	Pressure (MPa)	Yield (wt%)				Composition of oil phase (%)			
		Oil phase	Water phase	Total	Hydrocarbons	Alcohols	Esters	Ethers	Others
340	2	/	/	78.2	32.4	20.7	16.1	24.0	6.9
370	2	18.5	53.3	71.8	90.7	1.7	1.9	2.0	3.7
400	2	27.7	35.8	63.5	98.2	0.7	0.4	0.0	0.7
430	2	26.6	33.1	59.7	99.3	0.7	0.0	0.0	0.0
400	0.1	5.5	50.6	56.1	87.5	1.9	2.4	3.5	4.7
400	1	24.2	41.2	65.4	90.1	0.0	3.2	6.7	0.0
400	3	27.8	33.1	60.9	98.1	1.9	0.0	0.0	0.0

As mentioned above, in the main cracking reactions of oxygenated compounds in bio-oil, olefins were the predominant intermediate products, which are obtained from the deoxygenation (decarboxylation, decarbonylation, and dehydration) of oxygenated compounds. The comparison between the direct cracking of oxygenated compounds and the co-cracking of oxygenated compounds and ethanol illustrated that the conversion efficiency of bio-oil to hydrocarbons was apparently enhanced due to the fact that the existence of alcohols promoted the deoxygenation of oxygenated compounds. It is generally accepted that the cracking of methanol to hydrocarbons (MTG) is based on a “hydrocarbon pool” mechanism. In this mechanism, products are derived from the active intermediates in the “hydrocarbon pool” and this mechanism could also

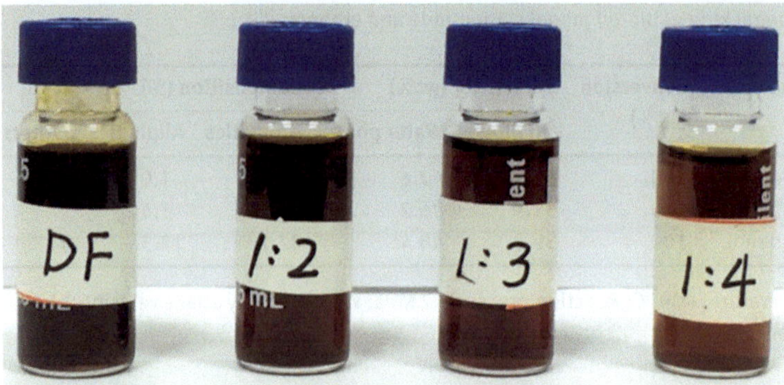

(a)

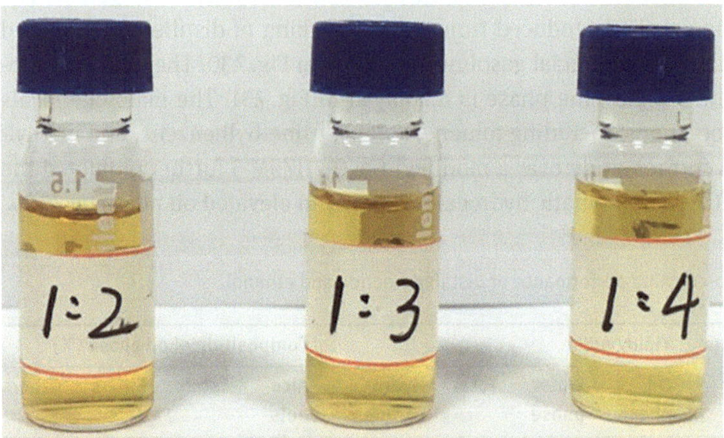

(b)

Fig. 7.30: The distilled fraction and the bio-gasoline

be applied in the conversion of ethanol to gasoline [90, 91]. Depending on the promotion of the co-cracking process by alcohols, a double-route reaction mechanism has been proposed, as shown in Fig. 7.32. The first route is the direct deoxygenation of oxygenated compounds, involving the release of CO, CO_2, and water by decarbonylation, decarboxylation, and dehydration, respectively. The second route is the conversion of oxygenated compounds through the "hydrocarbon pool". The bio-gasoline phase from cracking contains a large number of aromatic hydrocarbons, of which the methyl-substituted benzenes like toluene and xylene dominate. These chemicals, which are considered as typical active intermediates in the "hydrocarbon pool", can propagate the reaction of producing aromatics. During the cracking process, the existence of active intermediates can significantly promote the formation of olefins by the deoxygenation of oxygenated compounds in bio-oil. Subsequently, these olefins can be converted to compounds with a single benzene ring through aromatization,

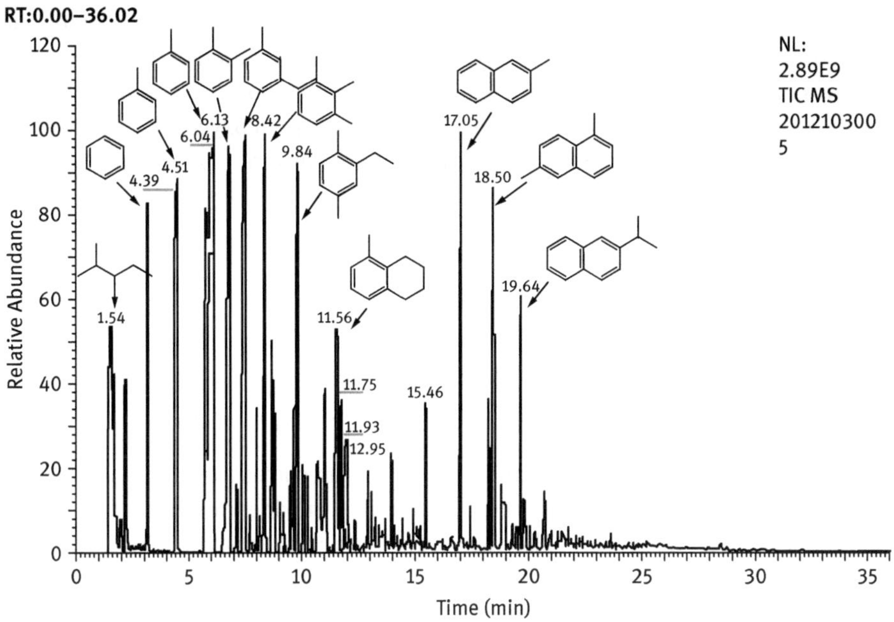

Fig. 7.31: GC-MS spectrum of the bio-gasoline.

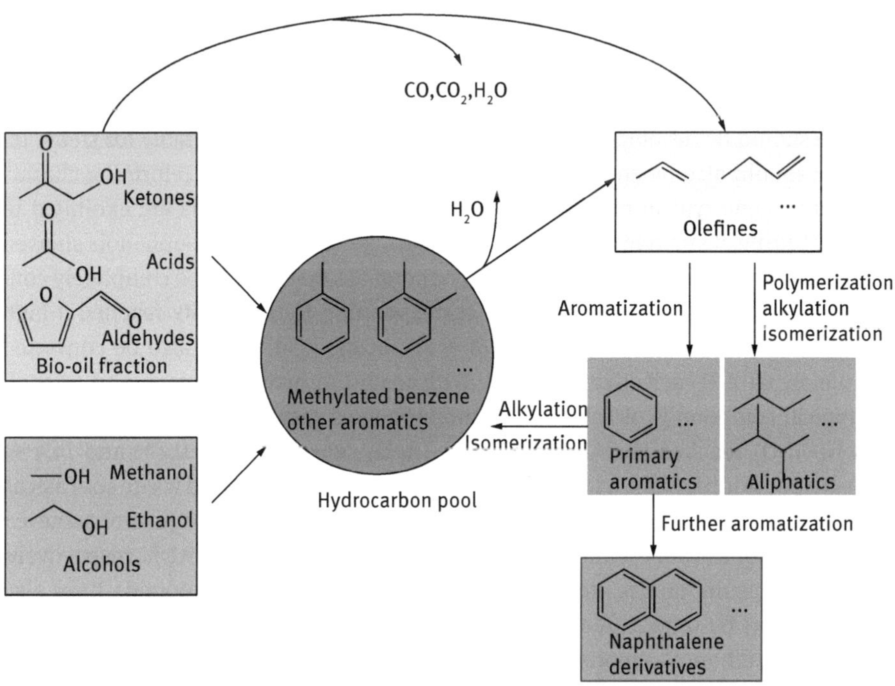

Fig. 7.32: Conversion mechanism of oxygenated compounds in bio-oil.

and C_4–C_6 isoparaffins through polymerization, alkylation, and isomerization. The compounds with a single benzene ring can promote the formation of the active intermediates like methyl-substituted benzene in the "hydrocarbon pool" through alkylation and isomerization, and then enhance the subsequent deoxygenation of oxygen-containing compounds and alcohols, or undergo aromatization to produce complex aromatic hydrocarbons.

7.4.2.3 Steam reforming of bio-oil fraction

Hydrogen can be produced by the catalytic reforming of bio-oil with steam under the effect of catalyst. Its overall reaction equation is listed as:

$$C_nH_mO_k + (2n - k)H_2O \rightarrow nCO_2 + (2n + m/2 - k)H_2 \,. \tag{7.1}$$

As can be seen from the above equation, the hydrogen comes from not only the bio-oil itself but also the water (the carbon in bio-oil is provided as the carbon source for the water-gas shift reaction). The main reactions during catalytic steam reforming include the following steps: (1) the reactants are adsorbed on the catalyst; (2) the adsorbed reactants are dissociated to produce CO and coke precursor and water is also decomposed into absorbed OH-species and H atom; (3) the hydroxyl and coke precursor can react to form CO and H (adjacent H can form H_2 by collision); (4) CO_2 and H_2 are formed from CO and H_2O through the water-gas shift reaction; (5) the active sites are available again after the desorption of CO_2 and H_2.

Coking is a major problem for the steam reforming process, which is similar to catalytic cracking. To overcome this drawback, catalysts with high efficiency and long lifetime should be developed. What's more, chemicals that are suitable for steam reforming should also be enriched in a specific fraction. The steam reforming characteristics of some typical model compounds over different catalysts are exhibited in Tab. 7.9. During steam reforming, alcohols, acids, ketones, and monophenols showed good reactivity. Ethanol, the typical compound of alcohols, could be completely converted over Ni-based catalyst, and the corresponding H_2 selectivity reached a high value of 98 %. The model compound of acids, acetic acid, could also be converted completely with H_2 selectivity of 96 %. With regard to hydroxyacetone and phenol, the typical compounds of ketones and monophenols respectively, their conversions over Ni/Al_2O_3 reached 98.7 % and 92.2 % with H_2 selectivities of 97.2 % and 74.4 %, respectively. Additionally, a low-cost Ni/Ash catalyst using the solid waste coal ash as the support was developed. The steam reforming of acetic acid and phenol achieved high conversions of 98.4 % and 83.5 %, and H_2 yields of 85.6 % and 79.1 %, respectively.

Similar results on the steam reforming of bio-oil model compounds have also been obtained by other researchers. Rioche et al. [92] selected acetic acid, phenol, acetone, and ethanol as model compounds to study their characteristics of steam reforming over noble metal based catalysts. The hydrogen yields of acetic acid, ethanol, acetone, and phenol reached 66 %, 70 %, 80 %, and 95 % over $Rh/CeZrO_2$, respec-

Tab. 7.9: Catalytic steam reforming of bio-oil model compounds.

Model compound	Catalyst	Temperature (°C)	Conversion (%)	H_2 yield (%)
Ethanol	Ni/Al_2O_3	600	99.0	88.0
	Ni/CeO_2	725	100.0	98.6
Acetic acid	Ni/Al_2O_3	700	98.2	87.0
	Co	400	100.0	96.0
	Pd/HZSM-5	600	95.7	60.2
Hydroxyacetone	Ni/Al_2O_3	700	98.7	97.2
Phenol	Ni/Al_2O_3	800	92.2	74.4

tively. Catalytic steam reforming of acetic acid over Ni/Co catalysts was also studied by Hu et al. [93]. Acetic acid was converted completely, and the H_2 selectivity reached up to 96.3 % at 400 °C. The Ni/Co catalysts displayed rather stable performances for the 70 h time-on-stream without any deactivation. The steam reforming of phenol reaction towards H_2 production was studied by Polychronopoulou et al. [94] over Rh catalysts. The conversion of phenol reached 87.4 % at 700 °C. In general, since acids, alcohols, ketones, and monophenols with low molecular weight can be effectively converted to obtain H_2, the enrichment of these chemicals favors overall reaction efficiency of steam reforming.

In addition to the model compounds, aqueous phase of bio-oil is also often selected to study its steam reforming behavior rather than crude bio-oil. The water-soluble chemicals, including carboxylic acids, ketones, aldehydes, sugars, and some monophenols, can be extracted efficiently from the bio-oil without phenolic oligomers having large molecular weight. Without the feeding of these oligomers into the catalytic bed, catalyst deactivation can be decelerated. In addition, since water is also one of the reactants, the extracted aqueous phase can be directly used as the feedstock of catalytic steam reforming. In general, the steam reforming of aqueous phase is able to maintain a stable run for 4–5 h. Nevertheless, the hydrogen yield decreased sharply because of coke formation on catalysts [95, 96]. Kechagiopoulos et al. [95] conducted an investigation on the steam reforming of model compounds of bio-oil (acetic acid, acetone, and ethylene glycol) and the aqueous phase of a real bio oil over commercial Ni-based catalyst. The results of the selected model compounds showed that all model compounds could be effectively reformed with hydrogen yields up to 90 %, while the reforming of the aqueous fraction of bio-oil proved to be more difficult, with the hydrogen yield fluctuating at about 60 %. The most serious problem encountered in these experiments is coking, which is attributed to the non-volatile sugars in aqueous phase. Monosaccharides and oligosaccharides, which are freely soluble in water and have low reactivity, tend to form carbonaceous deposits in the process of steam reforming. Marquevich et al. [97] investigated the production of hydrogen by the catalytic steam reforming of several model compounds of bio-oil (acetic acid, m-cresol, dibenzyl ether, glucose, xylose, and sucrose) using two commercial Ni-based catalysts. They found that except for sugars, other model chemicals were completely converted to produce hydrogen at high temperatures and high steam-to-carbon ratios, while sugars were decomposed to form char and gases before reaching the catalytic bed.

Acids and ketones with low molecular weights can be enriched in the distilled fraction through molecular distillation, while sugars and phenolic oligomers with high boiling points remained in the residual fraction. As a result, the distilled fraction by molecular distillation is feasible for steam reforming. Moreover, the water in bio-oil is almost evaporated into the distilled fraction, having a high steam-to-carbon ratio. Therefore, this fraction can be directly reformed without the addition of extra steam. A steam reforming investigation on the distilled fraction over Ni-based catalyst was carried out. The overall results indicated that the conversion of distilled fraction increased from 52.5 % to 95 % when the reaction temperature rose from 500 °C to 700 °C. The total yields of H_2 and carbon-containing gases also increased. A high H_2 yield reached 135 mg g^{-1} organics at 700 °C. In addition, the catalyst maintained its high activity for more than 11 h.

7.4.2.4 Catalytic hydrogenation of bio-oil fraction

Hydrogenation of bio-oil is an important upgrading technology of bio-oil and there are two main types. Mild hydrogenation is conducted under a low hydrogenation pressure to saturate C=C and C=O bonds, which can improve the stability of bio-oil. The other type is hydrodeoxygenation, which can convert the bio-oil into hydrocarbons by the removal of oxygen under a high hydrogen pressure.

A large portion of work addresses the catalytic hydrodeoxygenation of bio-oil, and the quality of upgraded bio-oil is improved remarkably due to its low oxygen content. Zhao et al. [98] investigated the selective hydrogenation of an n-hexane-soluble fraction from bio-oil over Ni/HZSM-5. The oxygenated organics (substituted furans, monophenols and phenol dimers) were almost completely converted into the liquid product containing less than 10 % of C_5–C_6 paraffins and more than 90 % of C_5–C_9 naphthenes and C_6–C_9 aromatics. Wildschut et al. [99] carried out bio-oil hydrogenation at temperatures of 250–350 °C and pressures of 10–20 MPa. The Ru/C catalyst was found to be superior with respect to oil yield (up to 60 wt%) and deoxygenation level (up to 90 wt%). The high heating value (HHV) of the upgraded products was about 40 MJ/kg, which is close to the value of commercial vehicle fuels. Catalytic hydrotreatment experiments (350 °C, 20 MPa) over mono- and bi-metallic metal catalysts of Rh, Pt, Pd on a zirconia support were performed by Ardiyanti et al. [100]. The highest yield of the upgraded oils was 47 wt% based on the feed, while the oxygen content declined from 40.1 wt% to 7 wt%. Because of the limitation of space velocity in a fixed bed reactor, multiple-stage hydroprocessing is often used. The first stage is low-temperature hydrogenation, which saturates the functional groups like the reactive carbonyl group. The second stage is hydrodeoxygenation, which removes the oxygen from alcoholic hydroxyl, phenolic hydroxyl and carboxyl groups in the form of water to obtain hydrocarbons. Venderbosch et al. [101] studied four-stage hydroprocessing of bio-oil, and found that the HHV of upgraded oil reached 40 MJ/kg when the operation pressure was higher than 20 MPa. The oxygen content in the upgraded oil decreased remarkably.

The hydrodeoxygenation of different chemical families in bio-oil shows that aldehydes and ketones have high reactivity. Therefore, the distilled fraction from molecular distillation is superior for converting into hydrocarbons through hydrodeoxygenation than crude bio-oil. Primary mild hydrogenation of distilled fraction over Ru/Al_2O_3 was conducted. The conversions of furfural and hydroxyacetone reached 75 % and 50 %, respectively, under the hydrogenation pressure of 3 MPa and temperature of 140 °C. Mild hydrogenation can convert the aldehydes and ketones in bio-oil distilled fraction into corresponding alcohols, which avoids the deterioration of bio-oil due to its polycondensation. The stability of distilled fraction is studied by the thermal treatment of upgraded bio-oil fraction, as shown in Tab. 7.10. The average molecular weight of upgraded bio-oil fraction varied little after thermal treatment at 120 °C, indicating that the thermal stability of bio-oil fraction was enhanced.

Tab. 7.10: Effect of mild hydrogenation on thermal stability of bio-oil.

Sample	Crude bio-oil	60 °C	80 °C	120 °C	140 °C
Before heating (M_n)	253	263	281	304	329
After heating (M_n)	308	319	307	310	332
Rate of rise (%)	21.7	21.3	9.3	2.0	0.9
Before heating (M_w)	259	274	299	325	365
After heating (M_w)	322	346	325	325	365
Rate of rise (%)	24.3	26.3	8.7	0.0	0.0

7.4.2.5 Emulsification of bio-oil distilled fraction and diesel

The emulsification of bio-oil fraction and diesel is a convenient upgrading technology. The stable emulsion can be obtained from the miscibility of bio oil and diesel with the assistance of surfactant, and it can substitute partially for diesel. The emulsion has lower acidity and viscosity but higher heating value than crude bio-oil. Therefore, it can be burned successfully without major modification of current diesel engines, which is considered to be a practical utilization of bio-oil in the short term.

The mixing ratio of bio-oil, diesel and emulsifier is quite important in the emulsification process. The appearance of emulsion fuels, the mixture of 5 % bio-oils, 92 % diesel, and 3 % surfactant, is displayed in Fig. 7.33. The results showed that the emulsions from crude bio-oil of rice husk, rice straw, rosewood, Mongolian Scots pine and

Fig. 7.33: Appearance of bio-oil emulsions from different species of biomass: rice husk, rice straw, rosewood, Mongolian Scots pine, bamboo, grassiness, and algae (from left to right).

bamboo all had stability times of around 15 days. However, the stability time of the emulsion from sea algae bio-oil was only about 5 days, while the bio-oil from grassiness was difficult to emulsify.

Other researchers have also done much work on the emulsification of bio-oil. Ikura et al. [102] produced emulsions of bio-oil and #2 diesel using mixtures of Hypermer B246SF and Hypermer 2234 as the surfactants. The optimal bio-oil and surfactant content ranged from 10–20 % and 0.8–1.5 %, respectively, and the longest stability time of emulsion was up to 15 days. Chiaramonti et al. [103] concluded that the optimal range of surfactant was between 0.5 % and 2 %, and the stable time at approximately 70 °C was about 3 days. Yu et al. [104] reported that when the bio-oil and surfactant content were 10 % and 4–5 % respectively, the stability of the emulsion fuel was longer than 120 h. The engine of a Taishen-25 tractor ran well when using this emulsion fuel.

Although emulsion fuel comprised of bio-oil and diesel can be directly applied in diesel engines, its heating value is lower and corrosiveness still exists compared to diesel, due to the introduction of water and carboxylic acids. The combustion characteristics of bio-oil emulsion in a diesel engine were evaluated by Chiaramonti et al. [105]. The results showed that the injector nozzle was corroded. Compared to crude bio-oil, the residual fraction from molecular distillation of bio-oil has a high heating value and a water content close to 0. Since most of the water is evaporated and enriched in the distilled fraction, the corrosiveness of the residual fraction is weak. Consequently, high-quality emulsion fuel can be obtained from the emulsification of residual fraction and diesel.

The emulsion characteristic of diesel and bio-oil fractions, including middle and heavy fractions produced from molecular distillation of pine bio-oil, was investigated. The emulsifier was a mixture of three surfactants (span 80, span 85, and tween 80) with a certain ratio. The stability time of the bio-oil fraction emulsions was relatively shorter than that of crude bio-oil emulsion, which was related to the compositions of bio-oil and its fractions. The bio-oil residual fraction was rich in sugars and viscous at room temperature, so alcohols were chosen as its diluents to increase its fluidity. However, in the emulsification process of alcohol-diluted heavy fraction with diesel, diesel was prone to emulsify directly with alcohols rather than with heavy fraction components. Therefore, improved emulsification needs to be developed to obtain more stable emulsions of bio-oil faction with diesel in the future.

References

[1] Hofbauer H. BIOBIB – a database for biofuels. Institute of Chemical Engineering, Vienna University of Technology, Vienna; 2004.

[2] Demirbas A. Relationships between heating value and lignin, moisture, ash and extractive contents of biomass fuels. Energy, Exploration & Exploitation. 2002;20(1):105–111.

[3] Miao XL, Wu QY, Yang CY. Fast pyrolysis of microalgae to produce renewable fuels. Journal of Analytical and Applied Pyrolysis. 2004;71(2):855–863.

[4] Zhang X, Bai X. Study on thermogravimetric analysis of some kinds of lignocellulosic biomass. Journal of Natural Science of Heilongjiang University. 2012;29(3):352–358.

[5] Zhao S, Tan WY, Wang SY, et al. Influence of temperature programmed rate on pyrolysis rate of the stalk of soybeans and white birch. Journal of Northeast Forestry University. 2003;31(2):75–77.

[6] Yang H, Yan R, Chen H, et al. Characteristics of hemicellulose, cellulose and lignin pyrolysis. Fuel. 2007;86(12):1781–1788.

[7] Wang S, Liu Q, Zheng Y, et al. Mechanism study of biomass pyrolysis by thermogravimetric analysis coupled with infrared spectroscopy. Journal of Engineering Thermophysics. 2006;27(2):351–353.

[8] Raveendran K, Ganesh A, Khilar KC. Pyrolysis characteristics of biomass and biomass components. Fuel. 1996;75(8):987–998.

[9] Haykiri-Acma H, Yaman S. Synergy in devolatilization characteristics of lignite and hazelnut shell during co-pyrolysis. Fuel. 2007;86(3):373–380.

[10] Vamvuka D, Kakaras E, Kastanaki E, et al. Pyrolysis characteristics and kinetics of biomass residuals mixtures with lignite. Fuel. 2003;82(15):1949–1960.

[11] Ren Q, Zhao C. Effect of heating rate on biomass pyrolysis. Journal of Fuel Chemistry and Technology. 2008;36(2):232–235.

[12] Yang S, Qiu K. Study on dynamic characteristics of pyrolysis of seven kinds of biomass in Yiyang area. Chemistry and Industry of Forest Products. 2009;2:10.

[13] Zhu X, Li G, Feng YT, et al. Thermogravimetric experiments and component analysis of biomass in Chongqing. Journal of Chongqing University(Natural Science Edition). 2006;29(8):44–48.

[14] Fu P, Hu S, Sun LS, et al. Release characteristics and formation mechanism of gas products during rice straw and maize stalk pyrolysis. Proceedings of the CSEE. 2009;29(2):113–118.

[15] Sun SB, Weng JJ, Jia YL, et al. Pyrolysis study of rice husk and rice straw by vacuum ultraviolet photoionization mass spectrometry. Journal of Chinese Mass Spectrometry Society. 2013;34(1):1–7.

[16] Li BS, Jiang EC, Wang MF, et al. Pyrolysis characteristic and kinetic analysis of elephant grass. Acta Energiae Solaris Sinica. 2011;32(12):1725–1729.

[17] Fu XF, Zhong ZP, Xiao G, et al. Comparative study on characteristics and dynamics of pyrolysis process of grass biomass. Boiler Technology. 2009;40(3):66–70.

[18] Li D, Chen L, Zhang X, et al. Pyrolytic characteristics and kinetic studies of three kinds of red algae. Biomass and Bioenergy. 2011;35(5):1765–1772.

[19] Wang J, Wang G, Zhang M, et al. A comparative study of thermolysis characteristics and kinetics of seaweeds and fir wood. Process Biochemistry. 2006;41(8):1883–1886.
[20] Wang J, Zhang M, Chen M, et al. Catalytic effects of six inorganic compounds on pyrolysis of three kinds of biomass. Thermochimica Acta. 2006;444(1):110–114.
[21] Zhao H, Yan H, Liu M, et al. Pyrolytic characteristics and kinetics of the marine green tide macroalgae, Enteromorpha prolifera. Chinese Journal of Oceanology and Limnology. 2011;29(5):996.
[22] Zhao H, Yan H, Zhang M. Pyrolysis characteristics and kinetics of Enteromorpha clathrata biomass: A potential way of converting ecological crisis "green tide" bioresource to bioenergy. Advanced Materials Research. 2010;113:114–170.
[23] Wang S. Pyrolysis and combustion experiments and mechanism research of seaweed biomass. Shanghai Jiao Tong University, Shanghai; 2010.
[24] Horne PA, Williams PT. Influence of temperature on the products from the flash pyrolysis of biomass. Fuel. 1996;75(9):1051–1059.
[25] Liu RH, Wang H. Effects of temperature of biomass fast pyrolysis on yield and properties of bio-oil. Transactions of the CSAE. 2006;22(6):138–144.
[26] Scott DS, Piskorz J. The flash pyrolysis of aspen-poplar wood. The Canadian Journal of Chemical Engineering. 1982;60(5):666–674.
[27] Scott DS, Piskorz J, Bergougnou MA, et al. The role of temperature in the fast pyrolysis of cellulose and wood. Industrial & Engineering Chemistry Research. 1988;27(1):8–15.
[28] Conti L, Scano G, Boufala J. Bio-oils from arid land plants: Flash pyrolysis of Euphorbia characias bagasse. Biomass and Bioenergy. 1994;7(1):291–296.
[29] Encinar JM, Beltran FJ, Ramiro A, et al. Pyrolysis/gasification of agricultural residues by carbon dioxide in the presence of different additives: influence of variables. Fuel Processing Technology. 1998;55(3):219–233.
[30] Zhang J, Fan ZL, Lin XF, et al. Online measurement of products during fast pyrolysis of biomass. Journal of Southeast University (Natural Science Edition). 2005;35(1):16–19.
[31] Maschio G, Koufopanos C, Lucchesi A. Pyrolysis, a promising route for biomass utilization. Bioresource Technology. 1992;42(3):219–231.
[32] Broido A, Kilzer FJ. A critique of the present state of knowledge of the mechanism of cellulose pyrolysis: Fire Res Abstr and Rev. 1963;5:157–161.
[33] Demirbas A. Determination of calorific values of bio-chars and pyro-oils from pyrolysis of beech trunkbarks. Journal of Analytical and Applied Pyrolysis. 2004;72(2):215–219.
[34] Onay O. Influence of pyrolysis temperature and heating rate on the production of bio-oil and char from safflower seed by pyrolysis, using a well-swept fixed-bed reactor. Fuel Processing Technology. 2007;88(5):523–531.
[35] Drummond AF, Drummond IW. Pyrolysis of sugar cane bagasse in a wire-mesh reactor. Industrial & Engineering Chemistry Research. 1996;35(4):1263–1268.
[36] Wang Q, Wang S, Wang L, et al. Experimental study of biomass flash pyrolysis for bio-oil production. Journal of Engineering Thermophysics. 2007;28(1): 173–176.
[37] Stiles HN, Kandiyoti R. Secondary reactions of flash pyrolysis tars measured in a fluidized bed pyrolysis reactor with some novel design features. Fuel. 1989;68(3):275–282.
[38] Gravitis J, Vedernikov N, Zandersons J, et al. Furfural and levoglucosan production from deciduous wood and agricultural wastes: ACS Symposium Series, ACS Publications, Washington DC; 2001.
[39] Müller-Hagedorn M, Bockhorn H, Krebs L, et al. A comparative kinetic study on the pyrolysis of three different wood species. Journal of Analytical and Applied Pyrolysis. 2003;68:231–249.

[40] Tan H, Wang S, Luo Z, et al. Influence of metallic salt on biomass flash pyrolysis characteristics. Journal of Engineering Thermophysics. 2005;26(5):742–744.
[41] Jensen A, Dam-Johansen K, Wójtowicz MA, et al. TG-FTIR study of the influence of potassium chloride on wheat straw pyrolysis. Energy & Fuels. 1998;12(5):929–938.
[42] Wang Z, Wang F, Cao J, et al. Pyrolysis of pine wood in a slowly heating fixed-bed reactor: Potassium carbonate versus calcium hydroxide as a catalyst. Fuel Processing Technology. 2010;91(8):942–950.
[43] Richards GN, Zheng GC. Influence of metal ions and of salts on products from pyrolysis of wood: applications to thermochemical processing of newsprint and biomass. Journal of Analytical and Applied Pyrolysis. 1991;21(1):133–146.
[44] Wang S, Liu Q, Wang K, et al. Study on catalytic pyrolysis of Manchurian ash for production of bio-oil. International Journal of Green Energy. 2010;7(3):300–309.
[45] Aho A, Kumar N, Eränen K, et al. Catalytic pyrolysis of woody biomass in a fluidized bed reactor: influence of the zeolite structure. Fuel. 2008;87(12):2493–2501.
[46] Aho A, Kumar N, Eränen K, et al. Catalytic pyrolysis of biomass in a fluidized bed reactor: influence of the acidity of H-beta zeolite. Process Safety and Environmental Protection. 2007;85(5):473–480.
[47] Uzun BAB, Sarioğlu N. Rapid and catalytic pyrolysis of corn stalks. Fuel Processing Technology. 2009;90(5):705–716.
[48] Williams PT, Nugranad N. Comparison of products from the pyrolysis and catalytic pyrolysis of rice husks. Energy. 2000;25(6):493–513.
[49] Qi WY, Hu CW, Li GY, et al. Catalytic pyrolysis of several kinds of bamboos over zeolite NaY. Green Chemistry. 2006;8(2):183–190.
[50] Song CC, Hu HQ, Zhu SW, et al. Biomass pyrolysis and its kinetic parameters with different methods . Journal of Fuel Chemistry and Technology. 2003;31(4):311–316.
[51] Yue JF, Ying H. Experimental study on industrial lignin pyrolysis. Transactions of the Chinese Society of Agricultural Engineering. 2006;22(1):125–128.
[52] Huang N. Study on pyrolysis characteristics and kinetics of three components of biomass. University of Chemical Technology, Beijing; 2007.
[53] Liu YQ, Long MN. Several different biomass fast pyrolysis. Chemical Industry and Engineering Progress. 2010;29(S1):126–132.
[54] Mohan D, Pittman CU, Steele PH. Pyrolysis of wood/biomass for bio-oil: a critical review. Energy & Fuels. 2006;20(3):848–889.
[55] Van der Stelt M, Gerhauser H, Kiel J, et al. Biomass upgrading by torrefaction for the production of biofuels: a review. Biomass and Bioenergy. 2011;35(9):3748–3762.
[56] Prins MJ, Ptasinski KJ, Janssen FJ. Torrefaction of wood: Part 2. Analysis of products. Journal of Analytical and Applied Pyrolysis. 2006;77(1):35–40.
[57] Ru B, Wang S, Dai G, et al. Effect of torrefaction on biomass physicochemical characteristics and the resulting pyrolysis behavior. Energy & Fuels. 2015;29:5865–5874.
[58] Harvey OR, Herbert BE, Kuo L, et al. Generalized two dimensional perturbation correlation infrared spectroscopy reveals mechanisms for the development of surface charge and recalcitrance in plant-derived biochars. Environmental science & technology. 2012;46(19):10641–10650.
[59] Pelaez-Samaniego MR, Yadama V, Garcia-Perez M, et al. Effect of temperature during wood torrefaction on the formation of lignin liquid intermediates. Journal of Analytical and Applied Pyrolysis. 2014;109:222–233.
[60] Neupane S, Adhikari S, Wang Z, et al. Effect of torrefaction on biomass structure and hydrocarbon production from fast pyrolysis. Green Chemistry. 2015;17(4):2406–2417.

[61] No S. Application of bio-oils from lignocellulosic biomass to transportation, heat and power generation—A review. Renewable and Sustainable Energy Reviews. 2014;40:1108–1125.
[62] Xiu S, Shahbazi A. Bio-oil production and upgrading research: A review. Renewable and Sustainable Energy Reviews. 2012;16(7):4406–4414.
[63] Bridgwater AV. Review of fast pyrolysis of biomass and product upgrading. Biomass and Bioenergy. 2012;38:68–94.
[64] Mortensen PML, Grunwaldt J, Jensen PA, et al. A review of catalytic upgrading of bio-oil to engine fuels. Applied Catalysis A: General. 2011;407(1):1–19.
[65] Trane R, Dahl S, Skjøth-Rasmussen MS, et al. Catalytic steam reforming of bio-oil. International Journal of Hydrogen Energy. 2012;37(8):6447–6472.
[66] Onay O, Gaines AF, Kockar OM, et al. Comparison of the generation of oil by the extraction and the hydropyrolysis of biomass. Fuel. 2006;85(3):382–392.
[67] Ertaş M, Hakk Alma M. Pyrolysis of laurel (Laurus nobilis L.) extraction residues in a fixed-bed reactor: Characterization of bio-oil and bio-char. Journal of Analytical and Applied Pyrolysis. 2010;88(1):22–29.
[68] Oasmaa A, Kuoppala E, Solantausta Y. Fast pyrolysis of forestry residue. 2. Physicochemical composition of product liquid. Energy & Fuels. 2003;17(2):433–443.
[69] Garcia-Perez M, Chaala A, Pakdel H, et al. Characterization of bio-oils in chemical families. Biomass and Bioenergy. 2007;31(4):222–242.
[70] Wang S, Wang Y, Cai Q, et al. Multi-step separation of monophenols and pyrolytic lignins from the water-insoluble phase of bio-oil. Separation and Purification Technology. 2014;122:248–255.
[71] Zhang QCJ, Wang TXY. Upgrading bio-oil over different solid catalysts. Energy and Fuels. 2006;20(6):2717–2720.
[72] Xiong WM, Fu Y, Lai DM, et al. Upgrading of bio-oil via esterification catalysed with acidic ion-exchange resin. Chemical Journal of Chinese Universities. 2009;30(9):1754–1758.
[73] Wang JJ, Chang J, Fan J. Experimental study on catalytic esterification of bio-oil by ion exchange resins. Journal of Fuel Chemistry and Technology. 2010;38(005):560–564.
[74] Xiong W, Zhu M, Deng L, et al. Esterification of organic acid in bio-oil using acidic ionic liquid catalysts. Energy & Fuels. 2009;23(4):2278–2283.
[75] Hu X, Gunawan R, Mourant D, et al. Esterification of bio-oil from mallee (Eucalyptus loxophleba ssp. gratiae) leaves with a solid acid catalyst: Conversion of the cyclic ether and terpenoids into hydrocarbons. Bioresource Technology. 2012;123:249–255.
[76] Lohitharn N, Shanks BH. Upgrading of bio-oil: Effect of light aldehydes on acetic acid removal via esterification. Catalysis Communications. 2009;11(2):96–99.
[77] Gunawan R, Li X, Larcher A, et al. Hydrolysis and glycosidation of sugars during the esterification of fast pyrolysis bio-oil. Fuel. 2012;95:146–151.
[78] Adjaye JD, Bakhshi NN. Catalytic conversion of a biomass-derived oil to fuels and chemicals I: model compound studies and reaction pathways. Biomass and Bioenergy. 1995;8(3):131–149.
[79] Adjaye JD, Bakhshi NN. Production of hydrocarbons by catalytic upgrading of a fast pyrolysis bio-oil. Part I: Conversion over various catalysts. Fuel Processing Technology. 1995;45(3):161–183.
[80] Gayubo AG, Valle B, Aguayo AT, et al. Olefin production by catalytic transformation of crude bio-oil in a two-step process. Industrial & Engineering Chemistry Research. 2009;49(1):123–131.
[81] Guo X, Zheng Y, Zhang B, et al. Analysis of coke precursor on catalyst and study on regeneration of catalyst in upgrading of bio-oil. Biomass and Bioenergy. 2009;33(10):1469–1473.

[82] Gayubo AG, Aguayo AT, Atutxa A, et al. Transformation of oxygenate components of biomass pyrolysis oil on a HZSM-5 zeolite. I. Alcohols and phenols. Industrial & Engineering Chemistry Research. 2004;43(11):2610–2618.
[83] Gayubo AG, Aguayo AT, Atutxa A, et al. Transformation of oxygenate components of biomass pyrolysis oil on a HZSM-5 zeolite. II. Aldehydes, ketones, and acids. Industrial & Engineering Chemistry Research. 2004;43(11):2619–2626.
[84] Valle B, Gayubo AG, Aguayo AST, et al. Selective production of aromatics by crude bio-oil valorization with a nickel-modified HZSM-5 zeolite catalyst. Energy & Fuels. 2010;24(3):2060–2070.
[85] Mentzel UV, Holm MS. Utilization of biomass: Conversion of model compounds to hydrocarbons over zeolite H-ZSM-5. Applied Catalysis A: General. 2011;396(1):59–67.
[86] Vispute TP, Zhang H, Sanna A, et al. Renewable chemical commodity feedstocks from integrated catalytic processing of pyrolysis oils. Science. 2010;330(6008):1222–1227.
[87] Ausavasukhi A, Sooknoi T, Resasco DE. Catalytic deoxygenation of benzaldehyde over gallium-modified ZSM-5 zeolite. Journal of Catalysis. 2009;268(1):68–78.
[88] Peralta MA, Sooknoi T, Danuthai T, et al. Deoxygenation of benzaldehyde over CsNaX zeolites. Journal of Molecular Catalysis A: Chemical. 2009;312(1):78–86.
[89] Graça I, Ribeiro FR, Cerqueira HS, et al. Catalytic cracking of mixtures of model bio-oil compounds and gasoil. Applied Catalysis B: Environmental. 2009;90(3):556–563.
[90] Mole T, Bett G, Seddon D. Conversion of methanol to hydrocarbons over ZSM-5 zeolite: An examination of the role of aromatic hydrocarbons using 13 carbon-and deuterium-labeled feeds. Journal of Catalysis. 1983;84(2):435–445.
[91] Haw JF, Song W, Marcus DM, et al. The mechanism of methanol to hydrocarbon catalysis. Accounts of Chemical Research. 2003;36(5):317–326.
[92] Rioche C, Kulkarni S, Meunier FC, et al. Steam reforming of model compounds and fast pyrolysis bio-oil on supported noble metal catalysts. Applied Catalysis B: Environmental. 2005;61(1):130–139.
[93] Hu X, Lu G. Investigation of steam reforming of acetic acid to hydrogen over Ni–Co metal catalyst. Journal of Molecular Catalysis A: Chemical. 2007;261(1):43–48.
[94] Polychronopoulou K, Costa CN, Efstathiou AM. The steam reforming of phenol reaction over supported-Rh catalysts. Applied Catalysis A: General. 2004;272(1):37–52.
[95] Kechagiopoulos PN, Voutetakis SS, Lemonidou AA, et al. Hydrogen production via steam reforming of the aqueous phase of bio-oil in a fixed bed reactor. Energy & Fuels. 2006;20(5):2155–2163.
[96] Wang D, Czernik S, Chornet E. Production of hydrogen from biomass by catalytic steam reforming of fast pyrolysis oils. Energy & Fuels. 1998;12(1):19–24.
[97] Marquevich M, Czernik S, Chornet E, et al. Hydrogen from biomass: steam reforming of model compounds of fast-pyrolysis oil. Energy & Fuels. 1999;13(6):1160–1166.
[98] Zhao C, Lercher JA. Upgrading Pyrolysis Oil over Ni/HZSM-5 by Cascade Reactions. Angewandte Chemie. 2012;124(24):6037 6042.
[99] Wildschut J, Mahfud FH, Venderbosch RH, et al. Hydrotreatment of fast pyrolysis oil using heterogeneous noble-metal catalysts. Industrial & Engineering Chemistry Research. 2009;48(23):10324–10334.
[100] Ardiyanti AR, Gutierrez A, Honkela ML, et al. Hydrotreatment of wood-based pyrolysis oil using zirconia-supported mono-and bimetallic (Pt, Pd, Rh) catalysts. Applied Catalysis A: General. 2011;407(1):56–66.
[101] Venderbosch RH, Ardiyanti AR, Wildschut J, et al. Stabilization of biomass-derived pyrolysis oils. Journal of Chemical Technology and Biotechnology. 2010;85(5):674–686.

[102] Ikura M, Stanciulescu M, Hogan E. Emulsification of pyrolysis derived bio-oil in diesel fuel. Biomass and Bioenergy. 2003;24(3):221–232.
[103] Chiaramonti D, Bonini M, Fratini E, et al. Development of emulsions from biomass pyrolysis liquid and diesel and their use in engines—Part 1: emulsion production. Biomass and Bioenergy. 2003;25(1):85–99.
[104] Yu JY, Peng YL, Li YF, et al. Stability of emulsion from bio-oil and diesel. Journal of Shandong University of Technology (Natural Science Edition). 2007;21(5):101–103.
[105] Chiaramonti D, Bonini M, Fratini E, et al. Development of emulsions from biomass pyrolysis liquid and diesel and their use in engines—Part 2: tests in diesel engines. Biomass and Bioenergy. 2003;25(1):101–111.

Abbreviations

ADF	Acid detergent fiber
ADS	Acid detergent solution
BC	Bacterial cellulose
^{13}C-CP/MAS-NMR	^{13}C-Cross polarization/Magic angle spinning- Nuclear magnetic resonance
CEL	Cellulolytic enzyme lignin
CPMD	Car-Parrinello molecular dynamics
CRTA	Constant rate thermal analysis
DAEM	Distributed activation energy model
DFT	Density functional theory
DGP	1,4:3,6-dianhydro-β-D-glucopyranose
DP	Degree of polymerization
EMAL	Enzymatic/mild acidolysis lignin
FF	Furfural
FTIR	Fourier Transform infrared
GC	Gas chromatography
GC-MS	Gas chromatography-Mass spectrometer
HF	Heavy fraction
HPLC	High performance liquid chromatography
ICP-AES	Inductive coupled plasma- Atomic emission spectrometer
IR	Infra-red
LCA	δ-lactone of 3-hydroxy-5-hydroxymethyltetrahydrofuran-3-carboxylic acid
LG	Levoglucosan
LPG	Liquefied petroleum gas
MCC	Microcrystalline cellulose
MF	Middle fraction
MFC	Microfibrillated cellulose
MTG	Methanol to gasoline
MWL	Milled wood lignin
NC	Networked cellulose
NCC	Nanocrystalline cellulose
NDF	Neutral detergent fiber
NDS	Neutral detergent solution
NMR	Nuclear magnetic resonance
Py-GC/MS	Pyrolysis-Gas chromatography/Mass spectrometry
SADF	Strong acid detergent fiber
SADS	Strong acid detergent solution
SEM	Scanning electron microscopy
TEM	Transmission electron microscopy
TG (TGA)	Thermal gravimetric analysis
TMS	Tetramethylsilane
TRS	Temperature of the radiation source
UPS	Ultraviolet photoelectron spectrometer
XPS	X-ray photoelectron spectroscopy

DOI 10.1515/9783110369632-010

Selected PhD theses supervised by the authors

Yanfen Liao. Mechanism study of cellulose pyrolysis. 2003.
Hong Tan. Mechanism study of biomass pyrolysis. 2005.
Xinshu Zhuang. Biomass hydrolysis under extremely low acids for fuel ethanol production. 2005.
Yan Yao. Separation and upgrading of bio-oil. 2008.
Qi Wang. Research on production and utilization of bio-oil from biomass fast pyrolysis. 2008.
Qian Liu. Biomass pyrolysis mechanism based on the multi-components. 2009.
Yingying Zhu. Experimental Research on dimethyl ether and ethylene glycol synthesis from syngas. 2010.
Xiujuan Guo. Mechanism research on the selective pyrolysis behavior of biomass. 2011.
Zuogang Guo. Bio-oil multiple upgrading research based on molecular distillation technology. 2012.
Xinbao Li. Researches on the bio-oil catalytic reforming for hydrogen production and dimethyl oxalate hydrogenation for ethylene glycol synthesis. 2013.
Qianqian Yin. Researches on the synthesis of hydrocarbon fuels. 2014.
Qinjie Cai. Study of bio-oil catalytic cracking based on the molecular distillation. 2015.
Wenwen Guo. Research on highly efficient synthesis of fuel ethanol from syngas. 2015.
Bin Ru. Research on biomass pyrolysis behavior and influence mechanism based on complex components. 2016.
Yurong Wang. Research on stepwise separation of bio-oil based on water extraction and molecular distillation. 2016.

DOI 10.1515/9783110369632-011

The Authors' representative academic papers published in this field

Luo Z, Wang S, Liao Y, et al. Research on biomass fast pyrolysis for liquid fuel. Biomass and Bioenergy. 2004;26(5):455–462.

Luo Z, Wang S, Liao Y, et al. Mechanism study of cellulose rapid pyrolysis. Industrial & Engineering Chemistry Research. 2004;43(18):5605–5610.

Luo Z, Wang S, Cen K. A model of wood flash pyrolysis in fluidized bed reactor. Renewable Energy. 2005;30(3):377–392.

Wang S, Liu Q, Liao Y, et al. A study on the mechanism research on cellulose pyrolysis under catalysis of metallic salts. Korean Journal of Chemical Engineering. 2007;24(2), 336–340.

Liu Q, Wang S, Zheng Y, et al. Mechanism study of wood lignin pyrolysis by using TG-FTIR analysis. Journal of Analytical and Applied Pyrolysis. 2008;82(1):170–177.

Liu Q, Wang S, Wang K, et al. Mechanism of formation and consequent evolution of active cellulose during cellulose pyrolysis. Acta Physico-Chimica Sinica. 2008;24(11):1957–1963.

Liu Q, Wang S, Luo Z, et al. Catalysis mechanism study of potassium salts on cellulose pyrolysis by using TGA-FTIR analysis. Journal of Chemical Engineering of Japan. 2008;41(12):1133–1142.

Liu Q, Wang S, Wang K, et al. Pyrolysis of wood species based on the compositional analysis. Korean Journal of Chemical Engineering. 2009;26(2):548–553.

Wang S, Gu Y, Liu Q, et al. Separation of bio-oil by molecular distillation. Fuel Processing Technology. 2009;90(5):738–745.

Wang S, Wang K, Liu Q, et al. Comparison of the pyrolysis behavior of lignins from different tree species. Biotechnology Advances. 2009;27(5):562–567.

Wang S, Liu Q, Wang K, et al. Study on Catalytic Pyrolysis of Manchurian Ash for Production of Bio-Oil. International Journal of Green Energy. 2010;7(3):300–309.

Guo X, Wang S, Guo Z, et al. Pyrolysis characteristics of bio-oil fractions separated by molecular distillation. Applied Energy. 2010;87(9):2892–2898.

Guo Z, Wang S, Gu Y, et al. Separation characteristics of biomass pyrolysis oil in molecular distillation. Separation and Purification Technology. 2010;76(1):52–57.

Guo X, Wang S, Wang Q, et al. Properties of bio-oil from fast pyrolysis of rice husk. Chinese Journal of Chemical Engineering. 2011;19(1):116–121.

Wang S, Guo X, Wang K, et al. Influence of the interaction of components on the pyrolysis behavior of biomass. Journal of Analytical and Applied Pyrolysis. 2011;91(1):183–189.

Guo X, Wang S, Wang K, et al. Experimental researches on milled wood lignin pyrolysis based on analysis of bio-oil. Chemical Research in Chinese Universities. 2011;27(3):426–430.

Guo Z, Wang S, Xu G, et al. Upgrading of bio-oil molecular distillation fraction with solid catalyst. BioResources. 2011;6(3):2539–2550.

Guo Z, Wang S, Guo L, et al. Catalytic steam reforming of ethanol for hydrogen production over Ni/CeO2-ZrO2 catalysts. BioResources. 2011;6(4):4092–4102.

Wang S, Guo X, Liang T, et al. Mechanism research on cellulose pyrolysis by Py-GC/MS and subsequent density functional theory studies. Bioresource Technology. 2012;104:722–728.

Luo Z, Wang S, Guo X. Selective pyrolysis of Organosolv lignin over zeolites with product analysis by TG-FTIR. Journal of Analytical and Applied Pyrolysis. 2012;95:112–117.

Wang S, Li X, Guo L, et al. Experimental research on acetic acid steam reforming over Co-Fe catalysts and subsequent density functional theory studies. International Journal of Hydrogen Energy. 2012;37(15):11122–11131.

DOI 10.1515/9783110369632-012

Wang S, Guo Z, Cai Q, et al. Catalytic conversion of carboxylic acids in bio-oil for liquid hydrocarbons production. Biomass & Bioenergy. 2012;45:138–143.

Wang S, Cai Q, Guo Z, et al. Renewable gasoline produced by co-cracking of methanol and ketones in bio-oil. BioResources. 2012;7(4):5019–5031.

Li X, Wang S, Cai Q, et al. Effects of preparation method on the performance of Ni/Al2O3 catalysts for hydrogen production by bio-oil steam reforming. Applied Biochemistry and Biotechnology. 2012;68 (1):10–20.

Wang S, Cai Q, Wang X, et al. Bio-gasoline Production from Co-cracking of Hydroxypropanone and Ethanol. Fuel Processing Technology. 2013;111:86–93.

Wang Q, Wang S, Li X, et al. Hydrogen production via acetic acid steam reforming over HZSM-5 and Pd/HZSM-5 catalysts and subsequent mechanism studies. BioResources, 2013:8(2):2897–2909.

Wang S, Ru B, Lin H, et al. Degradation mechanism of monosaccharides and xylan under pyrolytic conditions with theoretic modeling on the energy profiles. Bioresource Technology. 2013;143:378–383.

Wang S, Liang T, Ru B, et al. Mechanism study of xylan pyrolysis by Py-GC/MS. Chemical Research in Chinese Universities. 2013;29(4):782–787.

Wang S, Zhou Y, Liang T, et al. Catalytic pyrolysis of mannose as a model compound of hemicellulose over zeolites. Biomass & Bioenergy. 2013;57:106–112.

Wang S, Li X, Zhang F, et al. Bio-oil Catalytic Reforming without Steam Addition: Application to Hydrogen Production and Studies on its Mechanism. International Journal of Hydrogen Energy. 2013;38(36):16038–16047.

Wang S, Wang Y, Cai Q, et al. Multi-step separation of monophenols and pyrolytic lignins from the water-insoluble phase of bio-oil. Separation and Purification Technology. 2014;122(1):248–255.

Wang S, Zhang F, Cai Q, et al. Catalytic steam reforming of bio-oil model compounds for hydrogen production over coal ash supported Ni catalyst. International Journal of Hydrogen Energy. 2014;39(5):2018–2025.

Wang S, Wang Y, Cai Q, et al. Production of bio-gasoline by co-cracking of acetic acid in bio-oil and ethanol. Chinese Journal of Chemical Engineering. 2014;22(1):98–103.

Wang S, Cai Q, Wang X, et al. Bio-gasoline production from the co-cracking of the distilled fraction of bio-oil and ethanol. Energy & Fuels. 2014;28(1):115–122.

Guo Z, Wang S, Wang X. Stability mechanism investigation of emulsion fuels from biomass pyrolysis oil and diesel. Energy. 2014;66:250–255.

Wang S, Cai Q, Wang X, et al. Biogasoline production by co-cracking of model compound mixture of bio-oil and ethanol over HZSM-5. Chinese Journal of Catalysis. 2014;35(4):709–722.

Wang S, Lin H, Ru B, et al. Comparison of the pyrolysis behaviors of pyrolytic lignin and milled wood lignin by using TG-FTIR analysis. Journal of Analytical and Applied Pyrolysis. 2014;108:78–85.

Wang S, Ru B, Lin H, et al. Pyrolysis mechanism of hemicellulose monosaccharides in different catalytic processes. Chemical Research in Chinese Universities. 2014;30(5):848–854.

Wang S, Cai Q, Chen J, et al. Green aromatic hydrocarbon production from co-cracking of a bio-oil model compound mixture and ethanol over Ga_2O_3/HZSM-5. Industrial & Engineering Chemistry Research. 2014;53:13935–13944.

Wang S, Cai Q, Zhang F, et al. Hydrogen production via catalytic reforming of the bio-oil model compounds: acetic acid, phenol and hydroxyacetone. International Journal of Hydrogen Energy. 2014;39:18675–18687.

Li X, Wang S, Zhu Y, et al. DFT Study of Bio-Oil Decomposition Mechanism on a Co Stepped Surface: Acetic Acid as a Model Compound. International Journal of Hydrogen Energy. 2015;40:330–339.

Wang S, Ru B, Lin H, et al. Pyrolysis Behaviors of Four Lignin Polymers Isolated from the Same Pine Wood. Bioresource Technology, 2015,182:120–127.

Wang S, Guo W, Guo L, et al. Experimental and subsequent mechanism research on steam reforming of ethanol over Ni/CeO2 catalyst. International Journal of Green Energy. 2015;12:694–701.
Wang S, Ru B, Dai G, et al. Pyrolysis mechanism study of minimally damaged hemicellulose polymers isolated from agricultural waste straw samples. Bioresource Technology. 2015;190:211–218.
Wang S, Ru B, Lin H, et al. Pyrolysis Behaviors of Four O-Acetyl-preserved Hemicelluloses Isolated from Hardwoods and Softwoods. Fuel. 2015;150:243–251.
Wang Y, Wang S, Leng F, et al. Separation and Characterization of Pyrolytic Lignins from the Heavy Fraction of Bio-oil by Molecular Distillation. Separation and Purification Technology. 2015;152:123–132.
Ru B, Wang S, Dai G, et al. Effect of torrefaction on biomass physico-chemical characteristics and the resulting pyrolysis behavior. Energy & Fuels. 2015;29:5865–5874.
Wang S, Cai Q, Chen J, et al. Co-cracking of bio-oil model compound mixtures and ethanol over different metal oxide-modified HZSM-5 catalysts. Fuel. 2015;160:534–543.
Wang S, Zhang F, Cai Q, et al. Steam Reforming of Acetic Acid over Coal Ash Supported Fe and Ni Catalysts. International Journal of Hydrogen Energy. 2015;40:11406–11413.
Guo X, Wang S. Mechanism research on fast pyrolysis of Organosolv lignin by Py-GC/MS. Journal of Renewable and Sustainable Energy. 2015;7(023116):1–7.
Wang S, Wang Y, Leng F, et al. Separation and Enrichment of Catechol and Sugars from Bio-oil Aqueous Phase. Bioresources. 2016;11(1):1707–1720.
Wang S, Wang Y, Leng F, et al. Stepwise enrichment of sugars from the heavy fraction of bio-oil. Energy & Fuels. 2016;30:2233–2239.
Wang S, Chen J, Cai Q, et al. The Effect of Mild Hydrogenation on the Catalytic Cracking of Bio-oil for Aromatic Hydrocarbon Production. International Journal of Hydrogen Energy. 2016;41(37):16385–16393.
Wang S, Ru B, Dai G, et al. Influence mechanism of torrefaction on softwood pyrolysis based on structural analysis and kinetic modeling. International Journal of Hydrogen Energy. 2016;41(37):16428–16435.
Wang S, Ru B, Lin H, et al. Kinetic study on pyrolysis of biomass components: a critical review. Current Organic Chemistry. 2016;20(23):2489–2513.
Zhang F, Wang S, Chen J, et al. Effect of Coal Ash on the Steam Reforming of Simulated Bio-oil for Hydrogen Production over Ni/γ-Al_2O_3. Bioresources. 2016;11(3):6808–6821.
Wang S, Lin H, Zhang L, et al. Structural characterization and pyrolysis behavior of cellulose and hemicellulose isolated from softwood Pinus armandii Franch. Energy & Fuels. 2016;30 (7):5721–5728.
Wang S, Dai G, Ru B, et al. Effects of torrefaction on hemicellulose structural characteristics and pyrolysis behaviors. Bioresource Technology. 2016;218:1106–1114.
Wang S, Lin H, Ru B, et al. Kinetic modeling of biomass components pyrolysis using a sequential and coupling method. Fuel. 2016;185:763–771.

Wang S, Guo W, Guo L, et al. Experimental and subsequent mechanism research on steam reforming of ethanol over Ni/CeO2 catalyst. International Journal of Green Energy, 2015;12:694–701.

Wang S, Ru B, Dai G, et al. Pyrolysis mechanism study of minimally damaged hemicellulose polymers isolated from agricultural waste straw samples. Bioresource Technology, 2015;190:211–218.

Wang S, Ru B, Lin H, et al. Pyrolysis behaviors of four O-acetyl-preserved hemicelluloses isolated from hardwoods and softwoods. Fuel, 2015;150:243–251.

Index

Bei Fragen zur Produktsicherheit wenden Sie sich bitte an:
If you have any questions regarding product safety,
please contact:

Walter de Gruyter GmbH
Genthiner Straße 13
10785 Berlin
productsafety@degruyterbrill.com